The Springer Series in Materials Science covers the complete spectrum of materials physics, including fundamental principles, physical properties, materials theory and design. Recognizing the increasing importance of materials science in future device technologies, the book titles in this series reflect the state-of-the-art in understanding and controlling the structure and properties of all important classes of materials.

Springer Series in Materials Science

Volume 164

For further volumes:
http://www.springer.com/series/856

Alexander V. Kolobov · Junji Tominaga

Chalcogenides

Metastability and Phase Change Phenomena

Springer

Alexander V. Kolobov
Nanoelectronics Research Center
National Institute of Advanced Industrial Science and Technology
Tsukuba
Ibaraki
Japan

Junji Tominaga
Nanoelectronics Research Center
National Institute of Advanced Industrial Science and Technology
Tsukuba
Ibaraki
Japan

ISSN 0933-033X
ISBN 978-3-642-28704-6 ISBN 978-3-642-28705-3 (eBook)
DOI 10.1007/978-3-642-28705-3
Springer Heidelberg New York Dordrecht London

Library of Congress Control Number: 2012936758

Printed on acid-free paper

Springer is part of Springer Science+Business Media (www.springer.com)

To our daughters, Ilaria and Morie

Foreword

Alexander Kolobov and Junji Tominaga have in this book made a remarkable, indeed a monumental achievement. They cover the very large field of amorphous and glassy materials and phase change memories and their various manifestations, which is quite a task. Both authors are very important contributors to our field and are highly regarded. Their remarkable ability to cover such important basic science and technology and also understand that new concepts and mechanisms are continually expanding the field should be welcome to all who labor in this growing area. It is exemplary that this is the first book in our literature recognizing that the field of the chalcogenides is now playing an important part in the exciting field of topological insulators.

There are still broad and deep areas that our scientific efforts are addressing, and I believe there always will be. This book certainly provides a great and informative core which is a basis for new work. The outreach that the authors have given us is very impressive and comprehensive especially because the authors are strong participators and offer original insights of their own.

Their understanding and their own considerable talents make this a book that I believe will get the attention it deserves for amorphous materials generally and uniquely for the chalcogenides whose physics is now impacting various other applications where they are receiving great attention. I believe that the chalcogenides have entirely new physical, electronic, structural mechanisms effectively presented in this book that will be a source of knowledge for the readers who with enhanced understanding can make possible many new approaches and will interest many in the field. There should be a single book that covers the field of many investigations so that there is a commonality of knowledge for the reader. This is such a book.

Ovonic phase change and threshold switching devices are now being studied for important new functionality, cognitive computing, and multi-function thin film devices that open up a whole new group of potential applications including a provocative possibility that includes dynamic superconductivity.

The authors are personally wonderful models of the type of creativity that always seeks, explores, and contributes. I am certainly familiar with the important contributions of Junji Tominaga and I have long recognized the contributions, creativity, and excitement that Alex Kolobov provides to our field. We honor them for their leading roles and pay homage to those who preceded them. Japan has a leading role in development of phase change memory and Kolobov represents Russia with great pride as a student of Boris Kolomiets and pays respect to the Russian literature.

The authors have made a great contribution and the book lives up to its title. Metastability, chalcogenide glasses, and phase change alloys certainly deserve to be addressed in a scholarly, descriptive, and informative manner as they are indeed a rapidly expanding field with entirely new materials, physics, and electronic mechanisms. I am sure the book will be read and studied, and it will become the reference of choice by those who are involved and interested in what has become an area of not only such exciting research and development but a basis for expanding existing and valuable new industries.

The subjects of the book are expanding and important new fields of innovative science and technology are already being considered and proposed and certainly now one speaks of usages that will advance not our world's knowledge but the needed economic growth of our planet. Thus we are entering into a new advanced era and it promises a great time for its participants and builders.

This outstanding book is a welcomed contribution with the ambition and actualization to embrace the present field and lay the groundwork for future understanding. I heartily recommend this book by these very important authors for its publication is a landmark event.

Bloomeld Hills, MI, January 2012 Stanford R. Ovshinsky

Preface

The discovery in the mid-1950s by N. A. Goryunova and B. T. Kolomiets at A. F. Ioffe Physico-Technical Institute in Leningrad of the semiconducting properties of amorphous chalcogenide alloys opened up a new field in solid state physics, namely, amorphous semiconductors. The existence of semiconducting properties, i.e. of the forbidden energy gap in the density of states of disordered materials was very puzzling since the band structure of solids had been usually derived from the long-range order, or the periodic nature of the crystalline phase. Another important class of amorphous semiconductors includes hydrogenated amorphous silicon and related tetrahedrally bonded materials. While chalcogenide glasses typically possess p-type conductivity and cannot be doped, hydrogenated amorphous silicon can be doped which opened up completely new perspectives of the creation of amorphous semiconductor *devices* attracting further interest to this new field of research.

Possessing the typical properties of their crystalline counterparts, amorphous semiconductors additionally have characteristics unique to the amorphous phase. Intrinsically metastable, they can undergo transformations between different structural states with concomitant property changes. Such changes are known for both chalcogenide glasses and tetrahedrally bonded amorphous semiconductors. A special class of materials, the so-called phase-change materials exhibit a reversible transition between the crystalline and amorphous phases with an unusually large optical and electrical property contrast between the two phases. The proposal by S. R. Ovshinsky in the early 1960s to use these materials for memory applications marked the beginning of a vast and extremely fruitful field.

The reader may have never thought about this, but when one takes a laptop computer in one's hands, or watches a high-resolution TV one comes into contact with amorphous silicon: most of the flat panel displays contain thin film transistors that utilise amorphous silicon. Should one save data on a re-writable optical disc, one uses phase-change materials, another class of amorphous chalcogenides without which modern multimedia systems would have been impossible. In recent years phase-change materials have been attracting increasing attention due to their excellent scaling and cycling characteristics which made them the leading candidate to replace the currently ubiquitous FLASH memory. Samsung and Micron Technology Inc. have already started commercial production of phase-change random-access memory chips and Samsung has also started to use the latter in their mobile phones. Not only are amorphous semiconductors suitable for these applications, they are unique materials which ensured that the present computers, TVs, DVDs and quite a few other mobile devices became available to modern society.

The present monograph is devoted to structural metastability in both classes of chalcogenide alloys, i.e. chalcogenide glasses and phase-change alloys. While these two classes of materials are usually considered separately in the published literature, they possess numerous similar features and the authors believe that the inclusion of them under a single cover will close the gap and be enriching.

The structure of this book is as follows. Chapter 1 deals with the fundamentals of amorphous semiconductors whose properties are described in Chap. 2. Since material properties are determined by the structure, the latter is of significant importance and methods to investigate it are described in Chap. 3. Chapters 4–7 are dedicated to photo-induced metastability in chalcogenide glasses. This description starts with reversible photostructural changes (Chap. 4), followed by phenomena induced by polarised light (Chap. 5). Photo-induced crystallisation and amorphisation are the subject of Chap. 6 and phenomena occurring in chalcogenide glass–metal structures are described in Chap. 7. Chapters 8–11 concentrate on phase-change alloys. First, the structure of phase-change alloys in the crystalline (Chap. 8) and amorphous (Chap. 9) phases is described. Chapter 10 reviews pressure-induced transformations in phase-change alloys and this part of the book is concluded by Chap. 11 where atomistic mechanisms of the phase-change process are discussed. Finally, in Chap. 12 we discuss the present and future applications of these materials.

The authors are grateful to K. Shimakawa, K. Mitrofanov and especially P. Fons for critical reading of the manuscript and numerous useful suggestions. We sincerely hope that this monograph will be a useful reference not only for scientists and engineers working in the fields of solid-state physics, amorphous semiconductors and memory devices but also for graduate and post-graduate students specialising in solid-state physics and materials science.

Tsukuba, January 2012

Alexander V. Kolobov
Junji Tominaga

Contents

Part IV Applications

Part I
Amorphous Semiconductors

Chapter 1
Fundamentals of Amorphous Semiconductors

1.1 Definitions

Amorphous semiconductors can be largely divided into two classes. One main group of these materials is represented by chalcogenide glasses (also called chalcogenide vitreous semiconductors). Chalcogenides are materials containing one (or more) of the elements from Group VI of the periodic table, namely, S, Se or Te. While oxygen, strictly speaking, is also a chalcogen, oxide materials are usually treated separately. The other group includes hydrogenated amorphous silicon (a-Si:H) and related materials. The present monograph is dedicated to the first group of materials, i.e. chalcogenides.

Possessing typical semiconductor properties of crystalline materials, amorphous semiconductors also exhibit a number of effects that are unique to the amorphous phase: intrinsically metastable, they can undergo various transformations between different states. The metastability of chalcogenides is the main subject of this book and in this chapter we describe the most important attributes of amorphous semiconductors. The first step is to define the amorphous phase and to discuss what consequences this has on properties of amorphous materials. Numerous monographs have been devoted to this issue (e.g. Refs. [1–8]); here we shall discuss only those aspects that are relevant to the content of this book.

In order to introduce the concept of an amorphous material we start with the definition of a crystal. *A perfect crystal is a solid in which the atoms are arranged in a pattern that repeats periodically in three dimensions to an infinite extent.* This means that if one knows positions of just a few atoms that form the irreducible unit cell and the way their arrangement is repeated in space (the space group), one can unambiguously determine positions of all other atoms in the crystal. A crucial requirement for a solid to be in the crystalline phase is thus extended periodicity or long-range order.

A solid that does not possess the long-range order (or translational periodicity) characteristic of a crystal is called non-crystalline or amorphous. A few words should be said here about terminology related to the non-crystalline phase of solids. The

A. V. Kolobov and J. Tominaga, *Chalcogenides*, Springer Series in Materials Science 164, DOI: 10.1007/978-3-642-28705-3_1,

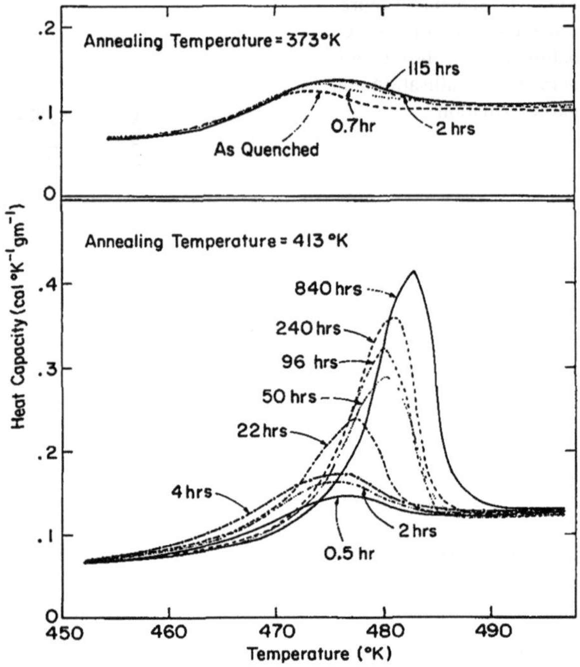

Fig. 1.3 Heat capacity versus increasing temperature in the region of the glass transition of amorphous $(1-x)As_2Se_3 \cdot xSb_2Se_3$ $(x = 0.3)$ after annealing for various times at 413 K. Reprinted from Das et al. [13] with permission from Elsevier

chalcogenides often require cooling rates in the order of 10^8 K/s in order to obtain the amorphous phase, a value that is rather difficult to achieve experimentally.

Despite a long history of glass science, there is no established theory that could predict whether or not a particular material is a good glass former. The first successful attempt to formulate the necessary conditions for glass formation was made by Zachariansen [15]. He argued that materials most likely to form glasses would have an internal energy in the amorphous state only slightly larger than in crystals. One of the most important issues in this approach was that *non-periodic arrangement of atoms is achieved mainly as a result of variations in the bond angles*.

It was suggested [16] that the ability of a material to form a (semiconducting) glass is related to directional, i.e. covalent bonds but it should be noted that covalent bonding is not a sufficient condition to form a glass, e.g. silicon is a typical semiconductor but cannot be prepared as a glass. At the same time, liquid silicon is a metal. Hence a requirement of the covalent nature of bonding in the melt may be a more appropriate formulation.

There have also been attempts to relate glass-forming ability to the electronic structure of the constituent elements. In particular, it was argued that in order to be a good glass former a material has to possess a concentration f of lone-pair electrons (i.e. pairs of valence electrons that do not participate in forming covalent bonds with the neighbouring atoms) relatively high compared to the total number of valence electrons ($f = 0.5 \div 0.66$) [17].

Fig. 1.4 S. R. Ovshinsky giving a talk at MRS Fall 2003 meeting (courtesy of E. Mytilineou)

1.2.3 Reversible Amorphous-to-Crystal Transition

As described earlier, a melt can be solidified in two different ways (Sect. 1.2.1; Fig. 1.1). If the melt is cooled slowly, an equilibrium crystalline phase is formed. If, on the other hand, the melt is cooled fast enough so that the structure cannot follow the change in temperature, the disordered phase is frozen; one obtains a supercooled liquid and subsequently a glass. When the obtained glass is annealed at temperatures between the glass-transition temperature and the melting point, it gradually transforms into the crystalline state. One can thus cycle a material between the ordered crystalline and disordered amorphous phases. With respect to chalcogenide alloys, the reversible crystal-to-amorphous transformation is called a phase-change process and materials undergoing such changes—usually Te-based alloys—are called phase-change materials. They are characterised by an unusually large optical and electrical property contrast and back in the 1960s Ovshinsky (Fig. 1.4) suggested to use these materials for memory devices [18]. The present section deals with this class of chalcogenides. The interested reader is also referred to a recent volume edited by Raoux and Wuttig [19].

From the description above it follows that almost any material can be prepared in either crystalline or amorphous phases. While this is generally true, there are several factors that determine practicality of such reversible changes between the ordered and disordered phases.

One of the major parameters is the characteristic time scale. Thus good glass formers possess a very high viscosity and crystallisation of such glasses can take years even at temperatures very close to the glass-transition temperature. Marginal glass formers, on the other hand, can crystallise in nanoseconds.

Another important parameter is cyclability, or the number of crystallisation–amorphisation cycles the material can withstand without degradation of properties. This is predominantly determined by the ability of a material to maintain the same single phase in the crystalline, amorphous and liquid phases. Materials that are subject to phase separation usually exhibit strong degradation of properties after several cycles.

R&D of phase-change optical storage has a long history starting with the discovery of the switching effect in some chalcogenide alloys [18, 20, 21]. In particular, in $Si_{12}Te_{48}As_{30}Ge_{10}$ it was found that upon application of a voltage that exceeds a critical value called threshold voltage, the chalcogenide alloy switched from the low-conductivity state to the high-conductivity state [18]. As the current in the high-conductivity state was reduced below the characteristic value termed the holding current the unit switched back to the original low-conductivity state. The observed switching process was reversible and could be repeated many times. The observed increase in the conductivity was attributed to the formation of a current filament, growing in diameter with increasing current flow. In the same paper [18] it was reported that by tuning the composition a memory effect was observed that was attributed to a change in the structure of the material, namely, "after switching from a highly resistive state, structural changes result in the preservation of a conductive state even when the current is totally removed. The material can be reversibly switched back to the highly resistive state by application of a current pulse of either polarity exceeding a threshold value". This finding marked the beginning of intensive—and extensive—research on phase-change materials.

A group of materials, typically (but not exclusively) Te-based alloys, exhibit very high amorphisation and crystallisation rates in the order of nanoseconds which, combined with large cyclability and a pronounced property contrast between the crystalline and amorphous phases, makes them ideal for memory applications. The most studied materials lie along the quasibinary GeTe-Sb_2Te_3 tie-line and are usually referred to as GST alloys. Other groups of phase change materials include Ag-In-Sb-Te (AIST) alloys based on Ag- and In-doped Sb_2Te. Ge-Sb alloys are yet another example of phase-change materials. It should be noted here that there is an important difference between these alloys: while the crystallisation process in nucleation limited in Ge-Sb-Te alloys, it is growth limited in AIST alloys. The detailed description of various phase-change materials is summarised in [22]. In this volume we concentrate on Ge-Sb-Te alloys as the most studied and best understood class of phase-change materials.

The description of the phase-change process given at the beginning of this section relies on a glass-transition diagram, but are phase-change materials glasses? Despite years of research in this area, it is only recently that the glass transition has been observed in phase-change materials subject to preannealing [23] which provides a justification for the use of glass-transition concept to describe the phase-change process.[1]

---

[1] Effect of thermal treatment on the intensity of the glass-transition features in the phase diagram is described in Sect. 1.2.

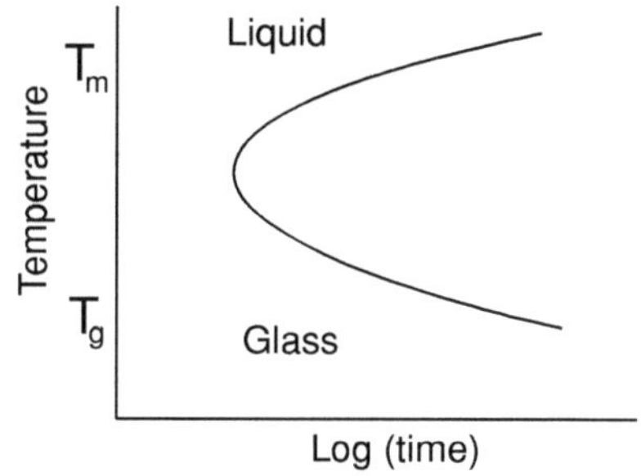

Fig. 1.5 Time-temperature-transformation diagram of an undercooled liquid

The crystallisation process takes place below the melting point (T_m) but above the glass-transition temperature (T_g) (we do not discuss any undesired recrystallisation of amorphous marks below the glass-transition temperature). A schematic plot of the crystallisation speed is illustrated by the time-temperature-transformation (T-T-T diagram) shown in Fig. 1.5. As the material is cooled below the melting temperature (undercooled liquid), the free-energy difference between liquid and crystal favours the formation of the crystalline state, i.e. the driving force to form crystalline nuclei increases with decreasing temperature. At the same time, atomic mobilities rapidly decrease because the viscosity becomes higher at lower temperatures. As a consequence, the highest crystallisation speeds are encountered at intermediate temperatures between T_m and T_g.

The phase diagram for quasibinary GeTe-Sb_2Te_3 phase-change alloys is shown in Fig. 1.6 [24, 25]. Figure 1.7 shows crystallisation times for different GST alloys [26]. It should be noted that, when talking about crystallisation times, one should be careful, since as-deposited films that are usually studied experimentally may require (significantly) larger times to crystallise than is required in subsequent cycles and the values reported in the literature do not always take this distinction into account. For example, while as-deposited Ge_xTe_{1-x} ($x = 29.5$–72.4) films need as long as 100–10^9 ns to crystallise (depending on composition), melt-quenched films (i.e. films that underwent at least one crystallisation–amorphisation cycles) require one to almost four orders of magnitude shorter (again depending on composition) times in order to crystallise [27].

The activation energy for crystallisation for various Ge-Sb-Te alloys has been determined by various authors to be around 2.3 eV. At the same time it should be noted that recently it was demonstrated that the liquid phase of phase-change alloys is 'fragile' (cf. Sect. 1.2.1), i.e. the temperature behaviour of viscocity does not obey the Arrhenius law [28]. This suggests that while the reported rather large value of the activation energy is appropriate to describe the retention properties at temperatures well below the glass-transition temperature, crystallisation in real devices, that takes place at much higher temperatures, may have a significantly smaller value of the activation energy.

In GeTe-Sb_2Te_3 phase-change alloys, the crystallisation temperature increases with the concentration of GeTe and can also be increased by doping with additional

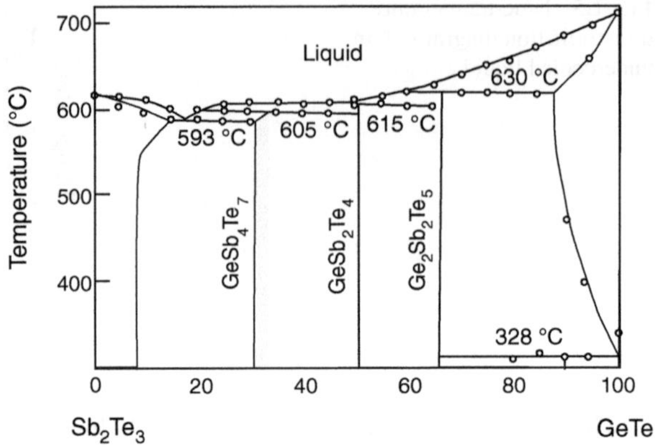

Fig. 1.6 Phase diagram of the GeTe – Sb_2Te_3 system after Abrikosov and Danilova-Dobryakova [24]

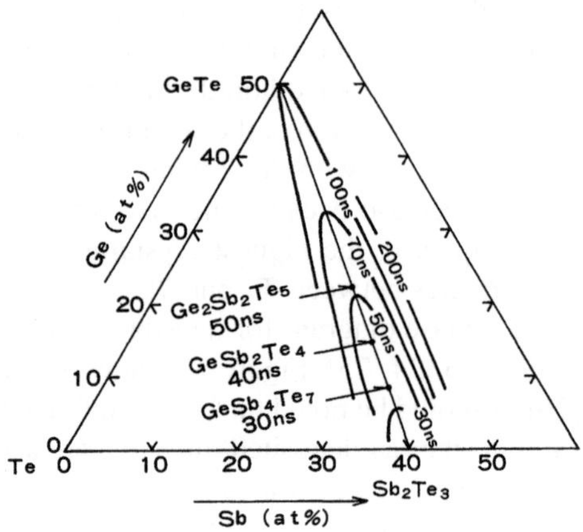

Fig. 1.7 Threshold crystallisation laser pulse durations for as-deposited Ge-Sb-Te films (laser power of 8 mW, active layer of 100 μm sandwiched between ZnS protective layers. Reprinted with permission by Japan Society of Applied Physics from Yamada et al. [26]

elements. For example, doping with nitrogen increases the crystallisation temperature from 180 °C in the undoped Ge-Sb-Te to 230 °C for films containing 3 at.% nitrogen [29]. Very importantly, doping with nitrogen concurrently increases the number of overwrite cycles [29].

Exposure to intense laser pulses or current pulses results in heating of the materials due to the non-radiative recombination of non-equilibrium charge carriers or Joule heating and consequently the phase-change process can be induced which is the basis of phase-change memory devices (cf. Chap. 12). Most investigators believe that the sole role of light or current pulses is to heat the material, the role of electronic excitation is usually neglected. The role of electronic excitation is discussed later in this volume (cf. Chap. 11).

Fig. 1.8 Professor B.T. Kolomiets (1908–1989), father of amorphous semiconductors

1.3 Atomic Structure

Amorphous chalcogenides, often also referred to as vitreous (or glassy) chalcogenides or chalcogenide glasses, are not new. What is relatively new is the discovery in 1954 by Kolomiets and Goryunova (Fig. 1.8) that these glasses possess semiconductor properties [30, 31]. This discovery opened a new field in semiconductor physics, namely, amorphous semiconductors. Following the pioneering work by Goryunova and Kolomiets, amorphous chalcogenides attracted a lot of attention due to their unique properties. First, it was very strange that amorphous, that is non-crystalline, materials could be semiconducting. The band structure of a crystalline solid is generally derived from translational symmetry in its structure that is absent in amorphous materials. How then can an amorphous material be a semiconductor? Second, amorphous materials possess a large number of defects that, in turn, should create a high concentration of states in the forbidden gap. Glasses, however, are transparent. How can a large number of defects coexist with such high transparency? In order to answer these questions, knowledge of the structure is needed and this issue is discussed below in this chapter.

1.3.1 Long-Range Order

As already mentioned, the main characteristic feature of crystals is their extended periodicity. If one plots the number of atoms as a function of distance for a given

1.3.3 Medium-Range Order

The existence of extended order in amorphous solids that extends above the short-range order was revealed in the mid-1970s by Grigirovici and Popescu who wrote that "many low-coordinated elements and compounds exhibit a more complex disorder or even a special kind or ordering at not too large a scale" [33]. A couple of years later Phillips [34] and Lucovsky and Galeener [35] introduced the concepts of medium-range order and intermediate-range order, respectively, with the same meaning. The medium-range order originates from the propagation of the order from the first coordination sphere and typically extends to about 1 nm.

1.3.4 Melting of Semiconductors

An important question is how do the structure and properties of an amorphous semiconductor change when the latter melts? Since both states are disordered, one might expect the average structure to remain essentially unchanged. This, however, is not always so. For a dedicated monograph an interested reader is referred to [36].

Generally, molten semiconductors can be largely divided into two classes. In the first class, semiconductor properties are preserved after melting, i.e. a $semiconductor \Longrightarrow semiconductor$ transition takes place and the molten semiconductor is a true liquid semiconductor. Sulphur- and Se-based chalcogenides generally belong to this group. The retention of semiconductor properties in the liquid phase has been attributed by Regel to preservation in the liquid state of the short-range order of the solid state [36].

The transition of the second type is a $semiconductor \Longrightarrow metal$ transition and such transitions are observed when silicon, germanium and most III–V compounds are molten [36]. The change in the conductivity type is attributed to a change in the short-range order upon melting, namely, to an increased coordination number in the melt.

An intermediate type is a transition when the melt initially preserves semiconducting properties but changes to a metal as the temperature is further increased; a typical representative of this type is tellurium [36].

To understand the different behaviour of different semiconductors upon melting, it is informative to remember that by quenching a melt only those materials that are good glass formers can be prepared in the amorphous state. Other materials, such as silicon, can be made amorphous (e.g. using sputtering or ion implantation) but cannot be prepared in the glassy state. This difference is fundamental and establishes a link between the structure (and properties) of a solid amorphous material with its melt.

A glass that has been derived from a melt and preserves essential elements of its structure naturally reverts to the same state upon melting. Good semiconducting glasses are thus expected to preserve semiconducting properties in the molten state.

On the other hand, the inability to freeze the disordered molten (metallic) phase of Si or GaAs suggests that the reverse transition is also not straightforward.

In this respect we would like to remind the reader that one of the criteria for a material to be a glass former is a relatively large concentration of lone-pair electrons [17] that act to form weaker interchain/interlayer bonds. The presence of bond energy hierarchy can be crucial for the melting behaviour. Thus, in a good glass former, melting is a result of rupture of weaker interchain/interlayer bonds preserving the covalent nature of the stronger covalent bonds and hence the semiconducting nature of the melt. The role of lone-pair electrons in this process has been discussed in detail in [17].

In materials such as Si or in III–V compounds, on the other hand, all bonds are characterised by the same energy. Melting thus results in the destruction of the covalent network proper and subsequent loss of semiconductor properties.

Both chalcogenide glasses [1] and, possibly, also phase-change materials [37, 38] retain their semiconductor properties upon melting.

1.4 Structure of Prototypical Chalcogenides

Many chalcogenides are molecular solids. In this class of materials, strong—primarily covalent—bonds coexist with weak—intermolecular, primarily Van-der-Waals bonds. As a result, the cohesive molecular grouping is formed when internally strongly bound fragments interact via weak forces. Depending on the extent of macroscopic extension of the covalently bonded unit one can introduce a concept of network dimensionality. Thus sulphur that consists of S_8 molecules is zero-dimensional, selenium that is built of extended Se chains is one-dimensional, and $As_2S(e)_3$ possesses a two-dimensional layered structure. Ge-based materials may possess three-dimensional networks. In what follows we provide the reader with more details about the structure of typical amorphous chalcogenides.

1.4.1 Elemental Chalcogens

We start this section with the structure of selenium. Selenium is a Group VI element possessing two *s*-electrons and four *p*-electrons. Two of the *p*-electrons are located at p_x and p_y orbitals and are used to form covalent bonds with the neighbouring atoms. The remaining two form a non-bonding pair located at p_z orbital and are usually called lone-pair electrons. Since *p* orbitals are orthogonal, the bonding angle subtended by Se atoms is rather close to 90 degrees with the lone-pair electrons occupying the orbital directed perpendicular to the plane formed by p_x and p_y orbitals. The structure of selenium can be thus viewed to zeroth approximation as nearly cubic (Fig. 1.12).

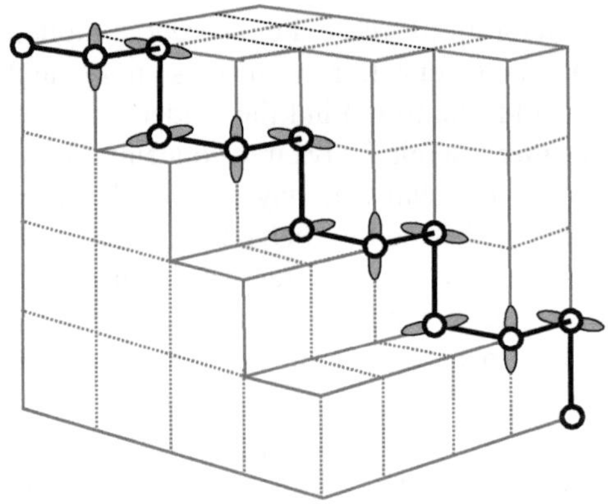

Fig. 1.12 Schematics of a selenium zig-zag chain in a simple cubic lattice. To obtain the real structure, intrachain atom distanced have to be decreased and interchain atom distances increased with a concomitant increase in the bond angle

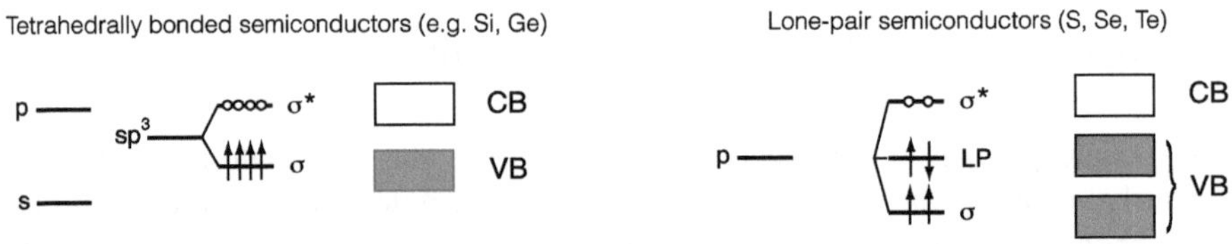

Fig. 1.13 Bonding in tetrahedrally bonded (*left*) and lone-pair (*right*) semiconductors

Kastner was the first to point out [39] the important role of lone-pair electrons in the chemistry of chalcogenides. In particular, he noted that while in tetrahedrally bonded semiconductors such as Si or Ge the hybridised sp^3 orbitals split into bonding (σ) and antibonding (σ^*) molecular states that are subsequently broadened into the valence and conduction bands, respectively in the solid, in chalcogens (S, Se, Te), the s states lie well below the p states and need not be considered. Because only two of the three p orbitals can be utilised for bonding, one normally finds chalcogens in twofold coordination. This leaves one non-bonding electron pair, termed lone pair. In a solid, these unshared, or lone-pair, electrons form a (LP) band near the original p-state energy. The bonding (σ) and anti-bonding (σ^*) bands are split symmetrically with respect to this reference energy. Both the (σ) and lone-pair bands are occupied. As a result, the bonding band is no longer at the top of the valence band; this role is played by the lone-pair band (Fig. 1.13). To account for this rather special situation Kastner suggested to use the term 'lone-pair semiconductors' when referring to chalcogenides.

Depending on the dihedral angle sign sequence (sense of rotation), the Se atoms can either form extended Se_N chains (for $-, -, -, -, -,$ etc. and $+, +, +, +, +, +, +, +,$ etc) or eight-membered rings (for $+, -, +, -, +, -, +, -$) as illustrated in Fig. 1.14.

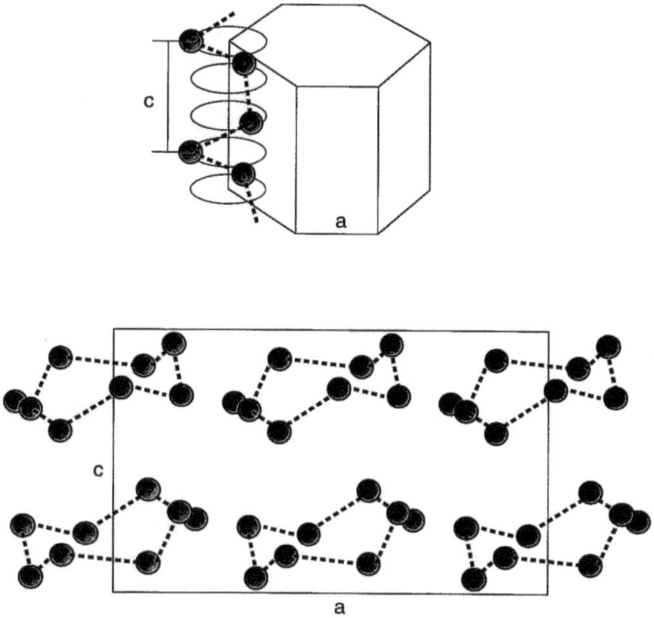

Fig. 1.14 Trigonal (*top*) and monoclinic (*bottom*) phases of selenium

The former modification is called trigonal (or hexagonal) selenium and the latter is monoclinic selenium. Bonding within a chain/ring is covalent. Bonding between the chains/rings is much weaker and is generally considered to be of van-der-Waals nature with a certain covalent component, although in several publications it has also been referred to as resonant [40, 41]. It should be noted that although the intermolecular interaction is significantly weaker, it is this interaction that is responsible for the long-range order in crystalline selenium. It may also be of interest to note that although sulphur and tellurium atoms have the same electronic configuration of the outer shell, the lighter sulphur atoms tend to form rings while the heavier tellurium atoms favour chains.

In amorphous chalcogens, intermolecular interaction is much weaker and the structure can be viewed as random arrangement of (possibly entangled) chains. The average coordination number in amorphous selenium slightly exceeds two [17, 42, 43] suggesting that a certain amount of Se atoms is more than twofold coordinated.

1.4.2 Chalcogenide Glasses

Binary chalcogenides are usually grouped into As-containing and Ge-containing materials. As-chalcogenides with an As:Chalcogen ratio of 2:3 such as $As_2S(e)_3$ possess the layered structure shown in Fig. 1.15. Crystals with 1:1 ratio of arsenic to

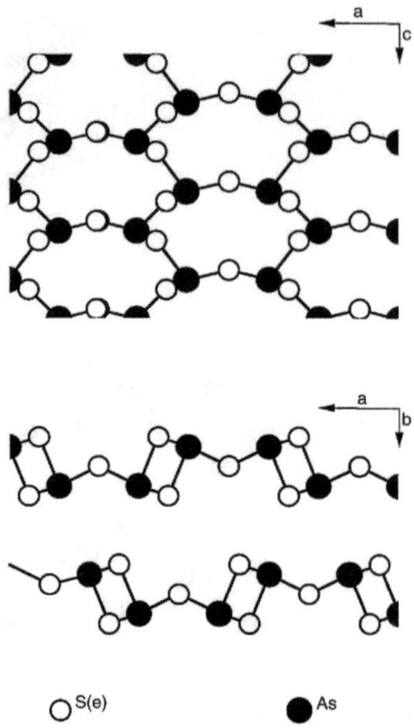

Fig. 1.15 The layered structure of As_2S_3

chalcogen are built of "spherical" $As_4S(e)_4$ molecules [44, 45] as shown in Fig. 1.16. Similar to the case of elemental chalcogens, bonding within layers (molecules) is covalent with the mean As-Se bond lengths of 2.38 Å and As-As bond lengths of 2.44–2.56 Å [44, 45] and the layers (molecules) are kept together by weaker interactions.

Ge-chalcogenides such as $GeS(e)_2$ possess a $GeS(e)_4$ tetrahedron as the main building block as shown in Fig. 1.17. The tetrahedra can be cross-linked either at corners or at edges, the former arrangement is called corner-sharing and the latter is called edge-sharing.

1.4.3 Phase-Change Materials

Structure of phase-change materials, such as Ge-Sb-Te alloys, in both crystalline and amorphous states have recently been very intensely investigated both experimentally and using ab initio simulations. This vast and rapidly developing subject is discussed in detail in Part III.

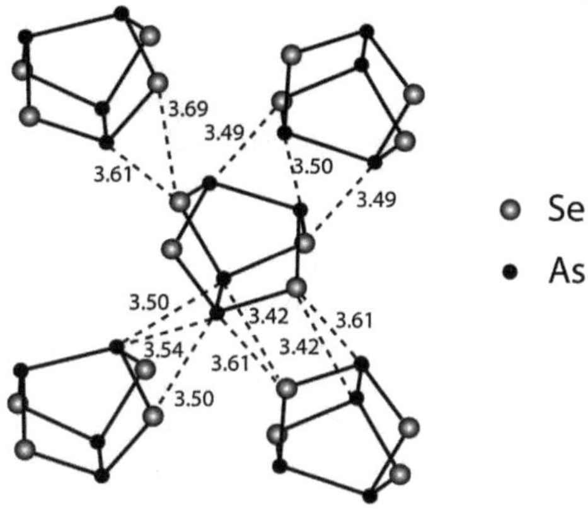

Fig. 1.16 The intermolecular arrangement in the low-temperature crystalline form of c-$As_{50}Se_{50}$ (As_4Se_4) showing close interatomic distances (in Å). Reprinted from Goldstein and Paton [44] by permission from IUCr

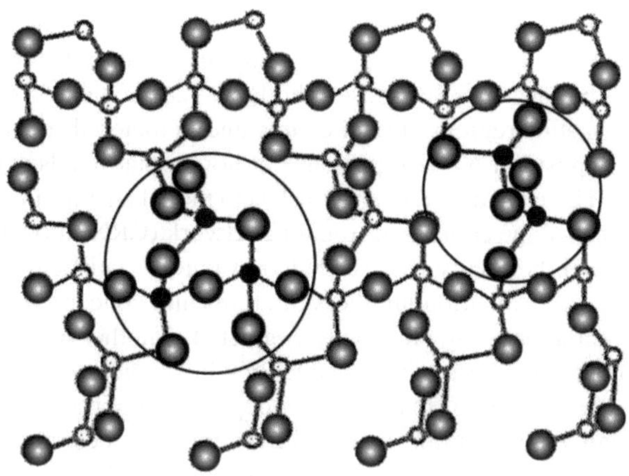

Fig. 1.17 Structure of crystalline $GeSe_2$ (*small circles* Ge, *large circles* Se). Both corner-sharing and edge-sharing $GeSe_4$ tetrahedra can be seen (*encircled*)

1.5 Defects

1.5.1 Ideal Glass and the "8–N" Rule

Talking about defects in amorphous solids which, in a sense, are totally defective may seem strange at the first glance. To establish a starting point, we should give a definition of a non-defective, ideal glass. We shall call *a glass in which all atoms satisfy their valence requirements an ideal glass*. Valence in this context is defined to mean the number of single (σ-type) covalent bonds necessary for the atom to complete, via the shared electrons comprising the bonds, its outer shell of s and p electrons. Since the complete shell contains 8 electrons, the atoms generally obey the "8–N" rule when $N > 4$, where N is the number of valence electrons and the coordination number is given by 8–N. For example, in the case of As_2Se_3 this means that all arsenic atoms are threefold coordinated and selenium atoms are twofold coordinated. The 8–N rule was proposed for chalcogenides glasses in order to account for the observation that chalcogenide glasses were insensitive to doping. This rule is often referred to as the Mott rule.[2]

We would like to note here that while the terms "Mott rule" and "8–N rule" are often used interchangeably, originally Mott proposed that in order to account for insensitivity of amorphous chalcogenides to doping [46], the atoms "use all their valence electrons in bonds with surrounding atoms or ions, modifying the local coordination to make this possible". The "8–N" formulation is a later attempt to quantify this requirement. Satisfaction of the covalent bonding requirements locally minimises the energy and thus allows one to obtain the total energy of the glass closest to that of the corresponding crystal in agreement with the Zachariansen's criteria (cf. Sect. 1.2.2) of glass formation.

In addition, in order to possess the short-range order similar to that of the corresponding crystal, local stoichiometry should be preserved; that is, arsenic species should only have selenium species as the first nearest neighbours and vice versa; in other words no "wrong" homopolar bonds, such as As-As or Se-Se bonds, exist in the ideal As_2Se_3 glass. The latter requirement is often loose because the glass-formation regions are quite large and in materials deviating from the stoichiometric composition As_2Se_3, presence of homopolar bonds is inevitable.

There are different approaches to model an amorphous structure. One such model is called the continuous random network (CRN) model. As the name implies, in this model the bonds are distributed in purely statistical terms, determined only by the local coordination numbers of atoms and concentrations of constituents. Any preferential ordering effects are neglected. Another often used model is the chemically ordered network (CON) model in which heteropolar bonds are favoured.

It may be fair to say that as-deposited amorphous films that are fabricated under highly non-equilibrium conditions are better described by the CRN model while for well-annealed films and bulk glasses the CON model may be more appropriate.

1.5.2 Topological Defects

Dangling Bonds

Deviations from the 8–N rule give rise to topological defects in the CRN network of the glass. The most straightforward kind of a topological defect would be atoms with unsatisfied, or broken, or dangling bonds. Thus any open S_8 ring will generate two dangling bond defects. Extended chains that are the building blocks of selenium also possess two dangling bonds at their ends. Each dangling bond contains one electron

[2] Sir Nevill F. Mott, P. W. Anderson and J. H. Van Vleck won the Nobel Prize in Physics in 1977 for their work on the electronic structure of magnetic and disordered systems, especially amorphous semiconductors.

and is electrically neutral. Such defects should be present in large quantities in an amorphous network consistent with the fact that the Fermi level in chalcogenide glasses is pinned near the middle of the gap [1]. At the same time, in amorphous chalcogenides a dark electron spin resonance (ESR) signal is usually not observed (with the exception of Ge-S glasses) whereas a large concentration of defect states has been demonstrated by techniques such as photoluminescence [47] and alternating current (ac) conductivity [48]. This inconsistency suggested that dangling bonds, while easy to imagine, do not form the major kind of topological defects in amorphous chalcogenides.

Negative-U Centres

To account for the above controversy, Anderson postulated [49] that, due to a strong electron–phonon coupling, electron pairing becomes energetically favourable despite the endothermic cost in energy required to place two electrons on one site. Such defects are called negative correlation energy, or negative-U, centres. This model was applied by Street and Mott to a case example of an elemental chalcogen [50]. They argued that the majority of defects in chalcogens are pairs of positively and negatively charged dangling bonds:

$$2D^0 \longrightarrow D^+ + D^- \tag{1.1}$$

where D stands for a dangling bond and the superscripts describe the charge. The resulting system is diamagnetic (i.e. all electrons are paired), which explains the absence of an ESR signal.

Valency Alternation Pairs

In the model subsequently put forward by Kastner et al. [51], usually known as the *valence-alternation pair* (VAP) model or Kastner–Adler–Fritzsche model, the formation of over-coordinated defects through involvement of lone-pair electrons was suggested. The essence of this idea is that an empty orbital of a positively charged dangling bond interacts with lone-pair electrons of a neighbouring chain, forming a threefold coordinated defect and lowering the energy of the system:

$$C_1^+ + C_2^0 \longrightarrow C_3^+ \tag{1.2}$$

where C stands for chalcogen, the subscripts describe the coordination, and the superscripts correspond to the charge. This process is demonstrated in Fig. 1.18.

The gain in energy due to the formation of an extra bond was considered to be the driving force for compensating the energy cost for creation of the doubly occupied site at the negatively charged dangling bond. The energy levels of various defect configurations that may exist in a-Se are schematically shown in Fig. 1.19. Within

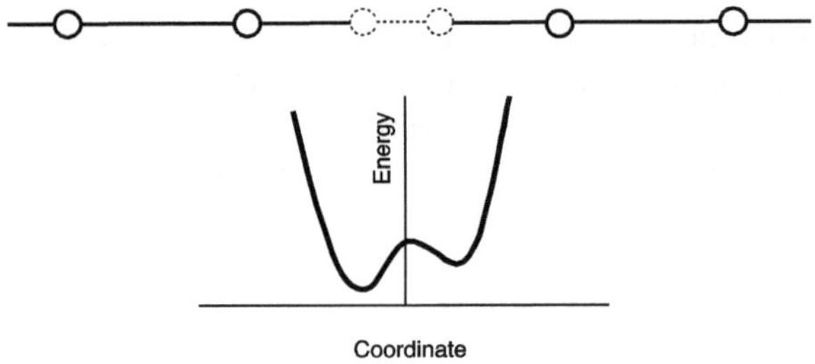

Fig. 1.20 Soft configurations that generate almost flat (double-well) potential reliefs

such soft configurations is characterised by double-well potentials when the central atom is usually closer to one of its neighbours (Fig. 1.20).

It has been argued that vibrational atomic motions in most of the soft configurations may also determine the dynamics of glasses at higher temperatures, e.g. interactions of electrons with soft configurations can result in the formation of negative-U centres [58]. The concept of soft configurations has also been applied to interpret the effect of pressure on chalcogenide glasses [59].

1.6 Electronic Structure

The existence of the bandgap in crystalline semiconductors is derived in most textbooks on solid-state physics on the grounds of perfect periodicity of crystals. In the presence of a perfectly periodic array of atoms, the eigenstates of the one-electron Hamiltonian can be chosen to have the form of a plane wave times a function with the periodicity of the Bravais lattice of the crystal. This statement is known as Bloch's theorem. The electrons propagating in crystals are sometimes referred to as Bloch electrons (in contrast to free electrons). The effect of a periodic potential can be described using the concept of the effective mass when the motion of an electron in a crystal can be treated identical to that of a free particle but with the effective mass, m^*, used instead of the free-electron mass. What happens in an amorphous solid where the atomic arrangement is not periodic? What is the origin of the bandgap in a glass? How does disorder effect the motion of charge carriers?

1.6.1 Origin of the Bandgap

On the basis of earlier studies of electronic transport in liquids, Ioffe and Regel [60] postulated that the "fundamental electronic property of a solid, i.e. whether it is a metal, semiconductor or insulator, is determined principally by the short-range order," that is, the nature of chemical bonding and the coordination number. This changed the paradigm that existed before and related the existence of the bandgap

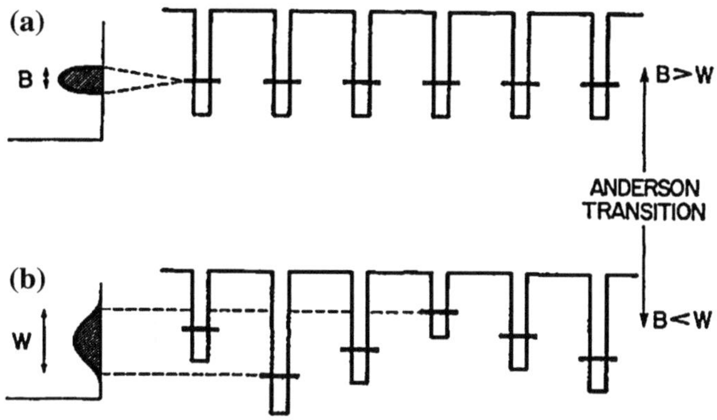

Fig. 1.21 Schematic illustration of the Anderson localisation. When the "width" of the disorder W exceeds the overlap bandwidth B, disorder-induced localisation takes place. Reproduced from Zallen [69] with permission by Wiley-VCH Verlag GmbH

to the periodicity of the crystal. It was later demonstrated [61] using a simple model Hamiltonian that a gap is, indeed, expected for an ideal tetrahedrally coordinated amorphous solid, provided that the interactions are of a certain magnitude.

Because the local short-range structure in an amorphous material is almost the same as in the corresponding crystal, the overall electronic structures are also similar. Small deviations in the bond lengths and angles as well as other structural imperfections result in broadening of the band edges making them not as abrupt as in crystals. An exception from this rule is phase-change materials that are characterised by pronounced differences between the local structure in the crystalline and amorphous phases.

1.6.2 Anderson Localisation

How does disorder effects transport properties? This issue was first addressed by Anderson in his famous paper "Absence of diffusion in certain random lattices" [62]. Anderson demonstrated that in a lattice with randomly distributed depths of potential wells (Fig. 1.21) the electron wave function becomes localised when the distribution of the depth of the random potential W becomes larger than the characteristic bandwidth B in the absence of disorder. Thus, while for small W/B the electronic states are delocalised (as in a crystal), for a sufficiently large disorder parameter W/B the states in the valence band are localised [2]. Such localisation of the electron wave function is called Anderson localisation.

The concept of disorder-induced localisation was subsequently extended [63, 64]. When Anderson's localisation criterium is satisfied, *all* of the states in the band are localised. In particular, it was pointed out that smaller degrees of disorder lead to

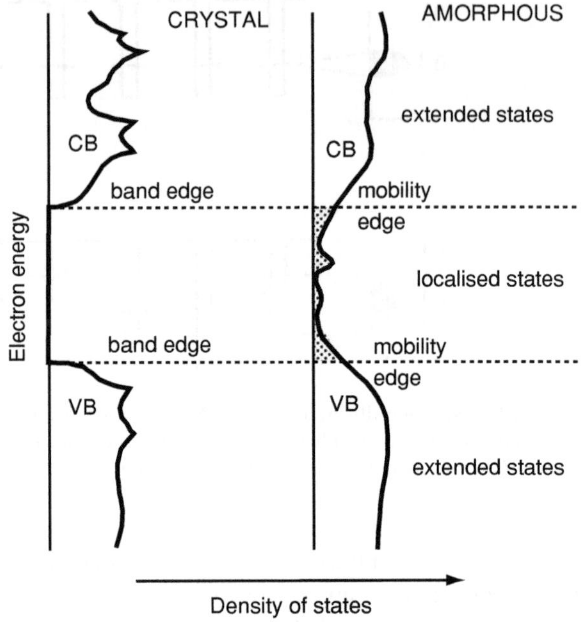

Fig. 1.22 Density of states in a crystal (*left*) and in the corresponding amorphous solid (glass) (*right*)

localisation of the states at edges, or tails, of the bands. (For historical fairness we would like to note that a couple of years earlier Gubanov conjectured [65] that states near the edges of conduction and valence bands are localised.) For energies on one side of the localisation threshold, the states are delocalised while for those on the other side of the threshold the states are localised. This means that deep in the valence and/or conduction bands the carriers are delocalised (extended) but the states in the band tails are localised [62] with a localisation radius of typically a few bond lengths. The boundary between the localised and delocalised states is usually referred to as a *mobility edge*. States at band edges are called tail states. The related value of mobility gap replaces the concept of bandgap used for crystals when transport properties are described. Comparison between the density of states (DOS) in an amorphous solid and the corresponding crystal is shown in Fig. 1.22.

Although the number of states in the tails is much smaller than the number of extended states, they play a very important role in transport and recombination of non-equilibrium charge carriers. Intensive efforts are therefore made to understand the nature and DOS in the gap of amorphous semiconductors.

It is interesting to note that Anderson localisation has also been proposed to account for the electrical properties of *crystalline* phase-change alloys [66].

1.6.3 Percolation

Another way to describe localisation of carriers is using the percolation approach. Disorder in atomic arrangement present in an amorphous solid causes local fluctuations in the electron potential. This results in creation of potential energy pockets of different heights (Fig. 1.23, upper panel). Electrons with a certain energy E can only reside in regions with an energy lower than E. At low energy, the allowed energy regions are isolated and small. As the energy is increased, the isolated regions grow larger and eventually merge. Finally, at a critical energy a continuous channel is formed that allows an electron to travel throughout the entire volume. This threshold energy is called *percolation level* or *percolation threshold.*

An easy way to visualise this concept is to consider water filling a landscape with hills and mountains of different height and shape. For the low water level (low electron density), water is localised in small reservoirs. As the water level rises, one can travel extended distances and finally as the water level rises above the percolation level one is free to go into the "open" sea (Fig. 1.23, lower panel).

One can also visualise the percolation threshold using the vandalised grid concept [2]. Imagine measuring conductivity of a metallic grid as shown in Fig. 1.24. If one cuts several segments interconnecting the grid nodes, the conductivity of the grid remains essentially unchanged. As this process continues, at a certain stage there will be no continuous conductive path between the electrodes and the conductivity drops to zero. Depending on whether nodes or bonds are eliminated from the grid, site and bond percolation can be introduced, respectively. The percolation thresholds are usually sharp and occur at $p_c \approx 0.5$ fraction of a conductive phase in two dimensions but only at $p_c \approx 0.15$ in three dimensions [68]. In one dimension it is, of course, 1.00.

1.7 Fabrication of Amorphous Semiconductors

The general idea of making an amorphous material is based on the fact that the amorphous state is thermodynamically less favourable than the crystalline state; that is, it is characterised by a greater free energy. There are various ways to fabricate amorphous materials.

1.7.1 Melt Quenching

The oldest technique to produce amorphous solids, especially bulk glasses, is melt quenching. In practice, using this method, a melt is rapidly cooled down by either simply turning off the oven, or by dropping an ampule with the melt into ice-cold water or liquid nitrogen depending on the required cooling rate. For most chalcogenides,

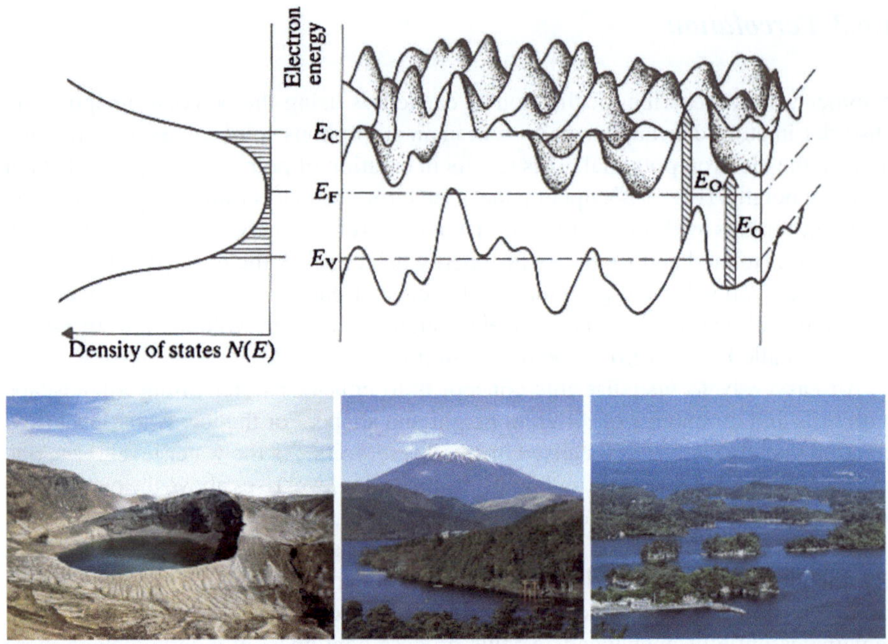

Fig. 1.23 *Upper panel* Symmetric part of long wavelength potential fluctuations. E_0 is an average optical gap, E_c and E_v are mobility edges or percolation thresholds. Reprinted from Fritzsche [67] with permission from Elsevier. *Lower panel* Visualisation of the percolation threshold: as the water level (electron energy) increases, individual reservoirs (*left* Okama crater lake, Japan) grow in size and merge (*centre* Fuji Five Lakes, Japan), eventually forming continuous pathways (*right* Matsushima Bay, Japan)

water or ice cooling are sufficient to generate the glassy phase. In extreme cases, the so-called splat cooling is used when a small droplet of the melt falls between two rotating drums that are cooled down using liquid nitrogen. This method allows one to obtain cooling rates in the order of 10^6–10^7 K/s and is usually used to produce amorphous metals.

1.7.2 Vapour Deposition

Thermal Evaporation

The melt quenching described above is a technique for obtaining bulk glasses. Thin amorphous films are usually prepared by vapour deposition. The simplest way to prepare a film is by thermal evaporation, in which a material to be evaporated is placed into a boat or crucible and heated in a vacuum. The produced melt evaporates and is collected on a substrate, forming a thin film. The parameter similar to the cooling rate

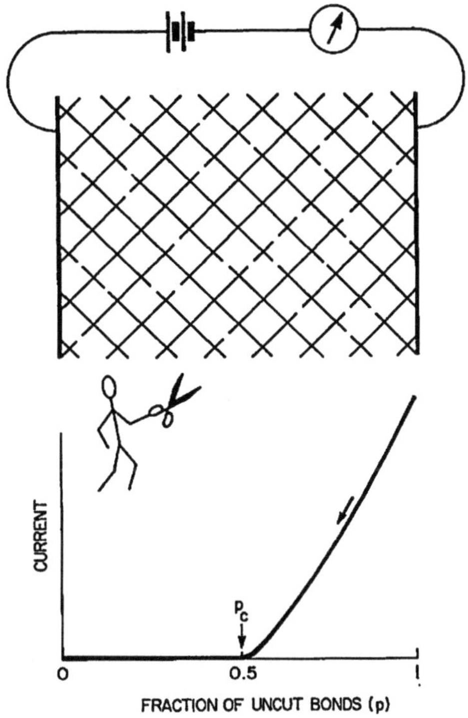

Fig. 1.24 Illustration of the percolation threshold concept. As one randomly cuts segments between nodes of a metallic grid, the conductivity drops to zero at the percolation threshold. Reproduced from Zallen [69] with permission by Wiley-VCH Verlag GmbH

in melt quenching is the deposition rate (whereas the substrate temperature is also important). The faster the film is deposited, the more "amorphous" it is. Sulphur- and Se-based chalcogenides, being good glass formers, usually do not impose strict requirements on the deposition rate.

Another parameter that makes these materials very suitable for this technique is their low melting points. The latter increases as one goes from elemental chalcogens to As- and Ge-based chalcogenides [70]. The method is ideal for preparing films of materials that melt congruently. If, however, it is used to evaporate a glass whose constituents have much different melting temperatures the resulting film has a composition (i) different from that of the bulk source and (ii) varying throughout its thickness.

Flash Evaporation

To ensure "simultaneous" evaporation of multicomponent glasses, a modification of this technique known as flash evaporation is used. In this method, a fine powder of the starting material is prepared and then dropped in small quantities onto a very hot filament. Pieces of powder "burst" and the material is transferred onto a substrate whereby the composition of the starting bulk is well preserved.

Since in the flash evaporation process the particle "explodes" into fragments that are large enough to preserve the composition of the starting bulk material it is not unreasonable to think that the structure of the initial materials is also partially preserved, i.e. the structure of films prepared by flash evaporation should be closer to that of the bulk glass that obtained by thermal evaporation when much smaller fragments of the materials are transferred through the gaseous phase.

At the same time it should be noted that the high temperature of the filament leads to radiation heating of the substrate and, for example, in the case of As-S glasses, results in partial re-evaporation of sulphur thus causing a change in the composition. It may also lead to partial crystallisation of films with a low crystallisation temperature.

Pulsed laser deposition is another technique that allows for regular stoichiometric transfer of target material to the films, and offers a possibility to fabricate films of unusual compositions [71].

Sputtering

Sputtering is a process whereby the target material is ablated by bombardment with energetic ions from an electrical plasma struck in a gas. Ejected material is transferred to the substrate in the form of ionised atoms or clusters of atoms.

If an inert gas is used the process is known as physical sputtering where the target material is simply physically transferred to the substrate. One of the main advantages of sputtering is that most elements have rather similar sputtering rates which allows one to obtain films with (approximately) the same composition as that of the target. The deposition rates, however, are usually quite slow, in the order of 1–10 Å/s. If a reactive gas is included in the sputtering gas, chemical sputtering may occur and the resulting films are a synthesis of the target and the gas. Chemical sputtering can be used, for example, to make Si_3N_4 films by sputtering a silicon target using a N_2-containing gas.

The simplest way to induce sputtering is to apply a large d.c. voltage to the target, thereby attracting positively charged ions from the plasma struck in the sputtering gas that sputter away the target surface. However, this technique can only be used to sputter metals or other conductive materials when the target can act as an electrode. D.c. sputtering does not work for insulators. In order to sputter poorly conductive materials, an a.c. (usually 13.56 MHz radio frequency) field is applied between a non-conductive target and an electrode that is connected to the Earth (ground).

Magnetron Sputtering

The technique was developed to improve the film deposition rate of dc- and rf-sputtering. In normal sputtering, plasma is created between two electrodes, i.e. a target and a counter electrode with a substrate. However, the concentration of ions and electrons generated from the gas (Ar or Kr) injected to hold the plasma is rather low. In order to increase their concentration, two magnets with a strong magnetic field are placed under the target. The target is usually 3–5 mm thick, the magnetic field penetrates the target and generates a field on the surface. Electrons generated by the plasma are accelerated by the oscillating electric field and have a circular momentum along the magnetic flux. Therefore, a large number of electrons is trapped at the target and effectively collide with neutral gas atoms, resulting in generation of more ions available for the target bombardment.

Helicon-Wave Sputtering

Helicon-wave sputtering has been specially designed for fabricating atomically flat films. In dc- and rf-sputtering, it is difficult to maintain a plasma at low pressure (<0.1 Pa). In addition, the distance between the target and the electrode is usually set at 40–50 mm to keep the plasma stable. Under such conditions, the surface of the deposited film and substrate is irradiated by hot electrons and recoiled ions with a high energy. Therefore, the film surface is not flat. To obtain flatness comparable to that achievable using epitaxial growth techniques, the target-substrate distance should be long enough. In helicon-wave sputtering, an additional generator is connected to a helical antenna, which is placed just above the target in order to stabilise the plasma under low pressure. Helicon-wave sputtering has been successfully used for deposition of atomically controlled phase-change layers [72, 73].

Chemical Vapour Deposition

Chemical vapour deposition (CVD) is a chemical process used to produce high-purity, high-performance solid materials. The process is often used in the semiconductor industry to produce thin films. In a typical CVD process, the wafer (substrate) is exposed to one or more volatile precursors, which react and/or decompose on the substrate surface to produce the desired deposit. In some cases, volatile by-products are also produced, which are removed by gas flow through the reaction chamber. CVD is especially useful for applications where conformal coverage of surfaces that contain features with high aspect ratio. It can be used for fabrication of both chalcogenide glasses and phase-change materials [74, 75].

1.7.3 Spin Coating

Spin coating is a cheap technique ideal for mass production and consists of the following. The composition to be deposited is dissolved in an appropriate solvent and a small amount of the solution is deposited onto a rapidly rotating (spinning) substrate. Due to centrifugal forces the solution is uniformly distributed on the surface and after evaporation of the solvent a thin amorphous film of the starting material is formed.

Various chalcogenide glasses have been successfully fabricated using this technique [76]. However, not all compositions can be deposited using this method, the main problem being a choice of an appropriate solvent. Thus as regards phase-change materials the available literature reports are limited to GeSbSe [77].

1.7.4 Ion Implantation

One can also fabricate amorphous layers by ion bombardment where exposure of crystals to energetic ions results in their disordering. By tuning the ion energy and/or using a combination of implantation energies one can generate rather uniform amorphous layers [78–80]. While this method is rather expensive and not commonly used, it has an advantage of being able to fabricate small localised amorphous regions with high spatial resolution and also to control precisely the doping level [81–83].

References

1. N.F. Mott, E.A. Davis, *Electronic Processes in Non-Crystalline Materials*, 2nd edn. (Clarendon Press, Oxford, 1979)
2. R. Zallen, *The Physics of Amorphous Solids* (Wiley, New York, 1983)
3. S.R. Elliott, *Physics of Amorphous Materials*, 2nd edn. (Longman, London, 1984)
4. K.D. Tsendin (ed.), *Electronic Phenomena in Chalcogenide Glassy Semiconductors* (in Russian) (Nauka, St. Petersburg, 1996)
5. J. Singh, K. Shimakawa, *Advances in Amorphous Semiconductors* (Taylor and Francis, London, 2003)
6. M. Popescu, *Non-Crystalline Chalcogenides* (Kluwer, Dordrecht, 2002)
7. K. Tanaka, K. Shimakawa, *Amorphous Chalcogenide Semiconductors and Related Materials* (Springer Science + Business Media, Berlin, 2011)
8. A. Popov, *Disordered Semiconductors* (Pan Stanford, Singapore, 2011)
9. W. Kauzmann, Chem. Rev. **43**, 219 (1948)
10. C.A. Angel, in *Insulating and Semiconducting Glasses*, ed. by P. Boolchand (World Scientific, Singapore, 2000), p.1
11. R. Blachnik, A. Hoppe, J. Non-Cryst. Solids **34**, 191 (1979)
12. N. Clavaguera, M.T. Clavaguera-Mora, S. Surinnach, M. Baro, J. Non-Cryst. Solids **104**, 283 (1988)
13. G. Das, N. Platakis, M. Bever, J. Non-Cryst. Solids **15**, 30 (1974)
14. A. Feltz, *Amorphous Inorganic Materials and Glasses* (VCH, Weinheim, 1993)
15. W. Zachariansen, J. Amer. Chem. Soc. **54**, 3841 (1932)
16. R.L. Muller, *Vitreous State* (in Russian) (Leningrad State University, St. Petersburg, 1960)
17. S.A. Dembovsky, E.A. Chechetkina, *Glass Formation* (in Russian) (Nauka, Moscow, 1990)
18. S.R. Ovshinsky, Phys. Rev. Lett. **21**, 1450 (1968)
19. S. Raoux, M. Wuttig (eds.), *Phase Change Materials: Science and Applications* (Springer, Berlin, 2008)
20. A.D. Pearson, W.R. Northover, J.F. Dewald, W.F. Peck, *Advances in Glass Technology* (Plenum Press, New York, 1962), p. 357
21. B.T. Kolomiets, E.A. Lebedev, Radiotechnika i Elektronika (in Russian) **8**, 2097 (1963)

22. M. Okuda (ed.), *Technology and Materials for Future Optical Memories* (in Japanese) (CMC, Tokyo, 2004)
23. M. Salinga, J. Kalb, M. Klein, T. Sontheimer, F. Spaepen, M. Wuttig, in *European Phase Change and Ovonics Science Symposium*, Zermatt, 2007
24. N.K. Abrikosov, G.T. Danilova-Dobryakova, Izv. Akad. Nauk SSSR, Neorg. Mater. **1**, 204 (1965)
25. N.K. Abrikosov, G.T. Danilova-Dobryakova, Izv. Akad. Nauk SSSR, Neorg. Mater. **6**, 475 (1970)
26. N. Yamada, E. Ohno, N. Akahira, K. Nishiuchi, K. Nagata, M. Takao, Jpn. J. Appl. Phys. **26** (Suppl. 26-4), 61 (1987)
27. S. Raoux, H.Y. Cheng, M.A. Caldwell, H.S.P. Wong, Appl. Phys. Lett. **95**, 071910 (2009)
28. J. Orava, A.L. Greer, B. Gholipour, D.W. Hewak, C.E. Smith, Nature Mater. **11**, 279 (2012)
29. R. Kojima, S. Okabayashi, T. Kashihara, K. Horai, T. Matsunaga, E. Ohno, N. Yamada, T. Ohta, Jpn. J. App. Phys. **37**, 2098 (1998)
30. B.T. Kolomiets, N.A. Goryunova, Zh. Tekh. Fiz. **25**, 2069 (1955)
31. B.T. Kolomiets, N.A. Goryunova, Zh. Tekh. Fiz. **25**, 984 (1955)
32. J.M. Ziman, *Models of Disorder* (Cambridge University Press, Cambridge, 1979)
33. M. Popescu, Ph.D. Thesis, Bucharest, 1975
34. J. Phillips, J. Non-Cryst. Solids **34**, 153 (1979)
35. G. Lucovsky, F.L. Galeener, J. Non-Cryst. Solids **37**, 53 (1980)
36. V.M. Glazov, S.N. Chizhevskaya, N.N. Glagoleva, *Liquid Semiconductors* (Plenium Press, New York, 1969)
37. A.V. Kolobov, P. Fons, M. Krbal, R.E. Simpson, S. Hosokawa, T. Uruga, H. Tanida, J. Tominaga, Appl. Phys. Lett. **95**, 241902 (2009)
38. R. Endo, S. Maeda, Y. Jinnai, R. Lan, M. Kuwahara, Y. Kobayashi, M. Susa, Jpn. J. Appl. Phys. **49**, 5802 (2010)
39. M. Kastner, Phys. Rev. Lett. **28**, 355 (1972)
40. G. Lucovsky, R. White, Phys. Rev. B **8**, 660 (1973)
41. K. Shportko, S. Kremers, M. Woda, D. Lencer, J. Robertson, M. Wuttig, Nature Mater. **7**, 653 (2008)
42. A.V. Kolobov, H. Oyanagi, Ka. Tanaka, K. Tanaka, Phys. Rev. B **55**, 726 (1997)
43. D. Hohl, R.O. Jones, Phys. Rev. B **43**, 3856 (1991)
44. P. Goldstein, A. Paton, Acta Cryst. B **30**, 915 (1974)
45. A.L. Renninger, B.L. Averbach, Acta Cryst. B **29**, 1583 (1973)
46. N.F. Mott, Adv. Phys. **16**, 49 (1967)
47. R.A. Street, Adv. Phys. **25**, 397 (1976)
48. S.R. Elliott, Philos. Mag. B **37**, 553 (1978)
49. P.W. Anderson, Phys. Rev. Lett. **34**, 953 (1975)
50. R.A. Street, N.F. Mott, Phys. Rev. Lett. **35**, 1293 (1975)
51. M. Kastner, D. Adler, H. Fritzsche, Phys. Rev. Lett. **37**, 1504 (1976)
52. M. Kastner, H. Fritzsche, Philos. Mag. B **37**, 199 (1978)
53. N.A. Popov, JETF Lett. **31**, 409 (1980)
54. S.A. Dembovsky, Mat. Res. Bull. **16**, 1331 (1981)
55. D.J. Treacy, U. Strom, P.B. Klein, P.C. Taylor, T.P. Martin, J. Non-Cryst. Solids **35**, 1035 (1980)
56. V.G. Karpov, M.I. Klinger, F.N. Ignatev, Sov. Phys. JETP **57**, 439 (1983)
57. M.I. Klinger, Solid State Commun. **45**, 949 (1983)
58. M.I. Klinger, S.N. Taraskin, Phys. Rev. B **52**, 2557 (1995)
59. M.I. Klinger, J. Non-Cryst. Solids **232**, 257 (1998)
60. A.F. Ioffe, A.R. Regel, *Progress in Semiconductors*, vol. 4 (Wiley, New York, 1960), pp. 237–291
61. D. Weaire, M.F. Thorpe, Phys. Rev. B **4**, 2508 (1971)
62. P.W. Anderson, Phys. Rev. **109**, 1492 (1958)
63. N.F. Mott, E.A. Davis, Philos. Mag. **17**, 1269 (1968)
64. M.H. Cohen, H. Fritzsche, S.R. Ovshinsky, Phys. Rev. Lett. **22**, 1065 (1969)

65. A.I. Gubanov, *Quantum Electron Theory of Amorphous Conductors* (Consultants Bureau, New York, 1965)
66. T. Siegrist, P. Jost, H. Volker, M. Woda, P. Merkelbach, C. Schlockermann, M. Wuttig, Nature Mater. **10**, 202 (2011)
67. H. Fritzsche, J. Non-Cryst. Solids **6**, 23 (1971)
68. H. Scher, R. Zallen, J. Chem. Phys. **53**, 3759 (1970)
69. R. Zallen, *The Physics of Amorphous Materials* (Wiley-VCH, Weinheim, 2004), pp. 137, 232
70. Z.U. Borisova, *Glassy Semiconductors* (Plenum Press, New York, 1981)
71. M. Frumar, B. Frumarova, P. Nemec, T. Wagner, J. Jedelsky, M. Hrdlicka, J. Non-Cryst. Solids **352**, 544 (2006)
72. J. Tominaga, R. Simpson, P. Fons, A. Kolobov, in *European Phase Change and Ovonics Science Symposium*, Milano, 2010
73. R.E. Simpson, P. Fons, A.V. Kolobov, T. Fukaya, M. Krbal, T. Yagi, J. Tominaga, Nature Nanotech. **6**, 501 (2011)
74. D. Lezal, J. Optoelectron. Adv. Mater. **5**, 23 (2003)
75. J. Lee, H. Park, S. Cho, Y. Park, B. Bae, J. Park, J. Park, H. An, J. Bae, D. Ahn, et al., in *VLSI Technology, 2007 IEEE Symposium on* (*IEEE*), pp. 102–103
76. G.C. Chern, I. Lauks, J. Appl. Phys. **53**, 6979 (1982)
77. D. Milliron, S. Raoux, R. Shelby, J. Jordan-Sweet, Nature Mater. **6**, 352 (2007)
78. G. Azevedo, M. Ridgway, K. Yu, C. Glover, G. Foran, Nucl. Instrum. Meth. B **190**, 851 (2002)
79. M. Ridgway, C. Glover, G. Foran, K. Yu, J. Appl. Phys. **83**, 4610 (1998)
80. M. Ridgway, C. Glover, K. Yu, G. Foran, C. Clerc, J. Hansen, A. Larsen, Phys. Rev. B **61**, 12586 (2000)
81. S. Privitera, E. Rimini, R. Zonca, Appl. Phys. Lett. **85**, 3044 (2004)
82. B. Liu, Z. Song, T. Zhang, J. Xia, S. Feng, B. Chen, Thin Solid Films **478**, 49 (2005)
83. B. Liu, Z. Song, T. Zhang, S. Feng, B. Chen, Appl. Surf. Sci. **242**, 62 (2005)

Chapter 2
Properties of Amorphous Chalcogenides

2.1 Electrical Properties

Electrical properties of amorphous semiconductors have been described in various monographs, e.g. in [1]. As in the previous chapter, we shall only discuss the issues related to the subject of this book.

2.1.1 *Pinning of the Fermi Level*

In contrast to crystals, amorphous chalcogenides do not exhibit any significant change in conductivity upon doping. This fact was first established by Kolomiets [2] and is usually explained by the Mott rule (cf. Sect. 1.5.1).

The temperature dependence of the dark conductivity of chalcogenide glasses follows an exponential dependence (Fig. 2.1) with the activation energy close to $E_g/2$ implying (i) intrinsic-like semiconducting behaviour and/or (ii) a large concentration of gap states that fix the Fermi level at midgap position [1]. This situation is usually called pinning of the Fermi level. Even when impurities are added in concentrations that do change the optical gap, the Fermi level remains pinned close to the centre of the gap. Similar results, i.e. the location of the Fermi level remaining in the middle of the gap independent of doping, have been reported for the amorphous phase of phase-change materials [3].

While under certain circumstances the chalcogenide glasses *can* be doped, the concentration of impurities (usually Bi) is usually larger than 10% [4]; this process is usually referred to as modification.

2.1.2 *P-Type Conductivity*

With just a few exceptions, amorphous chalcogenides are *p*-type semiconductors. The reason for this is not clear [6]. One of the explanations offered was that the

A. V. Kolobov and J. Tominaga, *Chalcogenides*, Springer Series in Materials Science 164, DOI: 10.1007/978-3-642-28705-3_2,

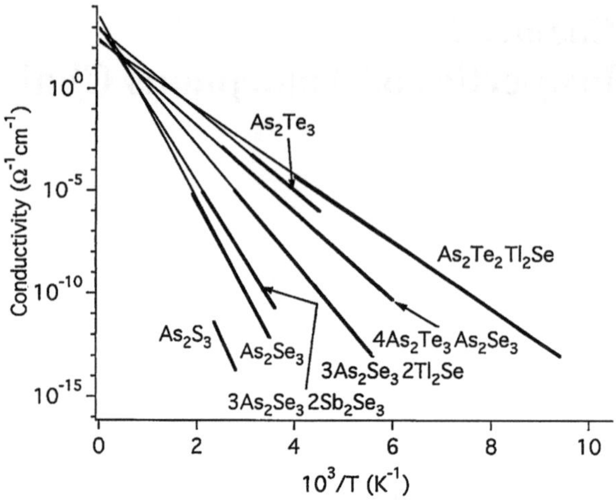

Fig. 2.1 Temperature dependence of dark conductivity of typical chalcogenide glasses (Reprinted with permission by Oxford University Press from Mott and Davis [5])

fluctuations in the conduction band tail are larger than those in the valence band, which leads to stronger localisation of electrons [6]. For crystalline GeTe that is an end point for the quasibinary $Ge_2Sb_2Te_5$-Sb_2Te_3 phase-change alloys it was argued that *p*-type conductivity is due to presence of vacancies on the Ge sublattice [7].

2.1.3 Hall Effect Anomaly

The conductivity type (*p*-type) of chacogenide glasses is determined by thermo-conductivity measurements using the Seebeck effect. Application of Hall effect measurements usually yields an opposite (n-) sign for the conductivity type. This phenomenon is referred to as the anomalous Hall effect and has been interpreted in terms of small polarons [8] or three-site interactions [9]. The Hall effect anomaly has also been observed for amorphous phase-change materials [10].

2.1.4 Phase-Change Materials

Phase-change materials of which $Ge_2Sb_2Te_5$ is a prototypical example, crystallise at temperatures around 160 °C. Crystallisation results in a drastic increase in their conductivity [11, 12] as illustrated by Fig. 2.2.

The temperature dependence of the dark conductivity of $Ge_2Sb_2Te_5$ in the amorphous and crystalline phases is shown in Fig. 2.3 [13]. In addition to the fact that conductivity in the crystalline phase is several orders of magnitude higher than in the amorphous phase, one can also see that while the amorphous and fcc phases are semiconducting (the conductivity increases with temperature), the hexagonal phase is believed to be metallic.

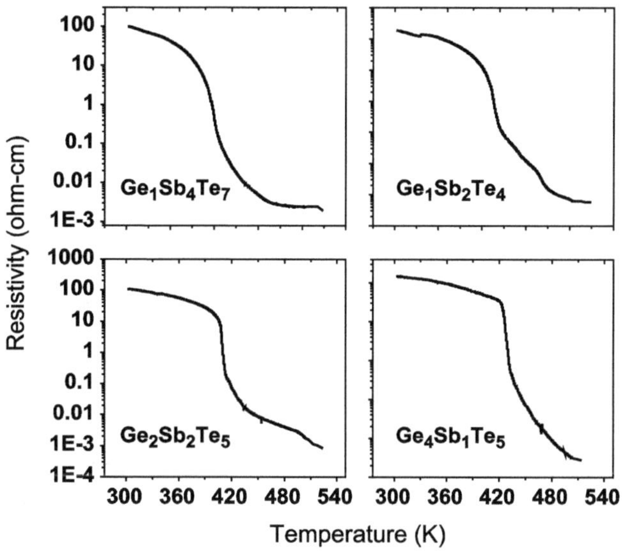

Fig. 2.2 Drastic change in dark conductivity accompanying the crystallisation of Ge-Se-Te alloys. Reprinted from Morales-Sánchez et al. [12] with permission from Elsevier

Fig. 2.3 Temperature dependence of the electrical dark conductivity of $Ge_2Sb_2Te_5$ in amorphous and crystalline (both the stable hexagonal and metastable face-centered cubic) phases. Reprinted with permission from Lee et al. [13]. Copyright 2005 by the American Institute of Physics (The label 'present work' in the figure refers to the cited work)

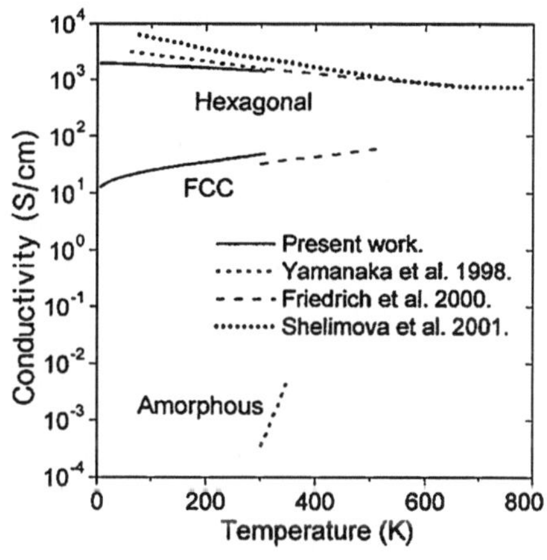

Semiconductors of Metals?

It is interesting to note that there is no general consensus on the conductivity type of the crystalline phases of GST. Thus, Lee et al. [13] proposed that the hexagonal phase of $Ge_2Sb_2Te_5$ is a narrow-gap degenerate semiconductor, where the Fermi level is within the valence band while based on the sign of the slope of temperature dependence of conductivity Wuttig argued that this phase is metallic [11, 14].

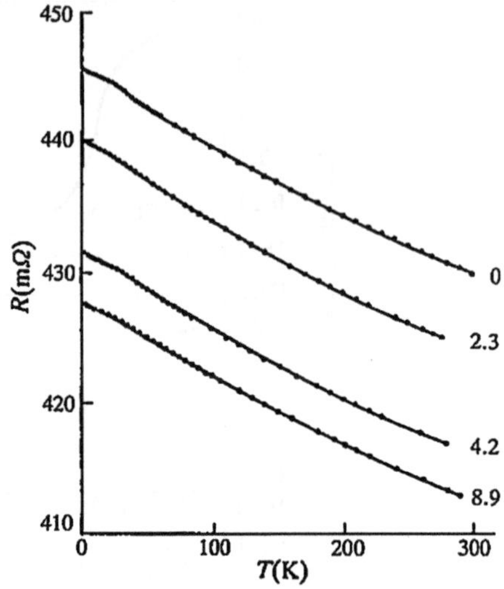

Fig. 2.4 Resistance of amorphous $Cu_{57}Zn_{43}$ as a function of temperature (with and without a magnetic field shown in Tesla). Reprinted from Fritsch et al. [15] with permission from IOP Publishing Ltd.

The fcc phase is generally described as a degenerate p-type semiconductor with a hole concentration of ca. 10^{20} cm^{-3} and a mobility of ca. 1 cm^2/V s although there are some results suggesting that the fcc phase is nondegenerate. At the same time, it should be noted that the experimentally observed variation in the temperature dependence of the conductivity is rather weak and similar to those observed in amorphous or 'dirty' metals [16] as illustrated in Fig. 2.4. The latter also refers to the liquid phase of $Ge_2Sb_2Te_5$.

It is interesting to note that the dielectric constant $\epsilon_1(\omega)$ of most phase-change materials (except GeTe) remains positive at all frequencies [17], which, again, is typical of 'dirty' metals where the plasma oscillations are overdamped [18].

2.1.5 Chalcogenides as Topological Insulators

Recently, a new class of materials called topological insulators [19] has been intensely investigated. Topological insulators are materials with a bulk insulating gap and protected conducting surface states arising from a combination of spin–orbit interactions and time-reversal symmetry which limits possible candidates to heavy elements and rather narrow bandgap materials. It was speculated that topological insulators may be promising materials for quantum computing [20]. So far, topological insulator behaviour has been observed in $Bi_{1-x}Sb_x$ alloys and some A_2B_3 crystals including Sb_2Te_3 [21–23].

It was also predicted that topological properties of $Ge_2Sb_2Te_5$ strongly depend on the stacking sequence along the $\langle 111 \rangle$ direction (cf. Sect. 8.1.2). Thus, $Ge_2Sb_2Te_5$ with the Petrov stacking sequence along the $\langle 111 \rangle$ direction was predicted be a

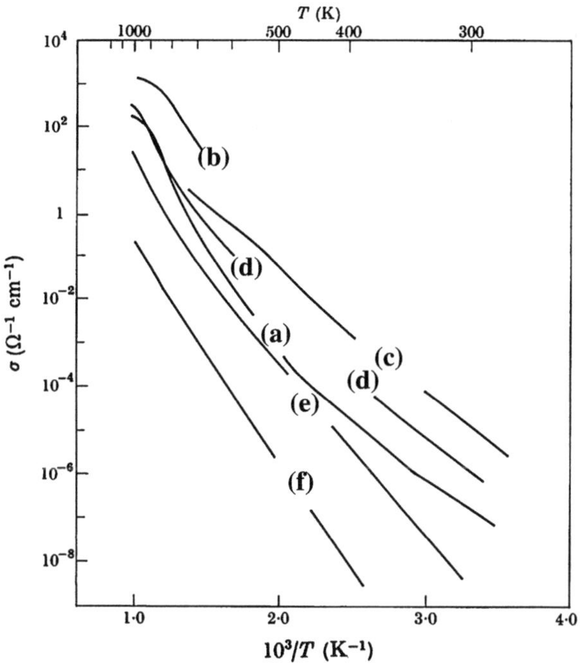

Fig. 2.5 Temperature variation of conductivity for several chalcogenide glasses in the solid and liquid states (*a*) $As_{30}Te_{48}Si_{12}Ge_{10}$; (*b*) As_2Te_3; (*c*) $As_2S_3Tl_2Te$; (*d*) As_2SeTe_2; (*e*) As_2Se_2Te; (*f*) As_2Se_3 (Reprinted with permission by Oxford University Press from Mott and Davis [5])

topological insulator while the Kooi sequence should exhibit surface-like conducting states [24]. A further prediction is that pressure modifies topological insulator properties of phase-change alloys. In particular, it was shown using DFT simulations that hexagonal $Ge_2Sb_2Te_5$ with the Kooi layer sequence transforms from a normal semiconductor to a topological insulator at a hydrostatic pressure of ca. 2.6 GPa, and reverts to a semiconductor at ca. 4.5 GPa, while $Ge_2Sb_2Te_5$ with Petrov sequence maintains a topological insulator state from zero pressure to 2.3 GPa.

2.1.6 Liquid Chalcogenides

Upon melting, chalcogenide glasses tend to preserve the semiconducting behaviour as illustrated by Fig. 2.5 for typical chalcogenide glasses [1]. This behaviour suggests that melting of chalcogenides proceeds without destruction of the covalent backbone of the structure.

A similar conclusion was made for liquid $Ge_2Sb_2Te_5$ based on the sign of the temperature dependence of conductivity (Fig. 2.6) [25] although the observed temperature dependence of conductivity was rather weak.

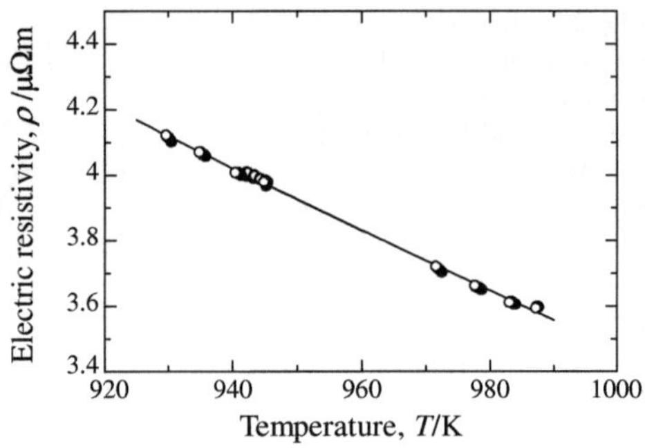

Fig. 2.6 Temperature dependence of conductivity of liquid $Ge_2Sb_2Te_5$ alloy. Reprinted with permission by Japan Society of Applied Physics from Endo et al. [25]

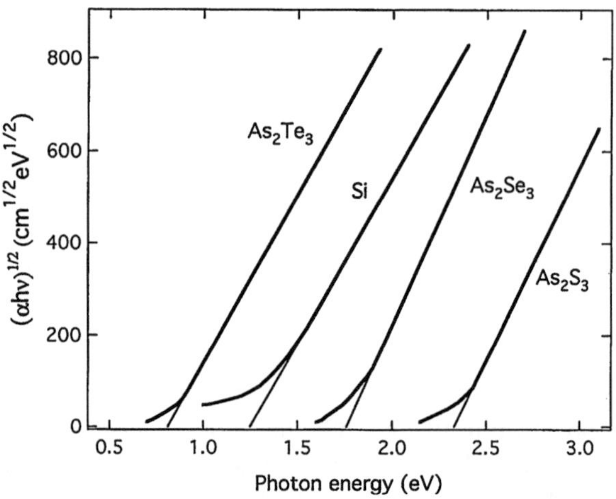

Fig. 2.7 Spectral dependence of optical absorption above the absorption edge for typical amorphous semiconductors (Reprinted with permission by Oxford University Press from Mott and Davis [5])

2.2 Optical Absorption

2.2.1 Interband Absorption

Figure 2.7 shows spectral dependencies of absorption coefficients for typical amorphous semiconductors [1]. As can be seen, this dependence can be described by the relation

$$\alpha h\nu = B(h\nu - E_0)^2 \tag{2.1}$$

This relationship is often called the Tauc law. The coefficient B is of the order of 10^5 $cm^{-1}eV^{-1}$. This kind of spectral behaviour is fairly easy to understand assuming parabolic dependencies for the density of valence and conduction band states.

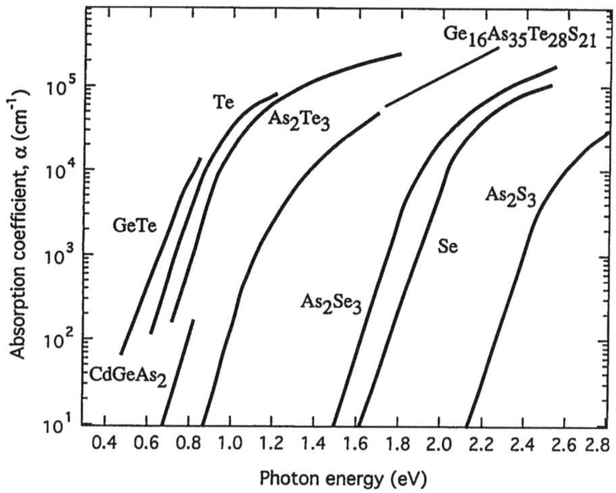

Fig. 2.8 Exponential absorption edge in amorphous semiconductors at room temperature (Reprinted with permission by Oxford University Press from Mott and Davis [5])

2.2.2 Absorption Edge (the Urbach Rule)

The presence of sharp band edges in crystals results in a sharp increase in the absorption coefficient when the photon energies exceed the optical gap. In amorphous chalcogenides, the presence of energy band tails results in significant absorption even at photon energies below the optical gap. Typical spectral dependencies for energies at the absorption edge [1] are shown in Fig. 2.8.

It can be seen that the absorption below the edge is described by an exponential dependence:

$$\alpha = \alpha_0 \exp[-\gamma(E_0 - h\nu)/kT] \tag{2.2}$$

which is usually called the Urbach rule. Here γ is a material constant and T is absolute temperature down to a critical value T_0 and equals T_0 for lower temperatures. It is interesting to note that, despite rather large differences in the bandgap, the slope of the Urbach edge, γ, is almost the same for various materials, typically ca. 50 meV. This constancy weakens a possible explanation that the Urbach rule results from the exponential distribution of the tail states that should vary from one material to another. Several possible mechanisms have been proposed [26–29]. In [27] it was suggested that the Urbach edge arises from an electric-field broadening of the exciton line. Although usually the exciton line has a Gaussian shape, in the presence of a uniform electric field it was shown to become exponential over a wide range of energies [27]. The responsible electric fields may arise, for example, from static spatial fluctuations, density variations and presence of charged defects.

An electric field was also considered to cause exponential absorption dependence through interaction of a bound exciton with lattice vibrations [26] leading to electric-field broadening of the absorption edge similar to the Franz–Keldysh effect.

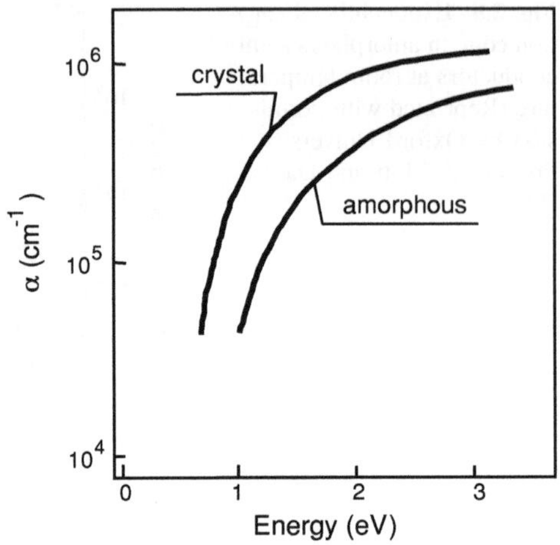

Fig. 2.9 Spectral dependence of absorption coefficients α of as-deposited amorphous and crystalline $Ge_2Sb_2Te_5$

Strong electron–phonon interaction was argued to be an alternative reason for the exponential dependence of the absorption coefficient [28, 30]. Within this approach, the slope of the Urbach edge ($u = kT/\gamma$) and absorption at larger energies were shown to be related through the following equation:

$$u = (4\alpha_0 E_0/3B)^{1/2} \tag{2.3}$$

where α_0 is the absorption coefficient at the bandgap energy E_0 and B is a parameter used for the interband absorption coefficient.

To summarise this discussion, it should be noted that such a variety of "successful" approaches to explain the exponential dependence is at the same time an indication that the question of its origin still remains open.

2.2.3 Phase-Change Materials

The spectral dependencies of optical constants for the prototypical phase-change material $Ge_2Sb_2Te_5$ in as-deposited amorphous and crystalline phases have been studied by various authors [13, 31–34] and there is a very good agreement between the results obtained by different groups. Fig. 2.9 shows generic absorption spectra for the amorphous and crystalline phases. The optical gaps have been determined as 0.7 eV for as-deposited amorphous $Ge_2Sb_2Te_5$ and 0.5 eV for the crystalline phase and the Urbach energy was found to be 80 meV for $Ge_2Sb_2Te_5$ [13], the latter value is slightly higher than that for chalcogenide glasses despite a smaller bandgap.

Since in optical devices using phase-change materials a change in optical reflectivity is employed, it is occasionally more convenient to use optical constants n and k to characterise the amorphous and crystalline phases [35]. The corresponding plots

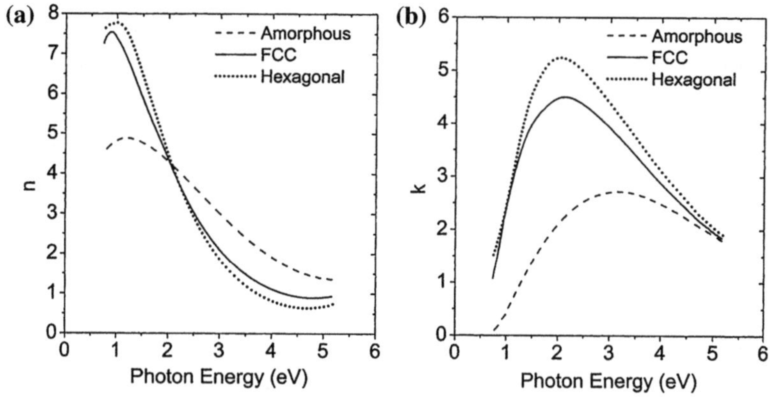

Fig. 2.10 Optical constants n (**a**) and k (**b**) of amorphous, fcc and hexagonal phases of $Ge_2Sb_2Te_5$, after Lee and Bishop [35]. Reprinted with permission from Springer

are shown in Fig. 2.10. It is interesting to note that computer simulations of the optical properties of Ge-Sb-Te phase-change alloys started in 1997 [36] and produced the results that are in very good agreement with experiment.

2.3 Photoexcitation and Recombination

Illumination of semiconductors with light having photon energies exceeding that of the optical gap generates electron–hole pairs. The photoexcited charge carriers of different signs can then either separate, giving rise to photoconductivity, or recombine, either radiatively (photoluminescence) or nonradiatively. Non-radiative recombination can, in principle, transfer the material into a state different from the one prior to irradiation. Such a possibility is especially likely in amorphous semiconductors, where the structure is more flexible than in the corresponding crystals and where structural disorder leads to localisation of charge carriers.

The recombination is called geminate if the very same electron and hole that were created recombine. This happens when the carriers do not drift apart, that is, at low temperatures, low mobilities or low light intensities. Below we briefly discuss specific features of photoluminescence in chalcogenide glasses.

2.3.1 Photoluminescence

Photoluminescence and excitation spectra for typical chalcogenide glasses, a-Se, As_2S_3 and As_2Se_3 [37, 38], are shown in Fig. 2.11. The main feature is a broad band centred at about half the bandgap energy. Interestingly, similar features are observed in crystalline chalcogenides [39].

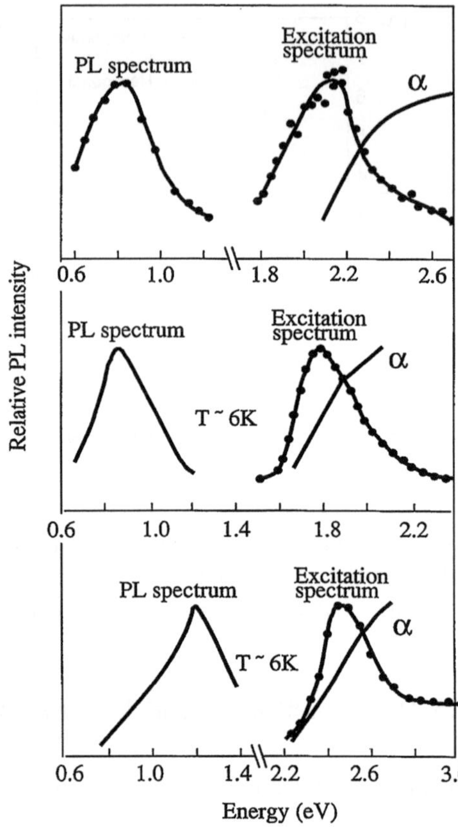

Fig. 2.11 *Top* Photoluminescence spectra, excitation spectra and absorption spectra for a-Se (reprinted from Street et al. [37] with permission from Taylor and Francis). *Middle and bottom* Photoluminescence spectra, excitation spectra and absorption spectra for As_2Se_3, and As_2S_3. Reprinted with permission from Bishop and Mitchell [38]. Copyright 1973 by the American Physical Society

The large difference between the luminescence energy and the bandgap implies that recombination occurs through deep states. Early models suggested that recombination takes place from the conduction band into deep states or between the tail states. These models tacitly assumed that the electron–phonon interaction is weak. It seems more likely, however, that electron–phonon coupling plays an important role in the recombination process [39].

The essence of this approach is illustrated by the coordination-coordinate diagram shown in Fig. 2.12. The adiabatic potentials are given by

$$E_1(q) = Aq^2 \tag{2.4}$$

$$E_2(q) = E_0 + Aq^2 - Bq \tag{2.5}$$

for the ground and excited states, respectively; A is related to the vibrational frequency ω through the equation $A = M\omega^2/2$, where M is the effective mass and B is the measure of electron–phonon coupling. The meaning of other parameters is

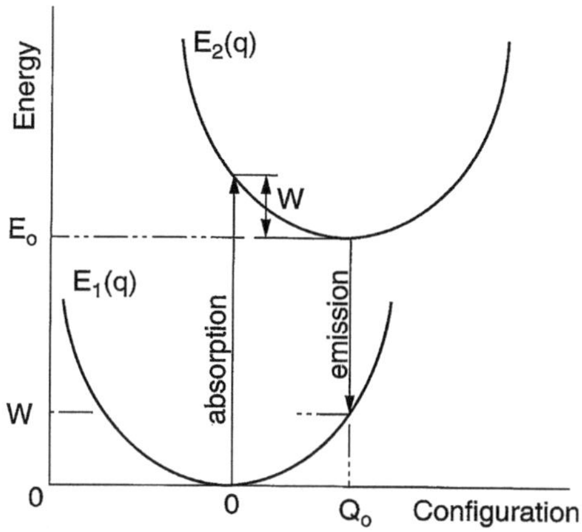

Fig. 2.12 Configuration-coordinate diagram illustrating optical absorption and emission in a system with strong electron–phonon coupling

obvious from the figure. It can be easily shown [39] that within this approximation the photoluminescence line shape is Gaussian and shifted by $2W$ to lower energies compared to the absorption edge.

The excitation spectrum measures the dependence of luminescence intensity on the excitation energy. In most chalcogenides this spectrum is presented by a broad peak close to the bandgap. The excitation mechanism is band-to-band absorption followed by the capture of carriers by recombination centres. A decrease in the intensity at lower energies is due to incomplete absorption whereas a decrease at higher energies most likely arises from the energy dependence of the quantum efficiency [39].

No data are available in the literature on photoluminescence in phase-change materials.

2.3.2 *Photo-Induced Metastability*

The phenomena described above do not change the structural state of the material. At the same time, being intrinsically metastable, amorphous semiconductors can be easily transferred between different metastable states by external stimuli such as light [40–42]. We would like to mention here that *meta*stable does not mean *un*stable. Thus diamonds are a metastable form of carbon with the stable phase being graphite. Despite this metastability diamonds are stable over a very large temperature and pressure range. Another example of a stable metastable state is stained-glass windows of medieval churches that remained essentially unchanged for centuries. Various aspects of metastability of chalcogenide glasses are described in detail in Chaps. 4 and 6 and structural transformations in phase-change materials are the subject of Chaps. 10 and 11.

References

1. N.F. Mott, E.A. Davis, *Electronic Processes in Non-Crystalline Materials*, 2nd edn. (Clarendon Press, Oxford, 1979)
2. B.T. Kolomiets, Phys. Stat. Sol. (b) **7**, 359 (1964)
3. S. Caravati, M. Bernasconi, T. Kühne, M. Krack, M. Parrinello, J. Phys. Condens. Matter **21**, 255501 (2009)
4. R. Rajesh, J. Philip, J. Appl. Phys. **93**, 9737 (2003)
5. N.F. Mott, E.A. Davis, *Electronic Properties of Non-Crystalline Materials*, 2nd edn. (Oxford University Press, Oxford, 1979)
6. A.V. Kolobov, J. Non-Cryst. Solids **198**, 728 (1996)
7. A.H. Edwards, A.C. Pineda, P.A. Schultz, M.G. Martin, A.P. Thompson, H.P. Hjalmarson, J. Phys. Condens. Matter **17**, L329 (2005)
8. D. Emin, Philos. Mag. **35**, 1189 (1977)
9. S.R. Elliott, *Physics of Amorphous Materials*, 2nd edn. (Longman, London, 1984)
10. S.A. Baily, D. Emin, H. Li, Solid State Commun. **139**, 161 (2006)
11. I. Friedrich, V. Weidenhof, W. Njoroge, P. Franz, M. Wuttig, J. Appl. Phys. **87**, 4130 (2000)
12. E. Morales-Sánchez, E.F. Prokhorov, J. Gonzalez-Hernandez, A. Mendoza-Galvan, Thin Solid Films **471**, 243 (2005)
13. B.S. Lee, J.R. Abelson, S.G. Bishop, D.H. Kang, B.K. Cheong, K.B. Kim, J. Appl. Phys. **97**, 093509 (2005)
14. T. Siegrist, P. Jost, H. Volker, M. Woda, P. Merkelbach, C. Schlockermann, M. Wuttig, Nature Mater. **10**, 202 (2011)
15. G. Fritsch et al., J. Phys. F Metal Phys. **12**, 2965 (1982)
16. N.F. Mott, *Conduction in Non-Crystalline Materials* (Clarendon Press, Oxford, 1987)
17. K. Shportko, S. Kremers, M. Woda, D. Lencer, J. Robertson, M. Wuttig, Nature Mater. **7**, 653 (2008)
18. R. Kiebooms, R. Menon, K. Lee, in *Handbook of Advanced Electronic and Photonic Materials and Devices*, vol. 8, ed. by H.S. Nalwa (Academic Press, San Diego, 2001), p. 1
19. M.Z. Hasan, C.L. Kane, Rev. Mod. Phys. **82**, 3045 (2010)
20. J. Moore, Nature Phys. **5**, 378 (2009)
21. H. Zhang, C.X. Liu, X.L. Qi, X. Dai, Z. Fang, S.C. Zhang, Nature Phys. **5**, 438 (2009)
22. G. Wang, X. Zhu, J. Wen, X. Chen, K. He, L. Wang, X. Ma, Y. Liu, X. Dai, Z. Fang, J. Jia, Q. Xue, Nano Res. **3**, 874 (2010)
23. D. Hsieh, Y. Xia, D. Qian, L. Wray, F. Meier, J.H. Dil, J. Osterwalder, L. Patthey, A.V. Fedorov, H. Lin, A. Bansil, D. Grauer, Y.S. Hor, R.J. Cava, M.Z. Hasan, Phys. Rev. Lett. **103**, 146401 (2009)
24. J. Kim, J. Kim, S.H. Jhi, Phys. Rev. B **82**, 201312 (2010)
25. R. Endo, S. Maeda, Y. Jinnai, R. Lan, M. Kuwahara, Y. Kobayashi, M. Susa, Jpn. J. Appl. Phys. **49**, 5802 (2010)
26. Y. Toyozawa, Progr. Theor. Phys. **22**, 455 (1959)
27. J. Dow, Phys. Rev. B **5**, 594 (1972)
28. A.V. Kolobov, O.V. Konstantinov, Philos. Mag. B **40**, 475 (1979)
29. B.L. Gelmont, V.I. Perel, I.N. Yassievich, Fiz. Tverd. Tela **25**, 727 (1983)
30. A.V. Kolobov, O.V. Konstantinov, Philos. Mag. B **47**, 1 (1983)
31. D.V. Tsu, J. Vac. Sci. Technol. A **17**, 1854 (1999)
32. S.Y. Kim, S.J. Kim, H. Seo, M.R. Kim, Jpn. J. Appl. Phys. **38**, 1713 (1999)
33. E. Garcia-Garcia, A. Mendoza-Galvan, Y. Vorobiev, E. Morales-Sanchez, J. Gonzalez-Hernández, G. Martinez, B.S. Chao, J. Vac. Sci. Technol. A **17**, 1805 (1999)
34. A. Pirovano, A.L. Lacaita, A. Benvenuti, F. Pellizzer, R. Bez, IEEE T. Electron Dev. **51**, 452 (2004)
35. B.S. Lee, S.G. Bishop, in *Phase Change Materials: Science and Applications*, ed. by S. Raoux, M. Wuttig (Springer, Berlin, 2009), p. 175

36. S. Ogawa, S. Yamanaka, Y. Ueshima, I. Morimoto, in *9th Symposium on Phase Change Optical Information Storage* (*PCOS' 97*), Numadu, 1997, p. 50
37. R.A. Street, T.M. Searle, I.G. Austin, Philos. Mag. **29**, 1157 (1974)
38. S.G. Bishop, D.L. Mitchell, Phys. Rev. B **8**, 5696 (1973)
39. R.A. Street, Adv. Phys. **25**, 397 (1976)
40. K. Shimakawa, A.V. Kolobov, S.R. Elliott, Adv. Phys. **44**, 475 (1995)
41. N. Itoh, M. Stoneham, *Materials Modification by Electronic Excitation* (Cambridge University Press, Cambridge, 2001)
42. A.V. Kolobov (ed.), *Photo-Induced Metastability in Amorphous Semiconductors* (Wiley-VCH, Winheim, 2003)

Chapter 3
Methods of Structure Analysis

3.1 Experimental Methods for Investigating Atomic Structure

3.1.1 Bragg Diffraction and Its Limitations

Bragg diffraction occurs when monochromatic electromagnetic radiation or subatomic particle waves with wavelength comparable to interatomic spacings is incident upon a crystalline sample, elastically scattered in a specular fashion by the atoms in the system and undergoes constructive interference in accordance with Bragg's law. For a crystalline solid, the waves are scattered from lattice planes separated by the interplanar distance d. Where the scattered waves interfere constructively, they remain in phase since the path length of each wave is equal to an integer multiple of the wavelength. The path difference between two waves undergoing constructive interference is given by $2d \sin \theta$, where θ is the scattering angle. This leads to Bragg's law that describes the condition for constructive interference from successive crystallographic planes of the crystalline lattice:

$$2d \sin \theta = n\lambda \tag{3.1}$$

where n is an integer and λ is the wavelength. A diffraction pattern is obtained by measuring the intensity of scattered waves as a function of scattering angle. Very strong intensities known as Bragg peaks are obtained in the diffraction pattern when scattered waves satisfy the Bragg condition. Powder Bragg diffraction data are usually analysed using the Rietveld refinement procedure [1] and allow one to obtain the atomic positions with 10^{-5} Å accuracy. At the same time, the requirement that a Bragg peaks form originates from a minimum spatial extent (vis-a-vis the Scherrer equation [2]) of typically more than 10 nm; this requirement inevitably leads to averaging over this or greater length scales.

A. V. Kolobov and J. Tominaga, *Chalcogenides*, Springer Series in Materials Science 164, DOI: 10.1007/978-3-642-28705-3_3,

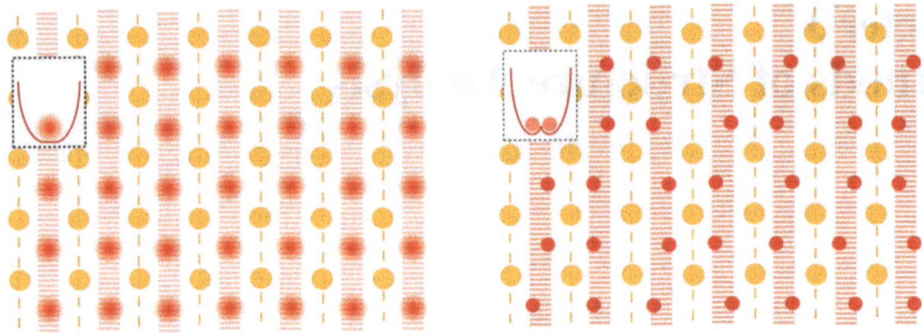

Fig. 3.1 Being a technique that probes an ensemble average structure, Bragg diffraction cannot distinguish between the structures with soft modes (*left*) and with stochastic local distortions (*right*). In both cases it "sees" a broad plane of atoms interpreted as a large isotropic thermal factor

In perfect crystals, the average structure is usually the same as the local structure. However, there are numerous cases when the average structure is different from the local structure even in crystals. One such example is crystals with stochastic local displacements. Consider two hypothetical two-dimensional crystals (Fig. 3.1), one with cation atoms loosely bound at the centre of the cell confined within a "soft" potential (cf. Sect. 1.5.3) so that the amplitude of thermal vibrations is rather large, and the other one with random local distortions. From the point of view of Bragg diffraction that probes periodic arrangement of atoms, the two structures are identical: in both cases Bragg diffraction will see a broad cation plane with a large isotropic thermal factor. Experimentally, this manifests as a decrease in the peak intensities. As a result, Bragg diffraction cannot distinguish between displacive and order-disorder transitions in solids as exemplified below (cf. Sect. 8.1.1) for the case examples of GeTe and $Ge_8Sb_2Te_{11}$.

Another somewhat related example when Bragg diffraction fails to obtain precise structural information is solid crystal solutions. While Vegard's law, derived from diffraction measurements, holds that a linear relation exists, at constant temperature, between the crystal lattice parameter of an alloy and the concentrations of the constituent elements, locally the bond lengths between the same elements remain almost constant. This is illustrated by Fig. 3.2 for the $Zn_{1-x}Mn_xTe$ system [3]. The same discrepancies between the average interatomic distance obtained through diffraction measurements and the actual bond lengths measured using extended X-ray absorption fine structure (EXAFS) (cf. Sect. 9.1.2) have been reported for various mixed crystals, e.g. $Si_{1-x}Ge_x$, $Ga_{1-x}In_xAs$, etc [3, 4].

A further example when use of the average bond length can lead to erroneous interpretation of the structure is silica when use of neutron powder diffraction detected a decrease in the average Si–O bond lengths while the pair-distribution-function (PDF) analysis (cf. Sect. 3.1.2) *of the same dataset* demonstrated that the actual bond length remained unchanged or even slightly increased [5]. The moral of the story is that the average structure may be quite different from the actual local structure.

A specific limitation of Bragg diffraction is that the Bragg peaks arise from diffraction on clouds of electrons that form crystal planes. If cation and anion possess a similar number of electrons, structural analysis based on X-ray diffraction (XRD)

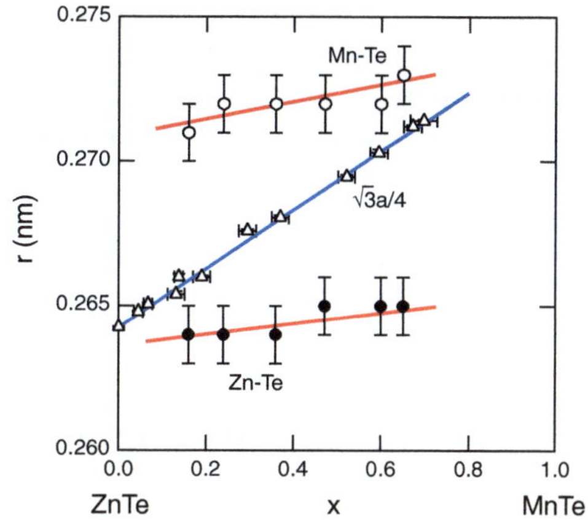

Fig. 3.2 The concentration dependence of the nearest neighbour distance obtained from Bragg diffraction (*triangles*) and XAFS (*circles*) results for mixed crystals of $Zn_{1-x}Mn_xTe$. Reprinted with permission from Hosokawa et al. [3]. Copyright 2009 by the American Physical Society

may lead to misinterpretation of the structure. Thus, in the case of KCl that possesses the rocksalt structure, XRD pattern is the same as that for a simple cubic structure since K^+ and Cl^- ions are indistinguishable from the point of view of XRD [6].

An additional complication for application of diffraction to amorphous materials is that the latter lack long-range order. As a consequence, Bragg diffraction, that is a very powerful tool to investigate crystals, yields broad features that are very difficult to interpret unambiguously.

3.1.2 Pair-Distribution-Function Analysis

A crystal structure is never perfectly periodic, even if the crystal is perfect, because atoms are vibrating due to either thermal or zero-point quantum vibrations. Deviations of atoms from their equilibrium positions due to vibrations are accounted for by the Debye–Waller factor. Contrary to intuitive expectations, when the atoms vibrate, the Bragg peaks do not broaden but their intensity is diminished. This "decreased" intensity gives rise to diffuse scattering between the Bragg peaks. The diffuse intensity is widely spread over Q-space and has an intensity much lower than that of the Bragg peaks. In powder diffraction, the diffuse scattering is usually considered as a broad background and discarded from analysis although it can provide important information regarding the local deviations from the average structure, be they due to dynamic or static disorder. Extraction of this information is done through a PDF analysis.

The PDF analysis method is a powerful tool for the study of glasses, liquids and amorphous materials, as well as crystalline or partly crystalline materials. The PDF method, which involves the direct model-free Fourier transformation of X-ray or neutron powder diffraction data, gives the probability of finding any two atoms at a given interatomic distance. Unlike other better known methods, PDF analysis can provide information not only about the long-range (>100 Å) atomic ordering but also about the short-range order in materials. This is because the technique takes into account both the Bragg as well as diffuse scattering (which is known to be related to short-range order effects). The interested reader is referred to the monograph by Egami and Billinge very appropriately titled "Underneath the Bragg peaks" [5]. Since PDF analysis does not depend on the periodicity of the lattice, it allows one to obtain accurate information about the local structure (which may be different from the average structure).

3.1.3 X-Ray Absorption Fine Structure Spectroscopy

A very powerful technique for determining local (short-range order) structure is X-ray absorption fine structure (XAFS) spectroscopy. A recent review of the technique can be found in [7].

XAFS is structured absorption on the high-energy side of an X-ray absorption edge. A simplified description of the underlying physics is the following. At the absorption edge of an element (when the energy of the incident X-ray matches the binding energy of a core electron) X-ray absorption abruptly increases and for energies above the edge energy, a core electron is ejected from the atom. The ejected core electron can be thought of as a spherical wave propagating outward from the absorbing (central) atom. The photoelectron wavelength is determined by its kinetic energy, which is in turn determined by the difference between the incident X-ray energy and the core-electron binding energy.

In a solid, the X-ray-excited outgoing photoelectron wave is scattered by the neighbouring atoms. If the outgoing and backscattered photoelectron waves are in phase, constructive interference leads to a local maximum in X-ray absorption; when they are out of phase, a local minimum in the X-ray absorption appears. As a result of this interference, the absorption above the absorption edge oscillates and such oscillations are called the fine structure. This is schematically illustrated in Fig. 3.3 (upper panel) that shows an X-ray absorption spectrum of copper, showing both X-ray absorption near-edge structure (XANES) and EXAFS regions. An artistic vision of such a process is shown in the lower panel of Fig. 3.3.

XANES is often used to refer to the structure in the near (around 50 eV) region of the edge. The sharp feature at the absorption edge is called the white line. EXAFS typically refers to structured absorption from approximately 50 to 1,000 eV or more above the absorption edge. XAFS's principal advantage is that it is a local structure probe and does not require the presence of long-range order. This means that it can also be used to determine the local structure in non-crystalline samples. The other advantage is that it is element selective, i.e. allows one to probe independently the local structure around different chemical species in a solid.

The magnitude of EXAFS oscillations $\chi(E)$ is obtained from the measured spectrum $\mu(E)$ by subtracting the atomic background $\mu_0(E)$ and normalising to the jump of the absorption at the edge energy $\mu(E_0)$:

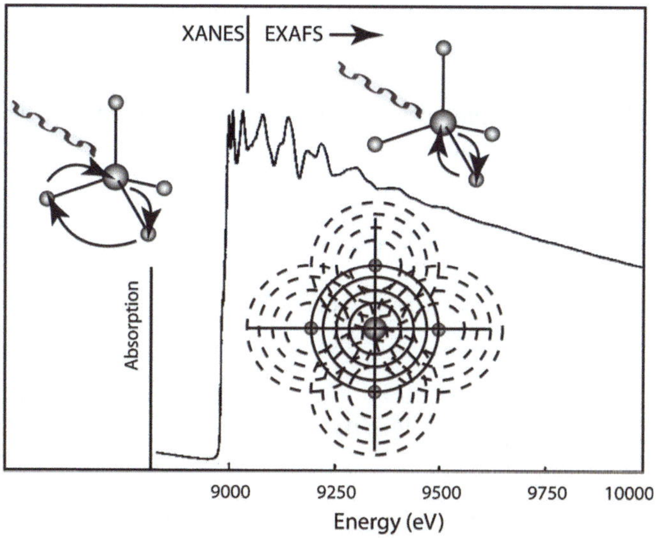

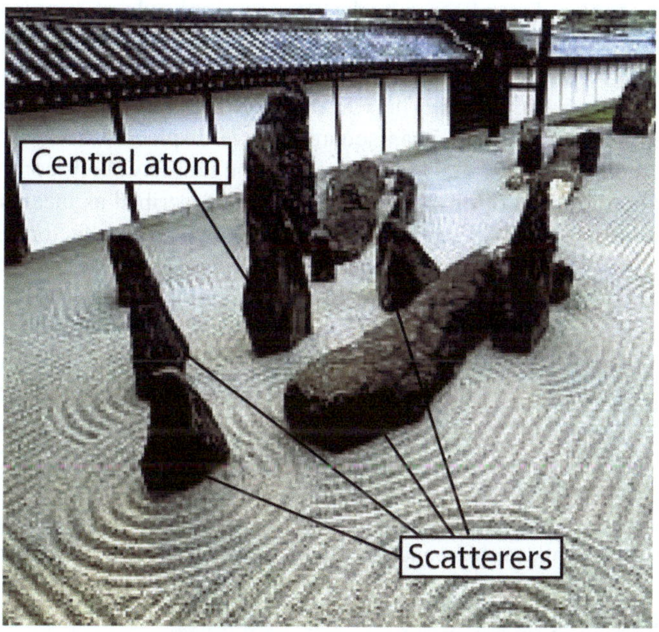

Fig. 3.3 *Upper panel* XAFS spectrum of copper. The *upper insets* show multiple scattering and single scattering processes responsible for XANES and EXAFS features, respectively; the *lower inset* illustrates the scattering interference between the on-going and back-scattered waves generating fine structures in the X-ray absorption spectrum. *Lower panel* An artistic illustration of XAFS seen in a Japanese zen garden

$$\chi(E) = \frac{\mu(E) - \mu_0(E)}{\mu(E_0)} \tag{3.2}$$

Usually, photoelectron momentum, k is used instead of energy, the two being related via the following formula:

$$k = \sqrt{\frac{2m(E - E_0)}{\hbar^2}} \tag{3.3}$$

Within a simple single scattering theory [7] the formula for EXAFS can be presented as:

$$\chi(k) = \sum_{i,j} \Im\left(\frac{N_i S_0^2 F_i^{\mathrm{eff}}(k)}{kR_i^2} \sin[2kR_i + 2\delta(k) + \varphi_i(k)] e^{-2\sigma_i^2 k^2} e^{-2R_i/\lambda(k)} \right) \tag{3.4}$$

The sum is taken over all shells of atoms j. The magnitude of oscillations is proportional to the coordination number N_i and decreases for larger interatomic distances R_i. It also depends on a backscattering amplitude of surrounding atoms F^{eff}. The fine structure is sinusoidally modulated by a function that, in addition to the distance R, involves the phase shift of the electrons, both at central, $2\delta(k)$, and surrounding, $\varphi_i(k)$, atoms. The EXAFS also depends on the mean free path of the electrons $\lambda(k)$, and the mean-square relative displacement σ_i^2. The EXAFS oscillations damp relatively quickly so that only the first few shells contribute to the fine structure. Structural disorder within EXAFS is described as mean-square relative displacement (MSRD) which, in contrast to Debye–Waller factor in diffraction studies, represents variations in the bond lengths for a pair of atoms (i.e. the interatomic distance) and not in positions of individual atoms.

The Fourier transformed data $\chi(R)$ contain information that is similar to PDF with three major differences. Firstly, $\chi(R)$ is obtained with respect to a particular kind of chemical species (an advantage). Secondly, the apparent distances are displaced to smaller values due to phase shifts (a disadvantage from the visualisation perspective). And thirdly, although $\chi(R)$ plots may look similar to radial distribution functions, they are Fourier transforms of an interference pattern between different scattering paths and as a result may contain features that do not represent any real interactomic distances. Both the phase shifts and interference effects are accounted for during the fitting [8].

The data are analysed using curve fitting analysis, for disordered materials some authors prefer to include the reverse Monte Carlo procedure [9, 10] to account for a broad distribution of bond lengths and bond angles. As the PDF analysis, EXAFS data also provide one-dimensional information such as distances to the first nearest neighbours and their chemical nature as well as the degree of bond disorder but do not allow direct unambiguous determination of the three-dimensional structure. While inclusion of multiple scattering paths may be sensitive to the bonding geometry, the effect of higher paths is rather weak and the success of the analysis, especially for amorphous solids, relies on the choice of the structural model used.

Different from EXAFS, XANES is strongly influenced by multiple scattering and contains information about spatial arrangement of neighbours within ca. 1 nm from the central atom, i.e. *XANES is a three-dimensional probe* [11]. This is a big advantage when one needs to distinguish between structures with similar interatomic distances but different spatial arrangement of atoms. On the other hand, XANES analysis is much more complicated than EXAFS (and not as widely used) and in most cases XANES spectra of the material under study are compared to those of reference materials with the known structures that contain fingerprints of certain local atomic arrangements.

Use of synchrotron radiation, especially at so-called third-generation sources, that allows one to obtain high quality data for amorphous films as thin as a monolayer and recent significant progress in theory [12–14] made XAFS the method of choice to investigate thin amorphous films.

The same principle is used by diffraction anomalous fine structure technique (DAFS) when intensity of Bragg peaks is measured as a function of the excitation energy across the absorption edge of constituent atoms [15]. Because DAFS combines the capabilities of diffraction and XAFS into a single technique, it has two enhanced sensitivities compared to the separate techniques, namely, DAFS contains EXAFS- and XANES-like information for the specific subset of atoms selected by the diffraction condition and, additionally, DAFS can provide site-specific absorption-like spectroscopic and structural information for inequivalent sites of a single atomic species within the unit cell. It has also been successfully used to analyse the structure of the amorphous phase in presence of the crystalline phase of the same composition [16]. The backside is a rather complicated data analysis procedure.

3.1.4 Nanobeam Electron Diffraction

Recently, nanobeam electron diffraction has been used to directly observe local atomic order in amorphous materials [17]. Nanobeam electron diffraction with a coherent electron beam smaller than 1 nm in diameter enabled the authors to acquire two-dimensional diffraction patterns from a nanoscale region to detect local atomic structure. While a selected area diffraction pattern for larger beam diameters is typical of an amorphous solid with a full halo ring without any detectable diffraction spots, upon reducing the beam diameter to 0.36 nm a diffraction pattern with a set of twofold symmetric spots, analogous to that of a single crystal can be observed (Fig. 3.4).

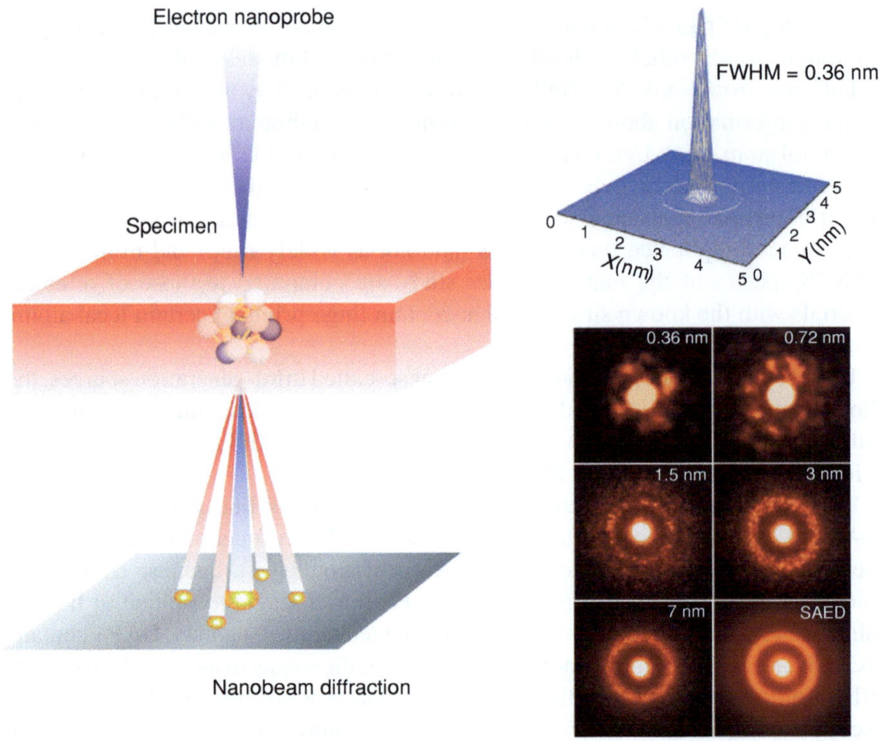

Fig. 3.4 Experimental scheme of nanobeam electron diffraction. The three-dimensional profile of a calculated electron nanoprobe with a FWHM beam size of 0.36 nm is shown in the *upper right inset* (spherical aberration coefficient Cs $= 0.002$ mm, defocus $\Delta f = 0$ mm, convergence angle $\alpha = 3.3$ mrad). The *lower right insets* show examples of the nanobeam size dependence of electron diffraction patterns. A large number of diffraction patterns from nanosized regions of a metallic glass were recorded by a television-rate CCD camera during nanoprobe scanning. Reprinted by permission from Macmillan Publishers Ltd.: Nature Mater. [17], copyright (2011)

3.2 Electronic Structure

3.2.1 Band Structure

The two major techniques that allow one to probe the gross features of the electronic structure are photoemission and ultraviolet (UV) and X-ray absorption.

Photoemission

Photoemission spectroscopy, also known as photoelectron spectroscopy, refers to the measurement of the energy of electrons emitted from solids, gases or liquids

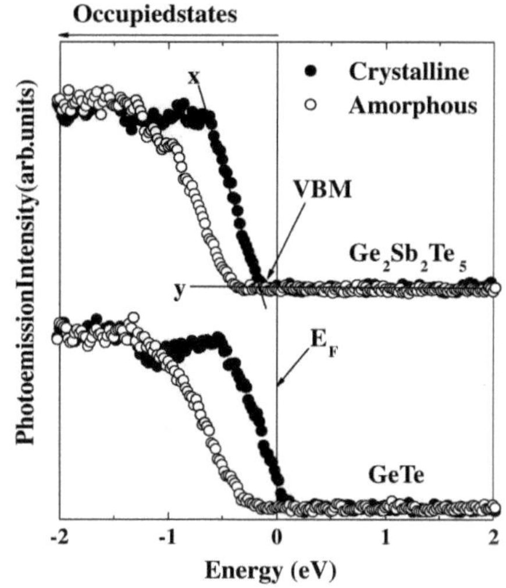

Fig. 3.5 Photoemission spectra near the valence-band top of the amorphous (*open circles*) and crystalline (*closed circles*) $(GeTe)_{1-x}(Sb_2Te_3)_x$ films for $x = 0$ (GeTe) and 1/3 ($Ge_2Sb_2Te_5$). The zero energy indicates the Fermi energy. Reprinted from Kim et al. [18]. Copyright 2009 by the American Physical Society

by the photoelectric effect, in order to determine the binding energies of electrons in a substance. The term refers to various techniques, whether the ionization energy is provided by an X-ray photon, an extreme ultraviolet (EUV) photon or an UV photon.

X-ray photoelectron spectroscopy (XPS) is used to study the energy levels of atomic core electrons, primarily in solids. Since the core levels have small chemical shifts depending on the chemical environment of the atom which is ionised, XPS measurements allow for chemical structure to be determined. An alternative name for this technique is electron spectroscopy for chemical analysis (ESCA).

While the escape length of electrons using typical laboratory equipment is on the order of a nanometer limiting the obtained information to surfaces, use of high-energy X-rays at synchrotron light sources allows one to probe the bulk properties. The latter is sometimes called hard X-ray photoelectron spectroscopy (HXPES).

In addition to core-level spectroscopy, due to the high brilliance of the third-generation synchrotron sources this method also offers high-energy resolution that enables one to obtain the valence-band structure. An example of such measurements is shown in Fig. 3.5 for crystalline and amorphous phases of quasibinary GeTe–Sb_2Te_3 alloys [18].

Other variations of photoelectron spectroscopy include UV photoelectron spectroscopy and EUV photoelectron spectroscopy.

X-ray and UV Absorption

In X-ray absorption the initial core state is a very narrow level and the absorption is primarily determined by the density of unoccupied (conduction band) states. For this reason, X-ray absorption near the corresponding absorption edge, the so-called

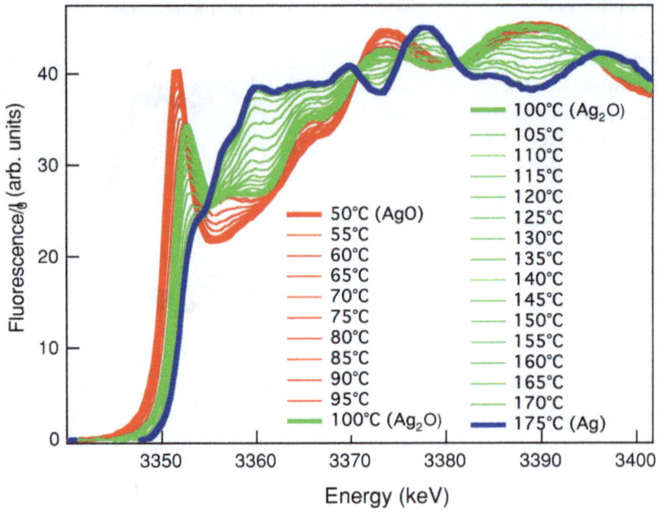

Fig. 3.6 Evolution of Ag L_2-edge during the reduction process from AgO to Ag_2O to metallic Ag. Reprinted with permission from Kolobov et al. [19]. Copyright 2004 by the American Institute of Physics

XANES, alternatively called near-edge X-ray absorption fine structure (NEXAFS), exhibits features that contain information about the local density of states as well as the chemical state of the element. Variation in XANES during the reduction process from AgO to Ag_2O to metallic Ag [19] is illustrated in Fig. 3.6.

For the UV range both the valence band and conduction band densities contribute to the absorption. A review of this technique with respect to amorphous materials is given in [20].

Although the two previous methods probe the band structure (including tails) they do not provide sufficient information about states whose density is orders of magnitude smaller. For this purpose other spectroscopic methods are used, the most important of which are briefly described below.

3.2.2 Defect Spectroscopy

Electron Spin Resonance

Electron spin resonance is a probe for unpaired electron spins. In tetrahedrally bonded amorphous materials neutral (paramagnetic) defects such as dangling bonds can be detected with a good sensitivity. In chalcogenide glasses, however, ESR signals are generally observed only after photoexcitation. The only exception are glasses of the Ge-S system that have a pronounced dark ESR signal. Although the reason for this exception is not well understood, a suggestion was made that it results from the fact that recombination between broken bonds in the melt proceeds much more slowly in Ge-S than in, say, As-S(e) glasses, so that broken bonds remain as paramagnetic defects in Ge-X glasses. Use of photo-induced ESR for defect analysis in chalcogenides is discussed in detail later in conjunction with photo-induced metastability (Sect. 4.3.1).

Photoluminescence

Photoluminescence probes the radiative recombination of optically created electron–hole pairs. After excitation, the created electron and hole thermalise rapidly until they are captured by localised states from which further thermalisation has a low probability. These can be either tail states or deep states. In photoluminescence experiments only information related to radiative centres can be accessible and a fraction of electron–hole pairs that recombine radiatively is called quantum efficiency.

The luminescence spectrum contains information about the distribution and properties of the radiative defects. The excitation spectrum, which represents the dependence of the luminescence intensity on the excitation energy, to a zeroth approximation resembles the spectral dependence of the absorption coefficient at the low-energy side.

Optical Absorption

Optical absorption in the sub-bandgap range photon energies can also provide indirect information about the density of states. The absorption coefficient can be measured either directly or, at very low absorption coefficients, using more sophisticated methods such as photoacoustic spectroscopy [21].

Time of Flight

In this technique, a sample is photoexcited by a short pulse of strongly absorbed light and a current across the sample thickness (in the "sandwich" geometry) is measured [22, 23]. The time at which the package of photoexcited carriers reaches the opposite side of the sample is called the time of flight. This technique usually measures the drift mobility, but because the mobility is determined by multiple trapping and excitation of carriers its temperature dependence can be used to obtain information on the density of localised states. It has been argued that investigation of the current change in the post-transit time range provides more accurate information [24].

3.3 Structural Modelling

The goal of structural modelling is to obtain a model that would account for the experimentally observed data and/or predict the structure of a material that is difficult to study experimentally.

3.3.1 Reverse Monte Carlo Modelling

Reverse Monte Carlo (RMC) modelling is a technique for generating structural configurations based on experimental data. The logic is very appealing: any model of a complex material worthy of further study should, at a minimum, agree with what is known (that is, experiment). In an ideal implementation, one should find a model agreeing with all known information. The approach was originally developed in [10, 25] for liquid and glassy materials for lack of alternative routes to explore experimental data, but in recent years progress has been made towards modelling crystalline systems as well. Starting with a suitable configuration, atoms are displaced randomly using the periodic boundary conditions until the input experimental data (usually the one-dimensional structure factor or the radial distribution function) match with the data obtained from the generated configuration. This is achieved by minimising a cost function which consists of either the structure factor or radial distribution function along with some appropriately chosen constraints to restrict the search space.

In order to avoid unphysical models, an appropriate choice of constraints is crucial. For example, in order to avoid the atoms getting too close to each other, a certain cut-off distance is imposed which is typically of the order of interatomic spacing. In RMC modelling, this value is usually obtained from the radial distribution function by Fourier transform of the measured structure factor. This is equivalent to adding a hard-sphere potential cut-off in the system which prevents the catastrophic build up of potential energy. In spite of the fact that RMC has been applied to many different types of system such as liquid, glasses, polymer and magnetic materials, questions are often raised about the reliability of results obtained from RMC simulation. The method has never been accepted without some degree of controversy, and the most popular criticism is the lack of the unique solution from RMC. This lack of uniqueness is not surprising, since usually only the pair correlation function or structure factor is used in modelling the structure. Additionallly, a large number of fitting parameters RMC that can produce multiple configurations having the same pair-correlation function. Use of multiple experimental data sets for RMC simulations is essential to generate a unique structural solution [26]. Another promising approach is to combine RMC and ab-initio simulations [27].

3.3.2 Ring Statistics

The analysis of topological networks (liquid, crystalline or amorphous systems) can be based on the part of the structural information which can be represented in the graph theory using nodes for the atoms and links for the bonds. The absence or existence of a link between two nodes in the structure is determined by the analysis of the total and partial radial distribution functions of the system. In such a network a series of nodes and links connected sequentially without overlap is called a path. A ring is a closed path and a N-membered ring is a ring containing N nodes.

There are different criteria to determine rings, e.g. a ring can be viewed as the shortest path which comes back to a given node (or atom) from one of its nearest neighbours [28]. Alternatively, a ring can be represented as the shortest path between two of the nearest neighbours of a given node [29]. It should be noted that use of different criteria may result in different ring statistics.

In binary (or quasibinary) materials without "wrong" bonds, only ABAB, i.e. the so-called even-membered rings, exist. On the other hand, the homopolar bond defects can have a significant influence on the ring statistics and, in particular such bonds often manifest themselves as odd-membered rings.

3.3.3 Ab-Initio Density-Functional Theory Simulations

In order to be able to reproduce the experimental results and possess predictive power, computer simulations should be based on first principles of quantum mechanics, i.e. be independent of any empirical parameters, hence the name ab-initio, or first principles, simulations.

An exact ab-initio solution of the Schrödinger equation for a many-body system comprising nuclei and ions is virtually impossible. In order to study such systems with realistic computational costs, a range of approximations has been introduced [30]. One of the most successful approaches based on the single particle approximation is the density functional theory (DFT). Using this theory, the properties of a many-electron system can be determined by using functionals, i.e. functions of another function, which in this case is the spatially dependent electron density. DFT is among the most popular and versatile methods currently available in condensed-matter computational physics and chemistry.

The major problem with DFT is that the exact functionals for exchange and correlation are not known except for the free electron gas. However, approximations exist which permit accurate calculations of certain physical quantities. In physics the most widely used approximation is the local-density approximation (LDA), where the functional depends only on the density at the coordinate where the functional is evaluated:

$$E_{XC}^{\mathrm{LDA}}[n] = \int \epsilon_{XC}(n) \cdot n(\boldsymbol{r}) d^3 r \qquad (3.5)$$

An extension of LDA is the generalised gradient approximation (GGA) which additionally takes into account the gradient of the density at the same coordinate:

$$E_{XC}^{\mathrm{GGA}}[n] = \int \epsilon_{XC}(n, \nabla n) \cdot n(\boldsymbol{r}) d^3 r \tag{3.6}$$

Using the latter (i.e. GGA), very good results for molecular geometries and ground-state energies have been achieved.

DFT simulations have recently been successfully applied to investigate phase-change materials as reviewed in [31]. The related results are discussed in detail later in this volume.

References

1. L.B. McCusker, R.B. Von Dreele, D.E. Cox, D. Louer, P. Scardi, J. Appl. Cryst. **32**, 36 (1999)
2. B. Cullity, S. Stock, *Elements of X-ray Diffraction* (Addison-Wesley, Reading, 1978)
3. S. Hosokawa, N. Happo, K. Hayashi, Phys. Rev. B **80**, 134123 (2009)
4. J.C. Mikkelsen, J.B. Boyce, Phys. Rev. Lett. **49**, 1412 (1982)
5. T. Egami, S.J.L. Billinge, *Underneath the Bragg Peaks: Structural Analysis of Complex Materials* (Pergamon, Oxford, 2003)
6. C. Kittel, *Introduction to Solid State Physics*, 7th edn. (Wiley, New York, 1996)
7. G. Bunker, *Introduction to XAFS: A Practical Guide to X-ray Absorption Fine Structure Spectroscopy* (Cambridge University Press, Cambridge, 2010)
8. B. Ravel, M. Newville, J. Synch. Rad. **12**, 537 (2005)
9. S.J. Gurman, R.L. McGreevy, J. Phys. Condens. Matter **2**, 9463 (1990)
10. R. McGreevy, J. Phys. Condens. Matter **13**, R877 (2001)
11. G. Smolentsev, A.V. Soldatov, M.C. Feiters, Phys. Rev. B **75**, 144106 (2007)
12. A.L. Ankudinov, B. Ravel, J.J. Rehr, S.D. Conradson, Phys. Rev. B **58**, 7565 (1998)
13. J.J. Rehr, R.C. Albers, Rev. Mod. Phys **72**, 621 (2000)
14. J.J. Rehr, J.J. Kas, F.D. Vila, M.P. Prange, K. Jorissen, Phys. Chem. Chem. Phys. **12**, 5503 (2010)
15. H. Stragier, J.O. Cross, J.J. Rehr, L.B. Sorensen, C.E. Bouldin, J.C. Woicik, Phys. Rev. Lett. **69**, 3064 (1992)
16. A.I. Frenkel, A.V. Kolobov, I.K. Robinson, J.O. Cross, Y. Maeda, C.E. Bouldin, Phys. Rev. Lett. **89**, 285503 (2002)
17. A. Hirata, P. Guan, T. Fujita, Y. Hirotsu, A. Inoue, A. Yavari, T. Sakurai, M. Chen, Nature Mater. **10**, 28 (2011)
18. J.J. Kim, K. Kobayashi, E. Ikenaga, M. Kobata, S. Ueda, T. Matsunaga, K. Kifune, R. Kojima, N. Yamada, Phys. Rev. B **76**, 115124 (2007)
19. A.V. Kolobov, A. Rogalev, F. Wilhelm, N. Jaouen, T. Shima, J. Tominaga, Appl. Phys. Lett. **84**, 1641 (2004)
20. S.R. Elliott, *Physics of Amorphous Materials*, 2nd edn. (Longman, London, 1984)
21. S. Yamasaki, Philos. Mag. B **56**, 79 (1987)
22. W.E. Spear, Proc. Phys. Soc. **76**, 826 (1960)
23. W.E. Spear, Proc. Phys. Soc. B **70**, 669 (1957)
24. A.V. Kolobov, J. Non-Cryst. Solids **198**, 728 (1996)
25. R.L. McGreevy, L. Pusztai, Mol. Simul. **1**, 359 (1988)
26. P. Jovari, I. Kaban, J. Steiner, B. Beuneu, A. Schops, M.A. Webb, Phys. Rev. B **77**, 035202 (2008)

27. J. Akola, R.O. Jones, S. Kohara, S. Kimura, K. Kobayashi, M. Takata, T. Matsunaga, R. Kojima, N. Yamada, Phys. Rev. B **80**, 020201 (2009)
28. S. King, Nature **213**, 1112 (1967)
29. L. Guttman, J. Non-Cryst. Solids **116**, 145 (1990)
30. R.M. Martin, *Electronic Structure*: *Basic Theory and Practical Methods* (Cambridge University Press, Cambridge, 2004)
31. W. Welnic, in *Phase Change Materials*: *Science and Applications*, ed. by S. Raoux, M. Wuttig (Springer Science + Business Media, Berlin, 2009), p. 17

Part II
Chalcogenide Glasses

Chapter 4
Reversible Photostructural Changes

Reversible photostructural change is a phenomenon unique to vitreous chalcogenides and is not observed in either amorphous group IV semiconductors, a-As, or in crystalline chalcogenides. Possible reasons for this are discussed later in this chapter and we start with a phenomenological description of this phenomenon.

4.1 Photodarkening

4.1.1 Reversible and Irreversible Photodarkening

The first reports of photodarkening in amorphous chalcogenides date back to the late 1960s–early 1970s [1–3]. The effect was first observed in thin films but later was reproduced for melt-quenched bulk glasses [4]. The effect manifests itself as a reversible change of optical absorption induced by the bandgap light and is represented in Fig. 4.1 for the typical chalcogenide glasses As_2S_3 and As_2Se_3 [5]. Upon exposure to the bandgap light, the absorption edge of an as-deposited film (solid curve) shifts to lower energies, or longer wave lengths (dotted curve). This effect is called photodarkening. Subsequent annealing near the glass-transition temperature results in a recovery of the initial parameters, which, however, is never complete: the curve corresponding to the annealed state occupies an intermediate position between those of the as-deposited and irradiated films. Should the annealed film be irradiated by the same light, one observes a completely reversible behaviour during the irradiation-annealing cycle.

The irreversible component of the photodarkening can be annealed out at temperatures close to the glass-transition temperature. It is believed that the irreversible component is related to thermal or photo-induced polymerisation of the as-evaporated film as well as disappearance of homopolar bonds. As mentioned in the previous chapter, in an ideal glass, the energetically favourable heteropolar bonds are predominantly formed. During the evaporation that is a non-equilibrium process, numerous

A. V. Kolobov and J. Tominaga, *Chalcogenides*, Springer Series in Materials Science 164, DOI: 10.1007/978-3-642-28705-3_4,

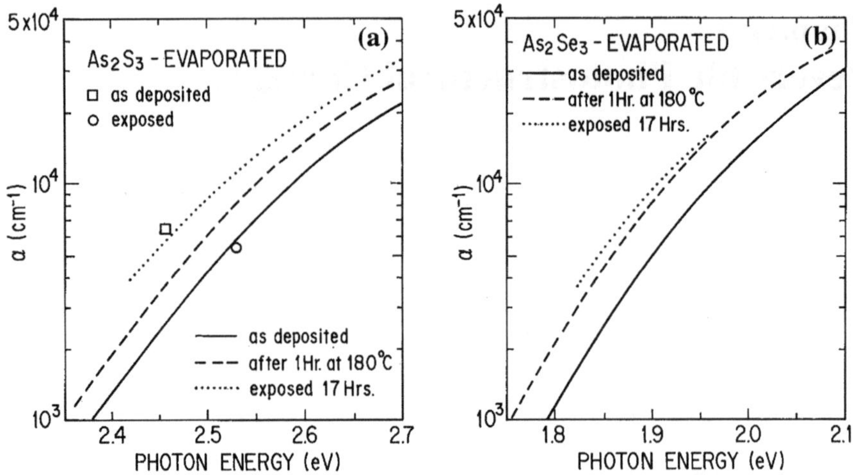

Fig. 4.1 Reversible shift of the absorption edge of typical chalcogenide glasses, As_2S_3 and As_2Se_3, under photoillumination and thermal annealing. Reprinted from De Neufville et al. [5] with permission from Elsevier

like-atom, or wrong, bonds can also be formed. Annealing results in a reorganisation of the amorphous network with the rupture of the wrong bonds and formation of energetically more stable heteropolar bonds.

Depending on the composition of the material, this process can lead to both an increase (e.g. in As-based glasses) and a decrease (e.g. in Ge-based glasses) of the material bandgap [6]. Accordingly, when the "wrong" bonds are broken and energetically more favourable heteropolar bonds are formed the bandgap either decreases or increases. Thus, while in as-deposited As-based films photodarkening is observed, in Ge-based films the absorption decreases upon illumination (photobleacing) [7]. At the same time, it should be pointed out that *in the annealed films illumination always results in photodarkening independent of the composition* [7]. Although the photo-induced changes in as-deposited films are crucial for certain applications, such as photoresists, they will not be further discussed here.

Usually the shift of the absorption edge is referred to as parallel, i.e. rigidly displaced, but in reality the slope also changes upon irradiation. This change was studied in detail for a-Se [8]. It was found that a change in the slope was equal to 629 and 316 $cm^{-1/2}\,eV^{-1/2}$ after illumination for 120 min at 13 and 77 K, respectively. This change in slope is correlated with a shift of the absorption edge by 0.157 eV (13 K) and 0.108 eV (77 K). It was suggested that the change in the slope results from the change in the width of band tails. The latter is claimed to arise from dihedral angle distortions. Although such a mechanism does sound reasonable, it should be noted that the observed changes correspond to energies in the range 2.1–2.3 eV; whereas the optical gap of a-Se equals 2.05 eV at room temperature [9]. The probed energy range thus corresponds not to the tail states but to the delocalised states above the mobility edge. Although it is quite likely that the same happens at lower energies,

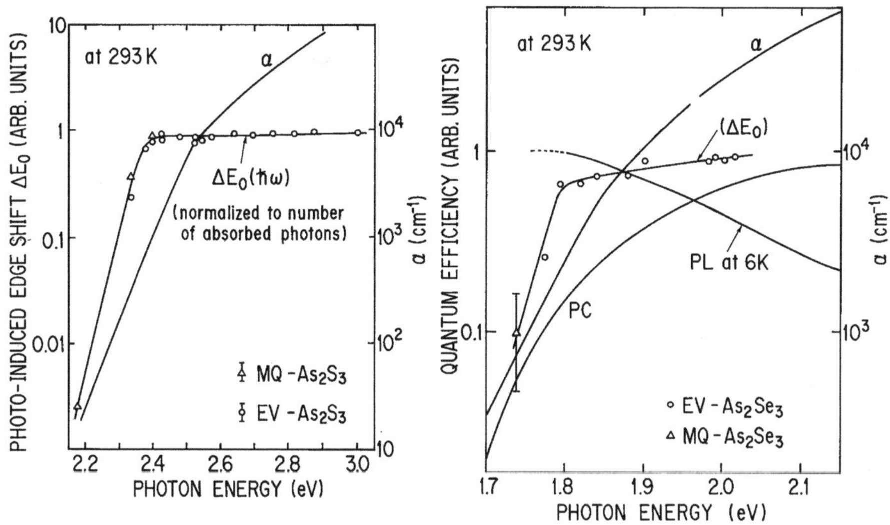

Fig. 4.2 Spectral response of photodarkening in As-based chalcogenide glases, As_2S_3 and As_2Se_3, Reprinted from Ka. Tanaka [10] with permission from Elsevier

an unambiguous conclusion cannot be made from the preceding data. The results of in situ EXAFS, IR absorption, Raman scattering and ESR studies described in detail later clearly show that bond switching plays a decisive role in this process. The bond-angle distortions most likely are a consequence of the bond switching.

The spectral dependence of the reversible photodarkening of As_2S_3 and As_2Se_3 [10] films is presented in Fig. 4.2. The long-wavelength edge of the spectral dependence of the photodarkening roughly coincides with the spectral dependence of the optical absorption coefficient. This fact suggests that for the photodarkening to occur bandgap excitation is necessary. It has been argued that the quantum efficiencies, normalised to an absorbed photon, are comparable for bandgap and sub-bandgap light [11]. At higher energies, the photodarkening either levels off (As_2S_3) or slightly increases (As_2Se_3) with photon energy [10]. In elemental vitreous chalcogens (Fig. 4.3) the spectral dependence of the photodarkening has a pronounced peak at the energy corresponding to the energy gap [12], which implies that excitation into localised band-tail states should be more efficient for the photodarkening than the excitation into higher lying extended states.

The kinetics of photo-induced structural change is described by a stretched logistic equation [13]. It has been further argued that creation of defects is initiated around pre-existing defects and that the created defects, in turn, induce weak bonds that further facilitate defect formation. The saturated value depends only on the total exposure at lower temperatures; that is, reciprocity holds between light intensity and exposure time. At higher temperatures, the saturated value of the photodarkening becomes light-intensity dependent [14] (Fig. 4.4).

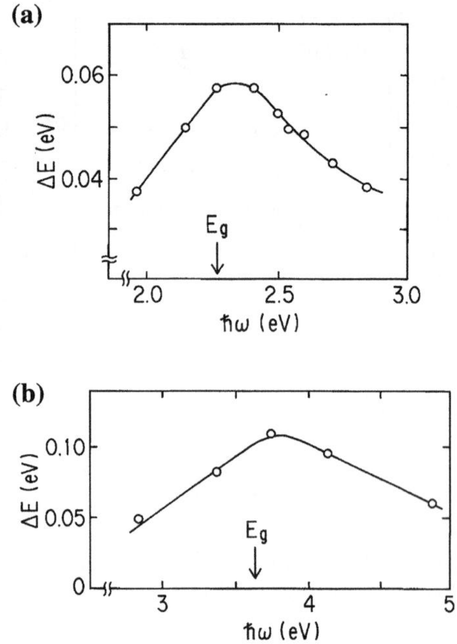

Fig. 4.3 Spectral response of photodarkening in elemental a-Se (*top*) and a-S (*bottom*). Reprinted with permission by Japan Society of Applied Physics from Tanaka [12]

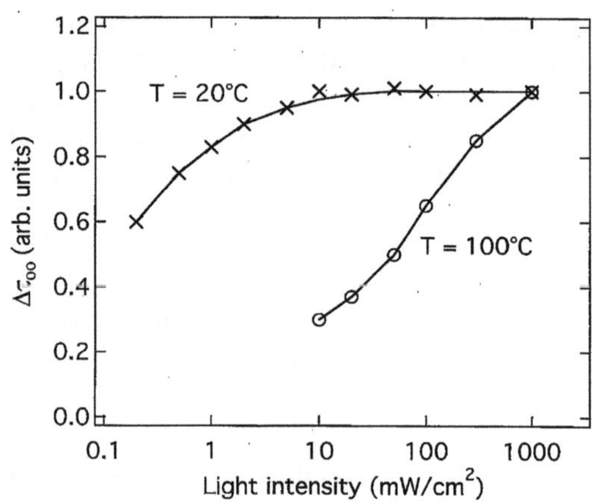

Fig. 4.4 Saturated photodarkening in an $As_{50}Se_{50}$ film at different temperatures as a function of light intensity. Reprinted from Averyanov et al. [14] with permission from Elsevier

The magnitude of photodarkening in the As–Se system increases in As-rich films, which has lead investigators [15] to conclude that As plays an important role in the process. Although this conclusion is certainly correct, the role of arsenic is not decisive: photodarkening is not observed in pure amorphous arsenic [16] whereas it does occur in pure chalcogens although this process can only be achieved at lower temperatures [1, 12, 17]. A selenium film photodarkened at 77 K has the initial absorption restored at room temperature within a few seconds. This result can be easily understood if one takes into account the fact that the glass-transition temperature of selenium is just above the room temperature [18].

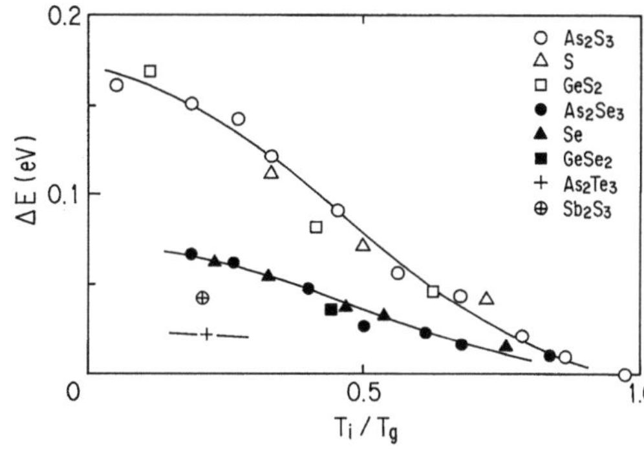

Fig. 4.5 Magnitude of photodarkening as a function of temperature (normalised to the glass-transition temperature). Reprinted from Tanaka [19] with permission from Elsevier

The above results indicate that there is a correlation between the temperature at which photodarkening is annealed out and the glass-transition temperature, that is, flexibility of the glassy network. The latter decreases as threefold coordinated arsenic or fourfold coordinated germanium atoms are added into the twofold coordinated selenium matrix, making the latter less reconstructable and thus requiring higher annealing temperatures. This correlation is clearly seen from Fig. 4.5, which shows that, for a fixed chalcogen species, the annealing temperature correlates very well with the glass-transition temperature [19], the latter being dependent on the flexibility of the glassy network.

4.1.2 Recovery of Initial Parameters

Annealing at temperatures higher than the temperature at which the irradiation was performed leads to recovery of initial transmission. The recovery process was studied by various authors [2, 17] who found that, in stoichiometric films, any increase of temperature above the photodarkening temperature leads to partial bleaching; that is, the recovery of the initial parameters is gradual. This result suggests that there is a broad distribution of different states responsible for the photodarkening. In contrast, in As-rich materials, for example for $As_{50}Se_{50}$, annealing at temperatures below 100 °C does not result in a noticeable recovery of parameters [17].

In As-rich selenides and in some sulphides, photoirradiation of previously darkened films results in partial bleaching [17, 20], provided that the second illumination is performed at a higher temperature than the initial one. This is illustrated in Fig. 4.6. Should the film photodarkened to saturation at room temperature (curve 1) be heated in the dark to a higher temperature and exposed to the same light as that which caused the photodarkening, its transmission increases (curve 2). The steady-state value of the absorption established on illumination of a previously photodarkened film is the same as that achieved by illumination of the annealed sample by the light with the same intensity and wavelength and at the same temper-

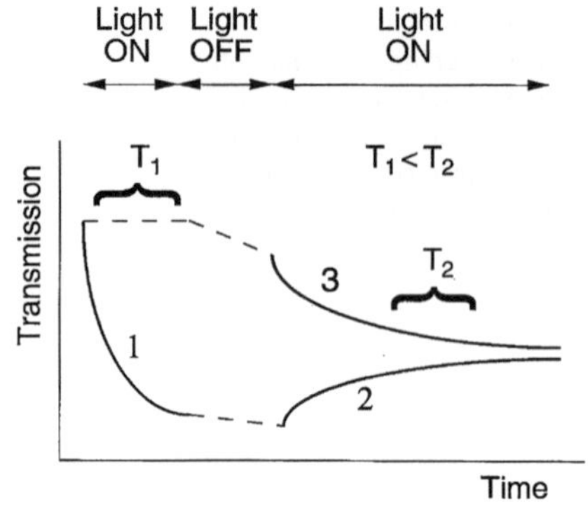

Fig. 4.6 Exposure of certain chalcogenides can lead to both photodarkening and photobleaching depending on the state of the film prior to the latest illumination. The final steady-state value is fully determined by temperature and the parameters of light during the latest illumination

ature (curve 3). This result demonstrates that the final state of the illuminated film is fully determined by the parameters (in this case, temperature) of the latest treatment and does not depend on the film's eartlier history.

The photo-induced shift of the absorption edge increases with decreasing temperature [21]. Besides, in contrast to the photo-induced change observed at room temperature, the spectrum caused by irradiation at low temperature (14 K) cannot be characterised by a simple shift of the absorption edge. This additional low-temperature component of the photodarkening can be annealed out by warming the sample to room temperature or by illuminating it with light for which $h\nu < E_g$. The presence of two components suggests that there can be at least two different mechanisms responsible for the photodarkening.

4.2 Reversible Structural Changes

4.2.1 Manifestations of Structural Changes

Reversible changes in the optical absorption are accompanied by (reversible) changes in the refractive index [15], electrical and photoelectric properties [22–25], volume [4], microhardness [26], glass-transition temperature [27, 28], dissolution rate in various solvents [29], etc. The totality of these changes has led investigators to the conclusion that the photo-induced changes in optical absorption are caused by changes in the structure. The first direct proof of the reversible structural changes was given by Ka. Tanaka [30], who demonstrated that the first peak (the so-called first sharp diffraction peak) in the X-ray diffraction pattern underwent reversible changes on illumination and subsequent annealing (Fig. 4.7).

Of special interest is the photo-induced volume change. A change in the volume (thickness) of a chalcogenide film (As_2S_3) was first reported in [4] and was one of the first indications that the structure of the material changed as a result of photoexcitation. The observed increase in the thickness of about 0.5% can be explained

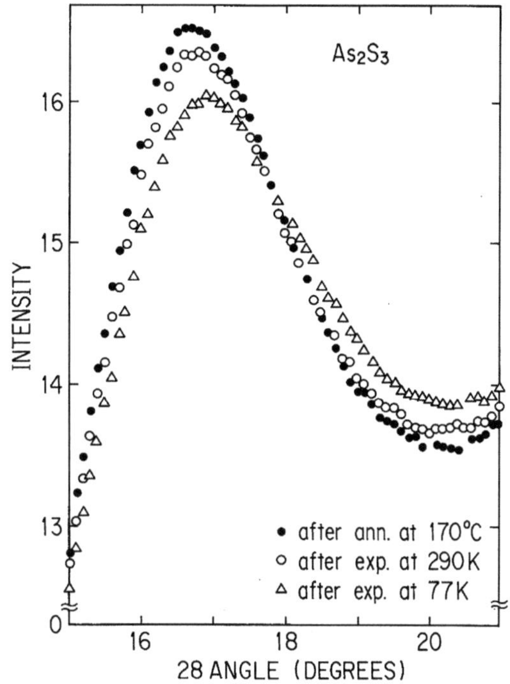

Fig. 4.7 Reversible change in the X-ray scattering pattern of As_2S_3. Reprinted with permission from Ka. Tanaka [30]. Copyright 1975 by the American Institute of Physics

by the assumption that light "pumps in" energy into the system, leading to its further disordering. The anharmonicity of the interatomic potential provides a simple explanation of the result. Illumination of As_2S_3 ($E_g = 2.4$ eV) by sub-bandgap light (He–Ne laser, ca. 2 eV 10^4 W/cm^2) results in much larger expansion reaching 6% (giant photoexpansion) [31].

Another possible explanation includes the photo-induced creation of "wrong" As-As bonds that are longer (2.57 Å) than heteropolar As-S bonds (2.24 Å). It should be noted, however, that S–S bonds (2.05 Å) are shorter than the heteropolar bonds. The heteropolar-to-homopolar bond conversion, therefore, does not necessarily result in the volume expansion.

For a long time it was believed that photodarkening and photoexpansion were two different sides of the same phenomenon. However, parallel studies of photodarkening and photoexpansion [33] have demonstrated that photoexpansion under bandgap irradiation increases faster than photodarkening. In contrast, under sub-bandgap irradiation photoexpansion increases more gradually than photodarkening. This result demonstrates that there is no one-to-one correlation between the two phenomena.

In Fig. 4.8 photoexpansion and photodarkening are shown as a function of exposure time for bandgap (top) and sub-bandgap (bottom) illumination. The absence of a direct correspondence between the processes is clearly seen. The independence of the two phenomena was further demonstrated by isochronal annealing characteristics [33]. A study of photoexpansion in As_2S_3 under intense light

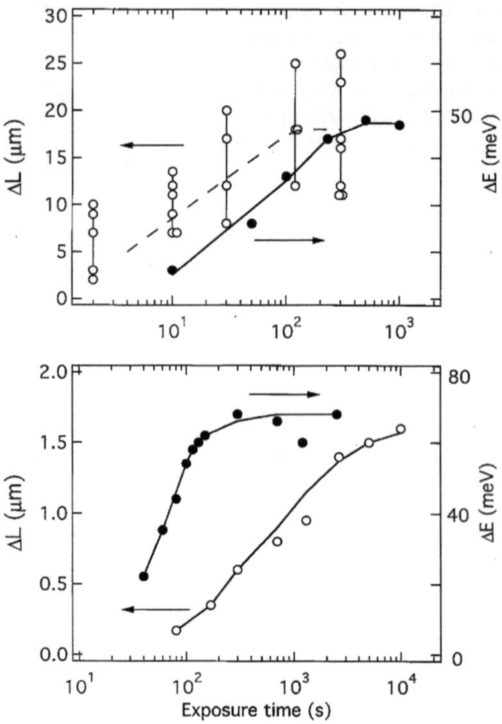

Fig. 4.8 Photoexpansion ΔL and photodarkening ΔE as a function of exposure time in As_2S_3 for the bandgap (*top*) and subbandgap (*bottom*) light. Reprinted with permission from Tanaka [33]. Copyright 1998 by the American Physical Society

(pulsed light was used to avoid heating) has shown that the expansion is maximal at about 2.0 eV, that is sub-bandgap illumination is more efficient than the bandgap light [34].

It is interesting to note that the photo-induced volume change has stable and transient components. Partial recovery of the volume change in an As_2Se_3 film is illustrated in Fig. 4.9. Similar results were obtained for a-Se, a-As_2S_3 and $GeSe_2$ implying that this is a general result.

4.2.2 Phenomenological Description: Configuration-Coordinate Model

Chalcogenide glasses are characterised by a strong electron–phonon coupling. As a result, the system in the excited state undergoes a significant lattice relaxation. The newly stabilised structure may be metastable and persist after cessation of the electronic excitation This metastable state has been described as a self-trapped exciton [35]. A phenomenological description of the photostructural changes has been given using a configuration-coordinate diagram approach (Fig. 4.10), originally developed for description of defects in solids.

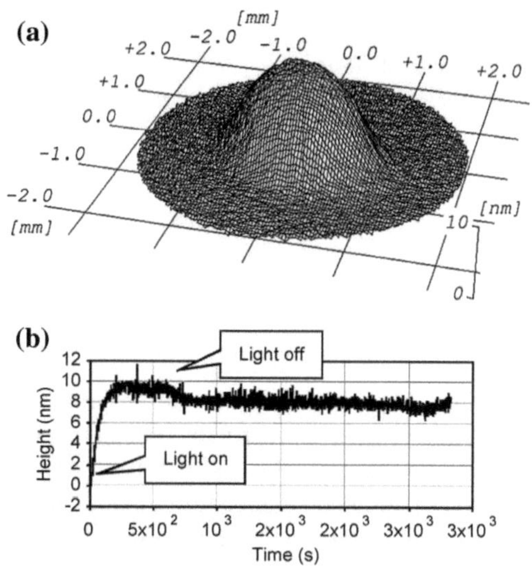

Fig. 4.9 **a** An example of surface height map for a-As_2Se_3 film ($\lambda = 532\,nm$, $W = 86\,mW/cm^2$). **b** Time evolution of the surface height for a-As_2Se_3 film. Reprinted from Ikeda and Shimakawa [32] with permission from Elsevier

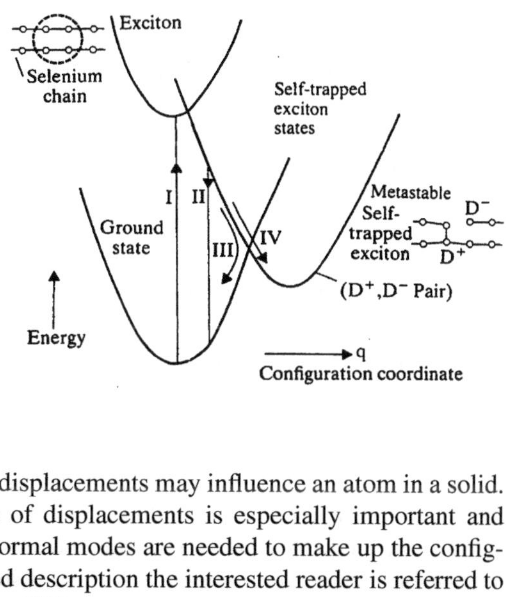

Fig. 4.10 Schematic illustration of the model for the self-trapping of excitons in chalcogenide glasses. Optical excitation (*I*) to an exciton state can be followed by two non-radiative decay channels [35], either directly back to the ground state (*III*) or to the metastable self-trapped exciton state (*IV*). Reprinted after Street [35] with permission from Elsevier

Any of a large number of atomic displacements may influence an atom in a solid. Often, a very specific combination of displacements is especially important and usually a combination of different normal modes are needed to make up the configurational coordinate Q. For a detailed description the interested reader is referred to the monograph by Itoh and Stoneham [36].

Depending on the details of the structural transformation, the models used to describe the photostructural change somewhat differ in detail [37–39] (Fig. 4.11), but all assume that the ground state of the system is characterized by a double-well potential with a barrier separating the ground and metastable states. Annealing returns the system to the ground state, overcoming the energy barrier.

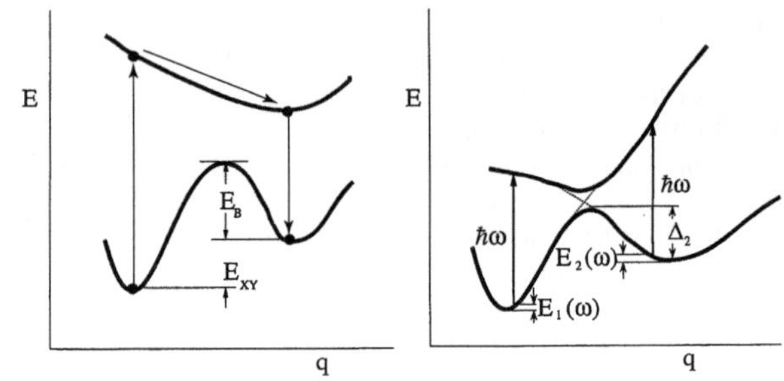

Fig. 4.11 Configuration-coordinate diagrams describing reversible photostructural change in chalcogenide glasses

In this chapter we consider the diagram shown on the right in Fig. 4.11 in more detail following [39]. "Vertical" optical transitions can occur between the two states, followed by thermal relaxation of the system to the position of new equilibrium. An equation of detailed balance for two optical transitions and a thermal transition from the metastable state to the ground state is

$$\mathrm{d}N_2/\mathrm{d}t = Q(\sigma_1 N_1 - \sigma_1 N_1) - N_2/\tau \quad (4.1)$$

where N_1 and N_2 are the number of sites in the ground and metastable states respectively, Q is the photon flux, σ_1 and σ_2 are the cross-sections for the photon absorption followed by the Frank–Condon transition from the ground state to the metastable state and vice versa, τ is the lifetime of the metastable state and depends exponentially on temperature:

$$\tau^{-1} = \nu \cdot e{-}\Delta_2/kT \quad (4.2)$$

where Δ_2 is the depth of the metastable state potential well, and ν is the "attempt-to-escape" frequency.

The optical cross-sections are given by

$$\sigma_{1,2} = \sigma_{1,2}(0)\{\hbar\omega/kT\}e^{-E_{1,2}(\omega)/kT} \quad (4.3)$$

where $\sigma_{1,2}(0) = |4\pi x_{12,21}|^2 e^2/hcn_0$ is the characteristic cross-section of the photoprocess depending on the coordinate matrix element x_{12} (x_{21}); n_0 is the refractive index at the frequency ω; $E_{1,2}(\omega)$ are energies of thermal excitations on a ground (metastable)-state curve when the vertical distance between the curves equals the photon energy (see Fig. 4.11).

Introducing the value $\gamma = N_2/(N_1 + N_2)$, that is, the fractional population of the metastable state, the balance equation takes the form

$$\mathrm{d}\gamma/\mathrm{d}t = -[1/\tau + Q(\sigma_1 + \sigma_2)]\gamma + Q\sigma_1 \quad (4.4)$$

The solution of the equation is

$$\gamma(T) = (\gamma_0 - \gamma_\infty)e^{-\alpha t} + \gamma_\infty \tag{4.5}$$

where γ_0 and γ_∞ are the starting and final values of the metastable state population, respectively, and $\alpha = Q(\sigma_1 + \sigma_2) + 1/\tau$.

The saturated value of the population of the metastable state γ_∞ is

$$\gamma_\infty = \{1 + [\sigma_2(0)/\sigma_1(0)]\exp\{[E_1(\omega) - E_2(\omega)]/kT\}^{-1} \tag{4.6}$$

If $E_1(\omega) < E_2(\omega)$, γ_∞ decreases with temperature as observed experimentally for the photodarkening. In the limit of high light intensities the saturated value of γ is determined only by temperature, in agreement with the experimental results. At lower light intensities (higher temperatures) the steady-state value of γ_∞ should become light-intensity dependent, which has been observed experimentally (Fig. 4.4).

Since probabilities of the transitions depend on $E_{1,2}(\omega)$, i.e. on the photon energy of excitation, the steady-state value, γ_∞ also depends on the photon energy (wave length) of the excitation light and through appropriate choice of the latter both photodarkening and photobleaching can be achieved.

Effect of Pressure

Because the probabilities of optical transitions depend on the mutual positions of atoms (configurational coordinate) they should also change with pressure. The pressure dependence of photodarkening has been confirmed by different groups although the details differ. Thus, in one experiment a film of As_2Se_3 was deposited on a substrate that was bent elastically in two dimensions by placing it between two coaxial tubes with different diameters and illuminating with light propagating along the tubes axis (Fig. 4.12). Depending on the location of the chalcogenide film (at the upper or lower surface of the substrate) the film was either stretched or compressed in two directions. Application or release of a load resulted in a reversible change of the stationary photodarkening value for both tensile and compressive stress [40] as shown in Fig. 4.13. It was found that compression of the film resulted in a decrease of the photodarkening with stretching giving rise to a larger effect [40] as shown in Fig. 4.14.

Application of hydrostatic pressure to As_2S_3 and Se results in an increase of the photo-induced shift of the [41, 42] as shown in Fig. 4.15. When the pressure is released, the state resulting from irradiation under pressure is stable in the dark but is reduced to 30 meV under irradiation. Investigation of the pressure dependence of photodarkening in a wide range of pressures has found that photodarkening in As_2S_3 reaches a maximum at about 5 kbar and then decreases gradually, vanishing at 57 kbar [43].

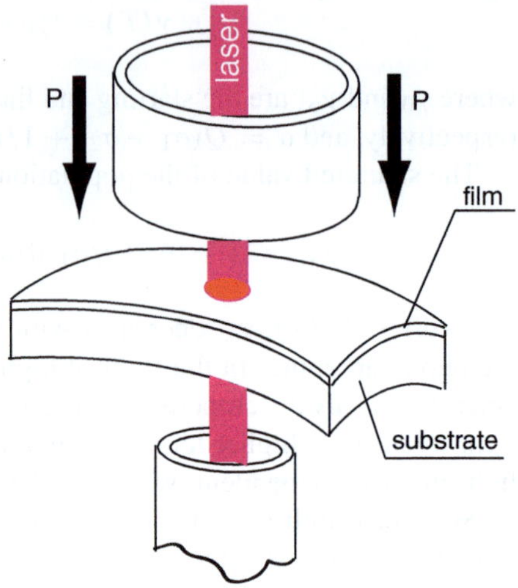

Fig. 4.12 Schematics of illumination of a chalcogenide film under two-dimensional deformation used in Ref. [40]

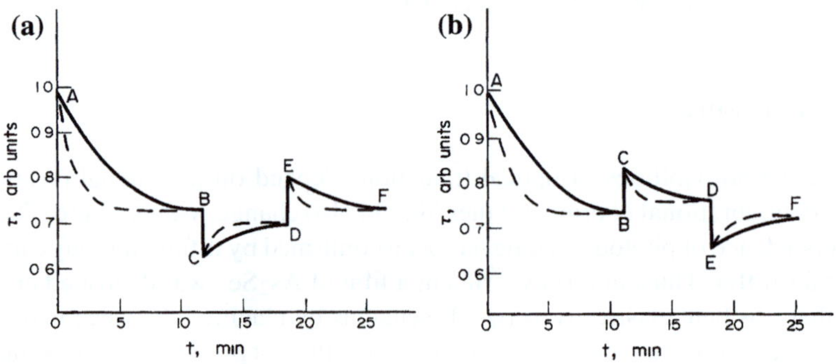

Fig. 4.13 Reversible change of photodarkening of a As_2Se_3 film exposed to light of different intensities (*solid* and *dashed curves*) under application and release of a load for the deformation conditions shown in Fig. 4.12 (panel **a** corresponds to compressing and panel **b** to stretching deformation). A load is applied at moment B and released at moment D. Drops (jumps) in transmission at these points are associated with the pressure dependence of the optical gap. Reprinted from Kolobov et al. [40] with permission from Elsevier

4.2.3 Microscopic Nature of Photostructural Changes

Although the configurational approach to photostructural changes proved to be very efficient in phenomenological description of the process, it was unable to explain its microscopic nature. Next we present the results of earlier works in which chalcogenide samples before and after photodarkening were studied using different structure-sensitive techniques.

A reversible change in the intensities of Raman peaks situated at $231\,cm^{-1}$ and $187\,cm^{-1}$ was detected in [44]. This change was attributed to an increase ($\sim$6%) in

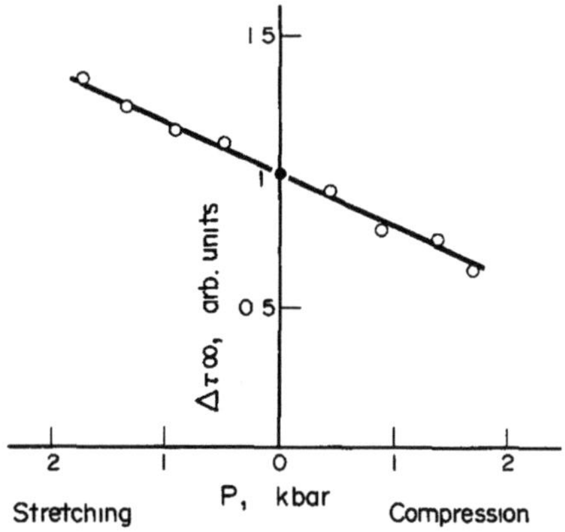

Fig. 4.14 Pressure dependence of (normalised) photodarkening of a As_2Se_3 film for the deformation conditions shown in Fig. 4.12. Reprinted from Kolobov et al. [40] with permission from Elsevier

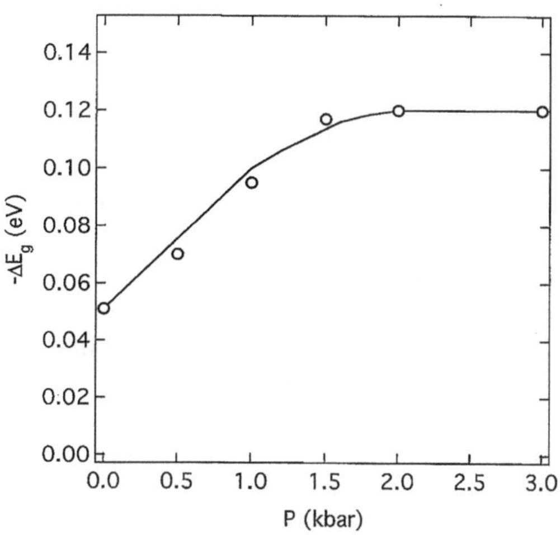

Fig. 4.15 Photodarkening as a function of applied hydrostatic pressure. Reprinted from Tsutsu et al. [41] with permission from Elsevier

the number of heteropolar As-As bonds in stoichiometric As_2S_3 on illumination. At the same time a reversible change in the Boson peak was detected [44]. Although an unambiguous interpretation of the Boson peak change is still a matter of controversy, it is usually agreed that this change represents a change in the medium range order.

This result shows the important role of As-As bonds in the process of photostructural changes and is in agreement with the conclusion based on the observation that an increase in arsenic content leads to a substantial increase of photo-induced changes of a viariety of properties of As-based chalcogenides [15]. The result, however, seemed to contradict the experimental fact that the optical gap of chalcogenides decreases on illumination, because any deviation from stoichiometry in the As-S(e) system should lead to an increase of the optical gap [6]. In addition, the formation of wrong bonds completely fails to explain photodarkening in elemental chalcogens.

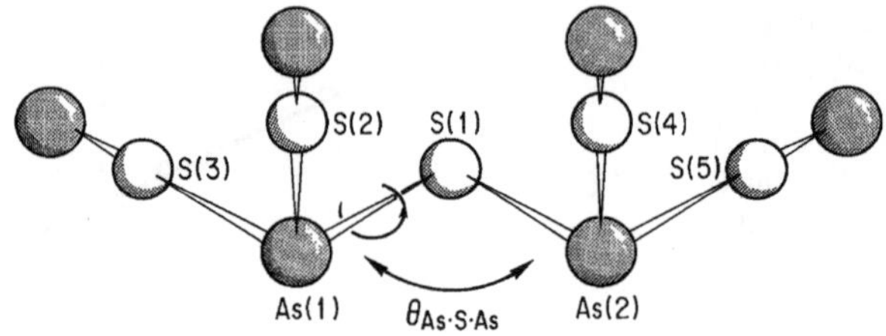

Fig. 4.16 A model of photostructural change in As_2S_3 proposed from EXAFS analysis and consisting of opening up of the valence angle subtended at a S atom with simultaneous change of the dihedral angle. Reprinted from Paesler and Pfeiffer [46] with permission from Elsevier

Studies using EXAFS (cf. 9.1.2), which is a local probe having the advantage of being able to probe independently the surroundings of different chemical constituents, is confirmed [45] that there is a small ($\sim$1%) increase in the concentration of As-As bonds but found no other changes in the first coordination sphere. The main changes occur in the second coordination sphere of arsenic. The interatomic distance between the two arsenic atoms bridged by a chalcogen atom reversibly increases, which means that the valence angle subtended by a chalcogen atom opens up on illumination. Interestingly, the distances in the third coordination sphere (As-S-As-S) remain unchanged, which led the authors to the conclusion [45] that the dihedral angle between two adjacent AsS_3 pyramids changed simultaneously with the increase of the valence angle at the bridging chalcogen atom (Fig. 4.16). (While the higher order shells are usually smeared out in amorphous materials, the authors observed a reversible change between the illuminated and annealed structures measured at low temperature.) Further studies using differential anomalous X-ray scattering [47] have led to a similar conclusion. It was suggested that excitation of lone-pair electrons and creation of an exciton change the electrical screening locally around the chalcogen atom. The atomic arrangement responds to this through fluctuations of the bond angles. The disorder in the bond angle strengthens the localised state. The exciton finally recombines either to the new (metastable) or to the old (ground state) geometry. This model is in essence very close to the one previously suggested in [42].

A unified model for photostructural changes has been suggested in [48]. It was suggested that not just the primary covalent (intramolecular) bonds but also weaker (intermolecular) bonds between non-bonded chalcogen atoms in neighbouring layers should be considered in this process. Within this approach, after absorption of a photon and subsequent excitation of an electron into the conduction band (anti-bonding states), the attractive interaction decreases. This changes the balance between the attractive and repulsive interactions and causes an increase in the intermolecular bond length. One consequence of this model is that a reversible increase in volume is predicted on illumination. An estimation of the expected expansion of the glass using experimental data for the photo-induced change in the elastic moduli gives a value for photo-induced expansion of 0.63% in As_2S_3 which compares well with the experimentally observed volume expansion of 0.6% [4]. Intramolecular bond breaking processes, such as homopolar-to-heteropolar bond conversion (and vice versa) and self-trapped excitons have also been considered in detail and it has been concluded that *intermolecular, rather than intramolecular, bond breaking processes are predominant* in the As-based chalcogenides.

A different approach was suggested in [49]. These authors argue that remnants of crystalline layers are preserved in the glass. Upon photoexcitation, electron–hole pairs are generated. The holes diffuse apart while electrons get trapped. The Coulomb repulsion between trapped electrons changes the interlayer interaction and, consequently, the geometry, resulting in repulsion of the layers (photoexpansion) and a slip motion along the layers. Such a change in the layer geometry results in the photodarkening through the mechanism proposed in [50], namely due to increased repulsive interaction between the lone-pair electrons.

4.2.4 In Situ Studies of Reversible Photostructural Changes

Earlier measurements were performed ex situ; that is, only annealed films and irradiated films were studied. As a result, information about the excited state of the semiconductor and hence about the mechanism of the structural change has been missing. Next we present results of in situ studies that provided direct information on the nanometer-scale mechanism of the photostructural change.

In Situ EXAFS

In this experiment, first an EXAFS spectrum of the starting (annealed) amorphous Se film was measured after cooling the sample down to 30 K. The sample was subsequently irradiated in a cryostat for several hours and an EXAFS spectrum of the photoexcited film kept under light irradiation was measured. Immediately after this, the light was switched off and the EXAFS spectrum of the irradiated sample (but no longer kept under light irradiation) was measured [51].

Raw EXAFS oscillations as a function of photoelectron wavenumber k for the annealed and photoexcited a-Se fiim and the corresponding Fourier transforms are shown in Fig. 4.17. A higher amplitude of the signal for the photoexcited sample is clearly seen. The data analysis reveals a reversible increase in the coordination number by about 5% in the photoexcited state (Fig. 4.18). Associated with the change in coordination, the disorder (mean-square relative displacement, or MSRD) also increases under light irradiation. After the cessation of illumination, the local change of coordination disappears whereas the light-enhanced structural disorder remains [51].

The observed light-induced increase in the coordination number indicates the formation of higher than twofold coordinated sites. This may look very unusual because normally one would expect the occurrence of photo-induced bond breaking,

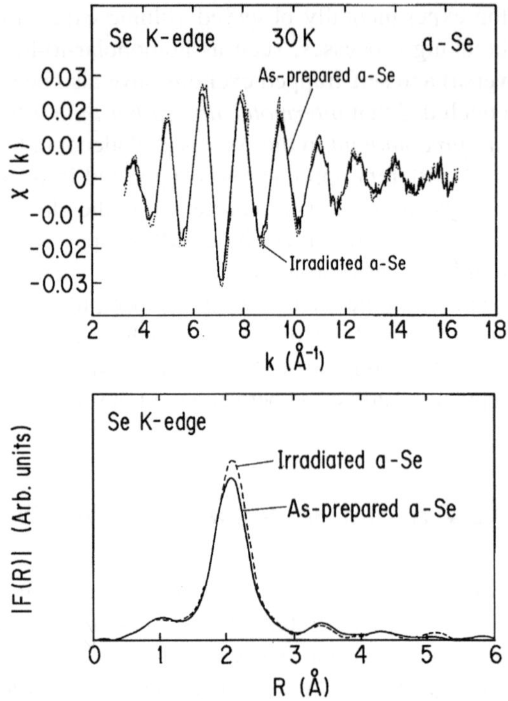

Fig. 4.17 Raw EXAFS oscillations (*top*) and magnitude of Fourier transforms (*bottom*) for the as-deposited and photoexcited a-Se film. Reprinted with permission from Kolobov et al. [51]. Copyright 1997 by the American Physical Society

and hence a decrease in the average coordination. This unusual behaviour observed in a-Se has been interpreted as follows.

Under light irradiation, electrons from the top of the valence band, formed in this case by lone-pair electrons, are excited into the conduction band, leaving behind unpaired electrons in the former lone-pair orbitals. If the excited atom finds itself close to the neighbouring chain, its unpaired electron in the formerly lone-pair orbital interacts with lone-pair electrons of the neighbouring chain, creating an interchain bond (Fig. 4.19). Comparison of total energies for the photoexcited atom close to a neighbouring chain and for two cross-linked chains, shown at the bottom of the same figure, clearly demonstrates that such a process is energetically favourable [52]. A pair of threefold neutral defect sites, $2C_3^0$, is therefore formed, introducing local distortion of atoms observed as an increase in MSRD. Subsequent evolution of the $2C_3^0$ defect pair is described in Sect. 4.3.1.

The formation of these new interchain bonds is expected to introduce a local distortion around the threefold coordinated sites, which would result in an increase in the MSRD. The formation of dynamical bonds would also cause a displacement of near-neighbour atoms. In situ EXAFS studies of photostructural changes in binary chancogenides were also performed [53].

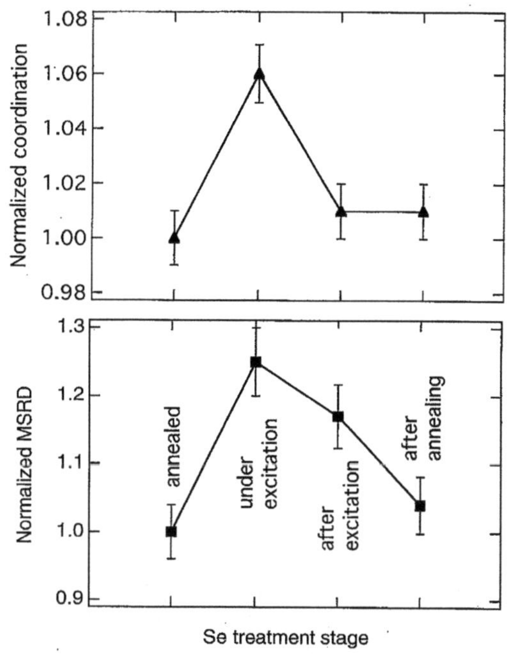

Fig. 4.18 Reversible increase in the average coordination number and disorder in the photoexcited state of a-Se. Reprinted with permission from Kolobov et al. [51]. Copyright 1997 by the American Physical Society

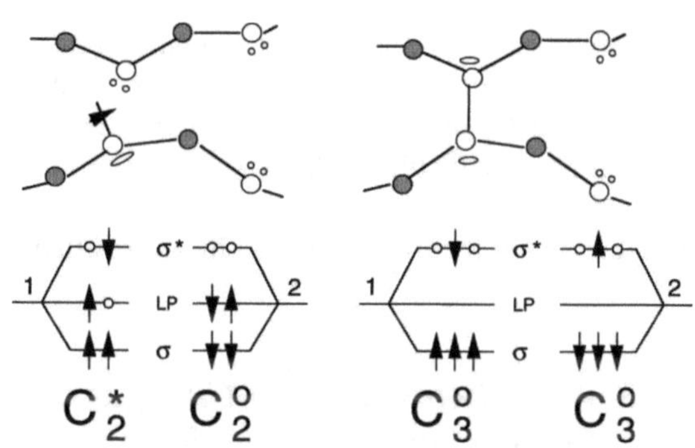

Fig. 4.19 Energy diagram of the formation of an interchain Se-Se bond in the photoexcited state. *Left-side panel* a photoexcited Se atom next to a neighbouring chain. *Right-side panel* a dynamical bond formed through interaction of the unpaired electron of the photoexcited Se atom and the lone-pair electrons of the neighbouring chain. Reprinted from Kolobov et al. [52] with permission from Elsevier

Raman Scattering

Further arguments for the formation of triply coordinated pairs in the photoexcited state come from Raman scattering studies. Figure 4.20 shows Raman spectra for an HV configuration (the electric-field vectors of the exciting and scattered light are orthogonal) taken at 20 K [52, 55] for the same three states of the sample: annealed, under photoexcitation and after photoexcitation.

Fig. 4.20 Polarised Raman scattering spectra demonstrating an appearance of new peaks in photoexcited Se, Reprinted from Kolobov et al. [54] with permission from Elsevier

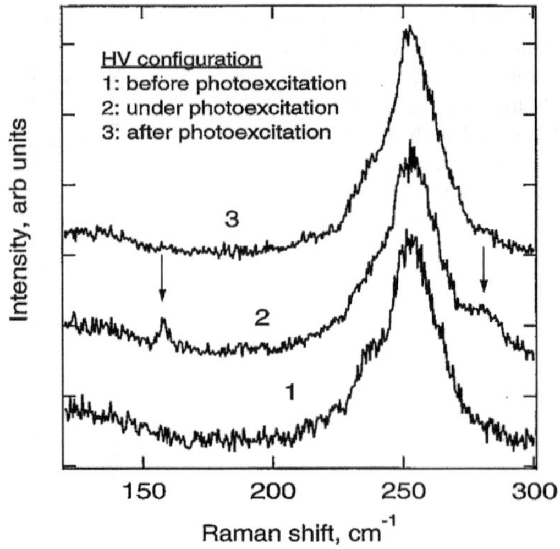

The spectrum for the photoexcited state is clearly different from those for the unexcited sample, both prior to irradiation and after the excitation (photodarkened but no longer photoexcited sample). New peaks at 280 and 158 cm^{-1} appear, none of which have been previously reported for selenium. These two peaks provide direct evidence of changes in the bonding configuration in the photoexcited state. A conclusion that "the nature of bonding during excitation is fundamentally different from that before and after" has also been made from structural modelling [56].

It is interesting to note that a similar result, that is, the appearance of a new peak shifted by about 25 cm^{-1} to higher energies as a result of photoexcitation was observed in the binary As_2Se_3 glass (Fig. 4.21) [57]. That study was performed ex situ but we believe that the very low temperature used (1.6 K) kept the photoexcited structure stable for some time and the comparison with the in situ data is justified.

4.2.5 Why Photostructural Changes are Restricted to Vitreous Chalcogenide Semiconductors

The top of the valence band in chalcogenides is formed by chalcogen lone-pair electrons and it is believed that a change in the interaction of lone-pair electrons resulting from the structural change is responsible for the photodarkening. The importance of the chalcogen lone-pair electrons is further emphasised by the fact that photodarkening disappears in metal-doped chalcogenides. This result was obtained for the incorporation of Cu, Ag and Zn [58, 59]. One of the reasons for the disappearance of photodarkening could be that, upon addition of the metals, the coordination of the chalcogen atoms increases. The chalcogen atoms become fourfold coordinated; that is, lone-pair electrons are no longer present in Cu-doped glasses containing

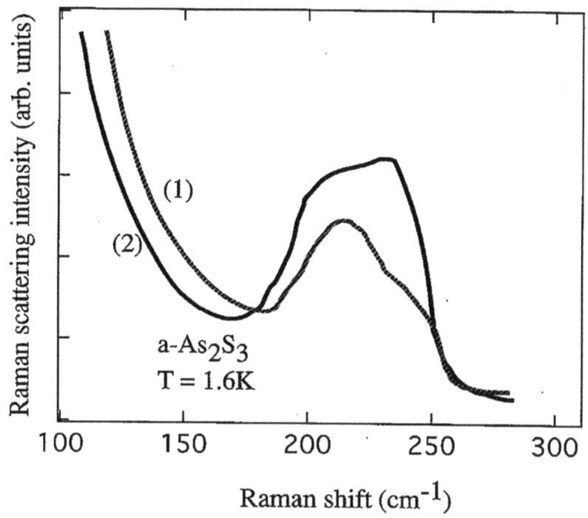

Fig. 4.21 Raman spectra of As_2Se_3 glass before (*1*) and after (*2*) illumination. Reprinted from Cernogora et al. [57] with permission from Elsevier

about 25 at.% of copper [60]. The nanoscale mechanism uncovered by the in situ EXAFS [51, 52] clarifies the role of lone-pair electrons: their presence is vital for the formation of dynamic interchain bonds through which the bond switching takes place.

Carrier localisation, which is intrinsic to disordered structures, is another important requirement. Upon absorption of a photon, electron–hole pairs are generated and the non-equilibrium carriers are localised in the tail states. Should they spend there a time comparable to—or longer than—the inverse phonon frequency, the lattice would be able to relax. This relaxation can lead to bond switching, that is, to photostructural changes. This process implies that geminate recombination is important for photodarkening. If the photon energy is larger than the optical gap, the carriers may diffuse apart during the thermalisation process, contributing to photoconductivity and lowering the probability of the structural relaxation. For example, a maximum in the quantum efficiency of photodarkening of a-S and a-Se (Fig. 4.3) is clearly observed, its position roughly corresponding to the optical gap.

Application of an external electric field can separate the carriers and this field-induced decrease of photodarkening in As_2S_3 has been observed experimentally [61]. This result suggests that geminate recombination is essential for both elemental and compound materials. It was also reported that the applied dc field changed the initial kinetics of photodarkening but did not have any effect on the saturated value [62]. At the same time, it should be noted that attempts of other investigators to reproduce the effect of external electric fields on photostructural changes have failed (V. M. Lyubin, unpublished).

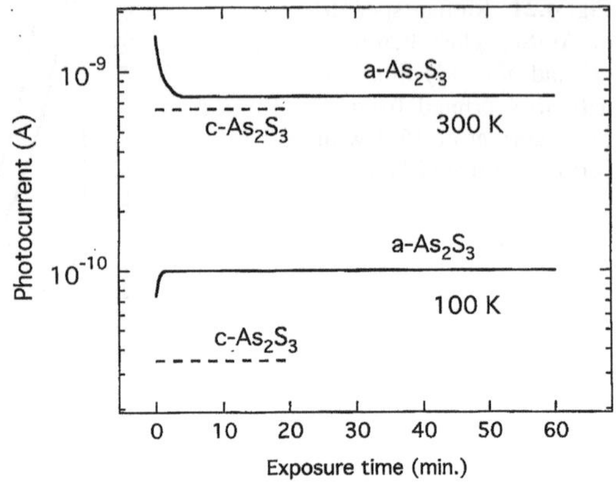

Fig. 4.22 Photocurrents of amorphous (*solid lines*) and crystalline (*dotted lines*) As_2S_3 as a function of exposure time at 100 and 300 K (illuminated by a Hg lamp, 50 mW/cm^2), Reprinted with permission from Toyosawa and Tanaka [64]. Copyright 1997 by the American Physical Society

4.3 Photo-Induced Defect Creation and Modification

4.3.1 Photo-Induced Electron Spin Resonance

Binary Chalcogenides

Following photo excitation, non-equilibrium charge carriers recombine. However, when the two excited C_2^* sites form a pair of C_3^0 defects, moving one of the lone-pair electrons to the bonding state the corresponding empty lone-pair state is pushed up to the antibonding state. Recombination in the excited state configuration is therefore impossible because there are no empty states in the valence band. Bond breaking is a necessary condition for the recombination, which means that the newly formed bonds are essentially dynamical and disappear after photoexcitation. The dynamical nature of the photoexcited state also has been stressed in [63].

Before the bonds break, there are non-equilibrium electrons in the conduction band. The observed photo-induced increase in the conductivity of amorphous chalcogenides under low-temperature excitation (Fig. 4.22) [64], found to be correlated with midgap absorption, is most likely caused by this mechanism. No such effect has been observed in the crystalline material. If, on the other hand, the excitation is performed at higher temperatures where bond breaking proceeds faster, the usual decrease [23] in the photocurrent is observed (Fig. 4.22).

After the bond breaking, unpaired electrons are created, which should result in the appearance of an ESR signal. In the first report by Bishop et al. [65], the photo-induced spin density was reported to be in the order of 10^{18} cm^{-3}. Later, employment of laser sources for the excitation resulted in an increase of the spin density up to 10^{20} cm^{-3} [66].

Biegelsen and Street [67] have found that photo-induced ESR kinetics in a typical material As_2S_3 had two components, a fast one and a slow one as can be seen from Fig. 4.23. They suggested that the fast component was due to excitation of

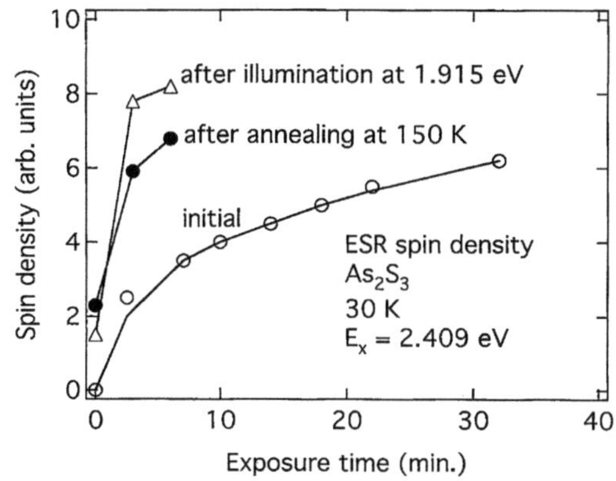

Fig. 4.23 Kinetics of photo-induced ESR in As_2S_3, Reprinted with permission from Biegelsen and Street [67]. Copyright 1980 by the American Physical Society

pre-existing defects. Most likely it was this component that was observed in the earlier study [65]. The slow component was associated with the photo-induced creation of metastable defects. The photo-induced ESR is annealed out at around 300 K or by illumination with long-wavelength light. An important finding of the authors [67] was that in the secondary illumination cycle, following annealing at 300 K, the kinetics changed drastically, namely, the fast component became much larger. The authors explained this increase in the fast component by the formation of new negative-U centres in As_2S_3 as a result of the photoexcitation. The number of defects in the secondary illumination cycle was therefore much larger than the number of the native "dark" defects, which manifested itself as an increase in the fast component. The authors further suggested that these defects, which they argued to be valence-alternation pairs, were also responsible for the photodarkening.

One further consequence of this work was an unambiguous proof that As_2S_3 had a negative correlation energy. The optically induced ESR observed previously [65], although providing strong support, could not serve as a final proof for the existance of negative-U because the observed spin density was less than $10^{18}\,cm^{-3}$ and chalcogenide glasses can easily have up to $10^{19}\,cm^{-3}$ impurities, such as oxygen, that can be responsible for the observed spin density. In addition to the possibility that the impurity itself can be ESR-active it is also possible that inclusion of an impurity perturbs the chalcogenide network, and the observed ESR signal, although coming from chalcogen species, may originate in the vicinity of impurities thus being caused by their presence [68].

In a binary glass, As_2S_3, the signal was shown to contain four different components. Two of the defects are annealed out at about 180 K whereas the other two are annealed out at about 300 K [69]. The two centres that anneal out at lower temperatures exhibit identical inducing and annealing behaviour. They were called type-I centres and represent about 15% of the total induced spins. One of these defects has been identified as a S_I centre that consists of a hole on a nonbonding $3p$ orbital of a sulphur atom. The other one (As_I) represents an electron on a non-bonding

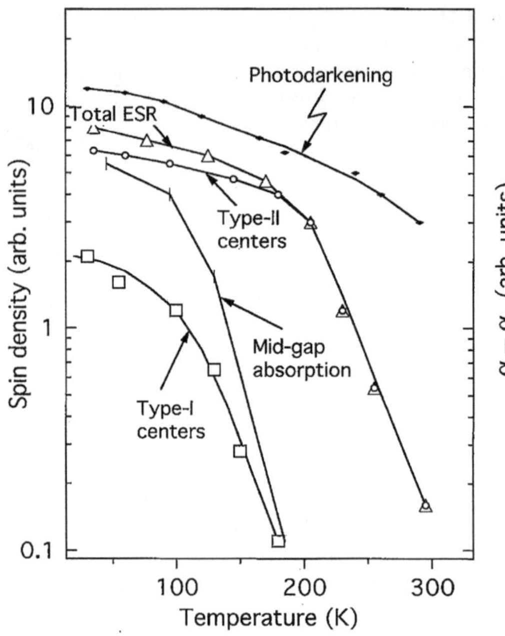

Fig. 4.24 Comparison of thermal decay of photo-ESR and photodarkening in As_2S_3. Reprinted with permission from Hautala et al. [69]. Copyright 1988 by the Americal Physical Society

sp-hybridised orbital of an arsenic atom. It was argued that these centres are generated simultaneously as a result of the rupture of an As-S bond.

The more stable (type II) centres represent 85% of the induced spins. They were identified as a hole on a non-bonding $3p$ orbital of a sulphur atom (S_{II}) and an electron on a nonbonding $4p$ orbital of a twofold coordinated arsenic atom (As_{II}). These centres were believed to result from rupture of homopolar S–S and As-As bonds, their relative concentration depending on the glass stoichiometry. A different assignment of the type-II centres, namely, to bonds broken in the immediate vicinity of homopolar bonds, was suggested in [70].

Later, from the analysis of the temperature dependence of the photodarkening and photo-induced ESR (Fig. 4.24) it was argued that because the annealing behaviours were quite different, the underlying mechanisms for the two processes were also different [69]. Below in this chapter it is suggested that, contrary to this conclusion, a relationship between photodarkening and photo-induced creation of valence-alternation pairs does exist.

Elemental Selenium

The preceding results were obtained on compound chalcogenides for which both experiment and theory [71] demonstrate negative correlation energy for the defects. In the case of elemental selenium the situation is more complicated. Calculations for

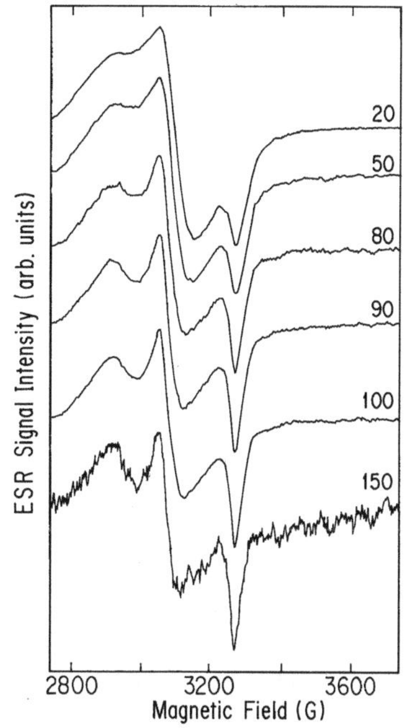

Fig. 4.25 Temperature evolution of photo-ESR in a-Se; the corresponding temperatures (in Kelvins) are indicated next to each curve. Reprinted with permission from Kolobov et al. [81]. Copyright 1998 by the American Physical Society

a-Se have, unexpectedly, found that in a-Se the correlation energy is positive [72]. Because a-Se is an important model material its response to irradiation is treated here in more detail following Ref. [73].

Although in the dark an ESR signal is not observed, photoexcitation results in an appearance of such a signal. An ESR spectrum measured at 20 K is shown in Fig. 4.25 (top curve). Annealing above 150 K for 20 min results in the disappearance of the ESR signal. Isochronal annealing experiments show that the line shape of the ESR signal changes at successive annealing steps for $T_a < 100$ K whereas no change occurs at higher temperatures ($T_a > 100$ K). This implies that at T < 100 K two (or more) centres coexist whereas at T > 100 K there is only one kind of centre. The line shape of this centre (Fig. 4.25, spectra taken after annealing between 100 and 150 K) reveals the characteristics of a triclinic centre with its principal values of g-tensors being 2.22, 2.099 and 1.983. The spin density of the triclinic (T) centre induced at 20 K is estimated to be 5.5×10^{19} cm^{-3}. The signal for lower temperatures can be deconvoluted into the triclinic component and another rather broad (B) and structureless component. The g-value and the linewidth for the B-centre are 2.09 and 150 G, respectively, and the spin density at 20 K is 5.3×10^{19} cm^{-3}. It is of interest to note that the starting concentrations of the two kinds of photo-induced defects are the same, indicating that the defects are created in pairs.

The annealing temperature dependencies of the spin density of the two centres (T and B) are shown in Fig. 4.26. In the region $T_a < 100$ K, it is found that these two components show complementary behaviour, namely, a decrease in concentration of

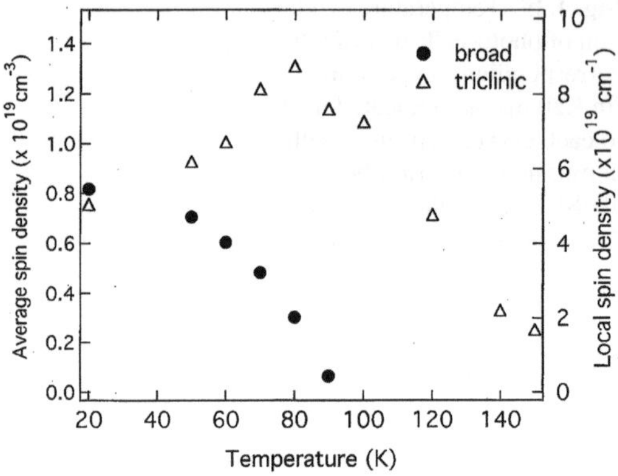

Fig. 4.26 Temperature dependence of concentrations of triclinic and isotropic (broad) ESR-active centres in a-Se. Reprinted with permission from Kolobov et al. [73]. Copyright 1997 by the American Physical Society

B-centres goes in parallel with an increase in concentration of T-centres. This result demonstrates that the *B-centres are converted into the T-centres with increasing temperature*, and that the T-centre is more stable than the B-centre. For $T_a > 100\,\text{K}$, only the T-centre is observed. Its intensity decreases with temperature increase. Finally, at $T_a > 150\,\text{K}$ the ESR signal disappears completely.

In the subsequent irradiation cycle the ESR signal saturates much more quickly whereas the line shape is identical to that observed in the first cycle. This result is similar to the case of As_2S_3 [67] and suggests that the number of defects existing in the semiconductor (although not ESR active any more) is larger after the primary irradiation process and only their excitation is needed to obtain an ESR signal.

The observed high density of spins ($\sim 10^{20}\,\text{cm}^{-3}$) clearly shows that the signal is not caused by impurities but is intrinsic to a-Se. This result provides direct evidence of negative-U in this material.

We now discuss the structure of the two kinds of defects. It was suggested [65] that the more stable triclinic centre arises from a localised hole on the Se site, because the g-shift is positive and large. If the T-centre is identical with the C_0^1 centre, the unpaired electron is located on one of the non-bonding orbitals while other orbitals are filled. Therefore, this centre is hole-like, which is consistent with the sign of the g-shift. On the assumption of sp-hybridization for single-bonded radicals with a fractional s-character of 0.4 the principal values of the g-tensor have been calculated to be $g_1 = 2.22$, $g_2 = 2.075$, and $g_3 = 2.0023$ [74]. These values are in good agreement with the experimental data and allow one to conclude that the T-centre is a neutral dangling bond, C_0^1. In addition, it is reasonable to expect that a singly coordinated atom will be more relaxed and thus the anisotropy of the centre should clearly appear. A triply coordinated centre will be more distorted due to the network strain and its structure should be smeared out. In the C_0^3 centre, the electron is in the anti-bonding orbital. The degeneracy of three bonding orbitals is lifted as a consequence of the lattice distortion. Therefore, it is expected that the C_0^3 centre shows less anisotropy (than C_0^1) and a more broadened spectrum due to the disorder.

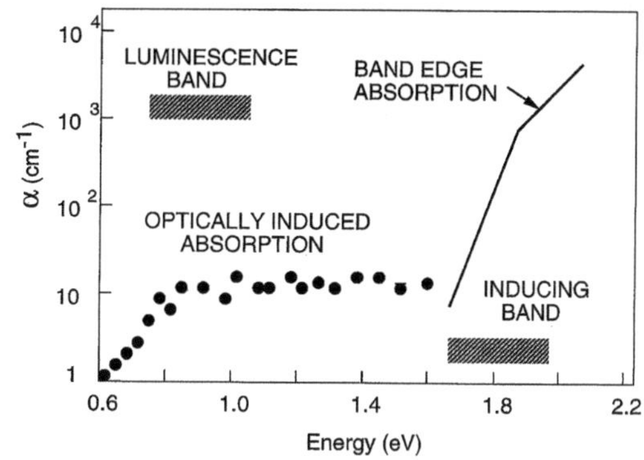

Fig. 4.27 Photo-induced midgap absorption in As_2Se_3 at 6 K. Reprinted with permission from Bishop et al. [75]. Copyright 1975 by the American Physical Society

4.3.2 Photo-Induced Midgap Absorption

The irradiation of chalcogenide glasses that results in the appearance of an ESR signal also leads to an increased absorption at about midgap energies (Fig. 4.27) [65]. The energy range of the midgap absorption and its very broad distribution correspond to excitation of tail states protruding deep into the gap. The induced absorption is stable at low temperature but can be annealed out in a way similar to ESR either by thermal annealing or by irradiation of the sample with photon energies corresponding to the induced absorption band.

Photo-induced ESR and midgap absorption are usually attributed to the same centres [65].

4.3.3 Photoluminescence Fatigue

Excitation of chalcogenides with continuous light causes a decrease in the photoluminescence intensity (the fatigue effect), which implies that after the excitation recombination centres obtain a new metastable state that is inefficient for photoluminescence [65, 76]. The change in the charge state of defects resulting in an ESR signal may also be the reason for the fatigue effect. The correlation between photo-induced ESR, midgap absorption and photoluminescence fatigue suggests that all these phenomena have the same underlying mechanism [67, 77].

4.3.4 Photo-Induced Change in AC-Conductivity

The photo-induced creation of neutral and charged defects causes a change in the ac conductivity and capacitance. The underlying mechanism for ac transport was suggested to be due to correlated two-electron hopping [78]. The kinetics of the

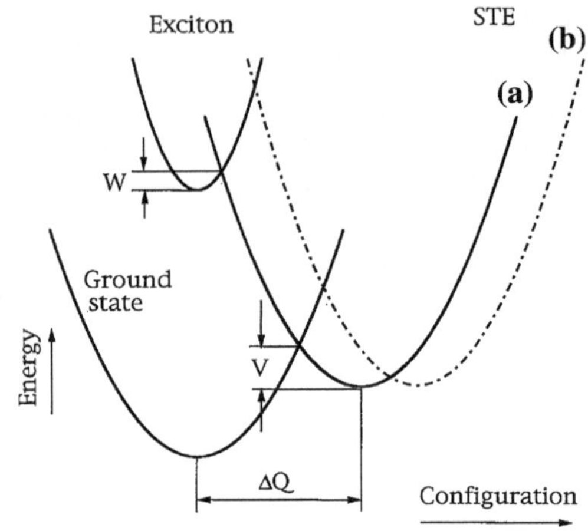

Fig. 4.28 Illustration of correlations between activation energy of creation, *W* and annealing, *V* of metastable defects. Reprinted with permission from Shimakawa and Elliott [24]. Copyright 1988 by the American Physical Society

photo-induced change in capacitance has two components: fast and slow. Similar to photo-induced ESR, the slow component is attributed to photo-induced creation of new defects and subsequent hopping between them [79].

Photo-induced creation of valence alternation pairs results in an increase of ac conductivity [24]. It was found that photoexcitation increases the ac conductivity, the degree of the increase being larger if the irradiation is performed at lower temperatures. Irradiation at 90 K increases the ac conductivity by almost an order of magnitude. Annealing results in recovery of the initial conductivity. An interesting finding was that the change in ac conductivity induced at lower temperature was also annealed out at a lower temperature (compared to the change induced at higher temperature). An explanation for this behaviour was suggested within the coordination-coordinate model shown in Fig. 4.28 that assumes a distribution of the electron–phonon coupling strengths [24]. States with larger electron–phonon coupling can only be optically populated at higher temperature and at the same time the barrier separating them from the corresponding ground state is also higher, resulting in higher annealing temperatures.

4.4 Relationship between Photo-Induced ESR and Reversible Photostructural Change (Photodarkening)

In this section we discuss the relationship between photo-induced ESR and reversible photostructural change that is characteristic of amorphous chalcogenides. It has already been mentioned that after the introduction of the valence alternation pairs model, it was suggested [92] that photo-induced charged defects created in large concentration were responsible for photodarkening. However, in a later study of

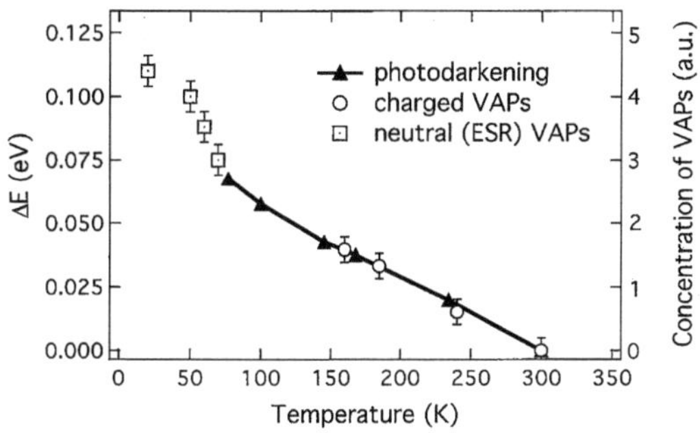

Fig. 4.29 Correlation between photodarkening and concentration of photo-induced VAPs. Reprinted with permission from Kolobov et al. [81]. Copyright 1998 by the American Physical Society

photo-induced ESR in As_2S_3 [69] it was argued that the origins of photodarkening and photo-induced ESR are different.

It is important to realise that (for the example of a-Se) after annealing at temperatures $T > 160$ K photocreated VAPs do not necessarily disappear. What disappears is the ESR signal. Its disappearance, however, is related not to the presence or absence of VAPs but to their charge state. The defects can become charged (e.g. $C_3^+C_1^-$ pairs) and hence ESR inactive, but they are nonetheless present in the sample. This fact is clearly demonstrated by the presence of the faster component in the ESR kinetics during the secondary irradiation process. The relative amplitude of the fast component is a measure of the concentration of charged VAPs. In Fig. 4.29 the temperature dependencies for photodarkening in a-Se (the data taken from [42]) and that for the concentration of charged VAPs are shown. The two curves are normalised. A very good agreement between the temperature behaviour of the two effects is obvious.

Furthermore, the concentrations of neutral (ESR-active) VAPs can be added on the same scale as the charged VAPs. One can easily see that this concentration also correlates rather well with the magnitude of the photodarkening.

The authors propose that photodarkening is closely correlated with the concentration of VAPs, either neutral or charged. The underlying mechanism can be the following. In the annealed state, the lone-pair orbitals of selenium, forming the top of the valence band, are oriented almost perpendicular to each other to minimise the energy of the system (Fig. 4.30, left) [80]. Any change in the atomic positions, such as the creation of VAP defects (either neutral and hence ESR active or charged and not detectable by ESR), results not only in displacements of the particular atoms involved but also in displacements of their neighbours from their original positions. This increases the repulsive interaction between the lone-pair electrons and results in a shift of the top of the valence band upward in energy thus leading to a decrease in the forbidden gap, that is, photodarkening. It has indeed been demonstrated that the change in the spatial orientation of the adjacent lone-pair orbitals leads to the formation of states at band edges [80]. The authors suggest that photo-induced VAPs

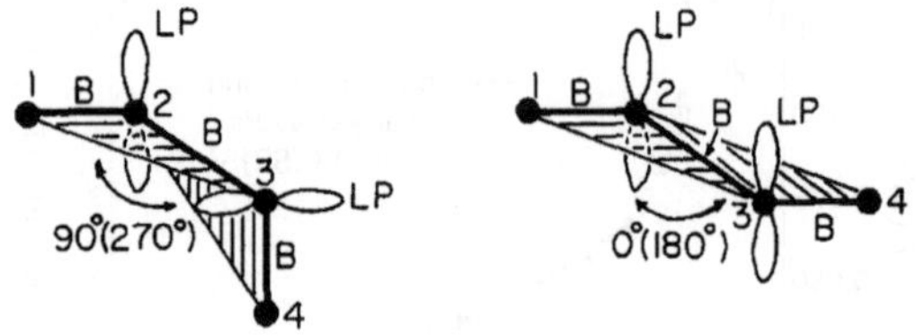

Fig. 4.30 Normal bonding arrangement for nearest neighbour Se atoms (*left*). *B* indicates the bonding directions, and *LP* the orientation of the lone-pair orbitals. For the normal bonding, the dihedral angle is either 90 or 270°. Upon structural rearrangement, the bonding arrangement can change. For the extreme case a dihedral angle distortion becomes 0 or 180° (*right*) which places the lone-pair orbitals on nearest neighbour Se atoms into a parallel alignment and results in the formation of localised states at band edges. Reprinted after Wong et al. [80] with permission from Elsevier

are a kind of seeds around which a structural relaxation takes place causing the photodarkening.

The results presented previously sum up to the following atomic-scale model of photo-induced structural change. After photoexcitation and creation of interchain dynamic bonds, bond breaking takes place at threefold coordinated atoms, resulting in a pair of ESR-active centres. One of the defects is singly coordinated and the other one is triply coordinated. This neutral VAP is not stabilised by direct charge transfer. Instead, the triply coordinated neutral defect is converted into another singly coordinated defect at temperatures around 90 K and a pair of dangling bonds is formed. At higher temperatures, pairs of singly coordinated neutral defects are transformed into pairs of charged defects one of which becomes threefold coordinated; that is, a charged VAP is formed. At yet higher temperatures charged defects recombine and the initial structure of the material is restored [81]. The whole process is summarised in Fig. 4.31 (top). A configuration diagram illustrating the structural changes during the reversible photodarkening cycle is shown at the bottom of the same figure. Direct calculations of light-induced changes in selenium and binary chalcogenides [82, 83] are in agreement with this description.

At the same time, it should be noted that alternative description of the photo-structural change has also been suggested. Thus, using ab initio molecular orbital calculations on clusters of atoms modelling the local structure of the amorphous system it was suggested that trigonal AsS_3 pyramidal units can be transformed into a fivefold coordinated As site having four As-S bonds and one As-As bond via a photoionization process [84].

While structural changes in chalcogenide glasses induced by light have been the most studied, it should be noted that rather similar changes can also be induced by other stimuli [85, 86].

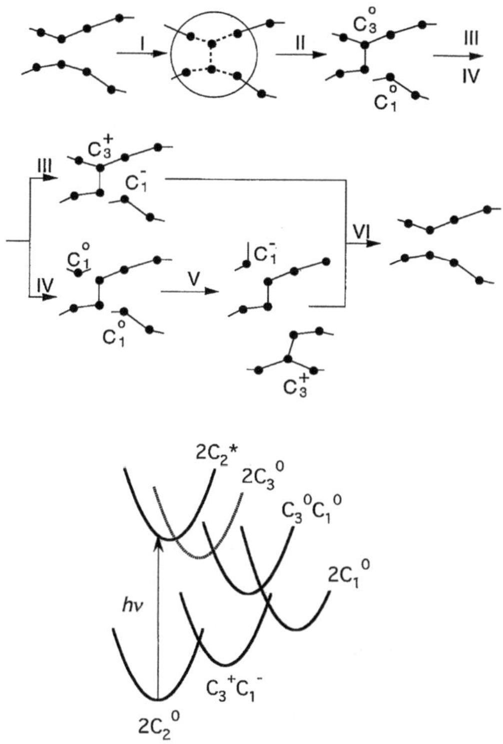

Fig. 4.31 Schematic presentation of a complete cycle of reversible photostructural change in a-Se. Reprinted with permission from Kolobov et al. [81]. Copyright 1998 by the American Physical Society

4.5 Photo-Induced Phenomena in Liquid Chalcogens

It is interesting to note that elemental liquid sulphur and selenium also exhibit significant photo-induced changes [87]. Here we shall briefly describe the main results following Ref. [87].

4.5.1 Liquid Sulphur

Transient Absorption Measurements

Figure 4.32 shows a logarithmic plot of the change in transmission, $\Delta T/T$, versus time for liquid sulphur at 130 °C, where T is the transmittance in the thermal equilibrium state and ΔT the photo-induced change of the transmittance [88]. The wavelength of the transmitted probe light is 390 nm, i.e. corresponds to the energy that is slightly larger than the optical gap of liquid sulphur [89] below the polymerisation temperature T_p in thermal equilibrium. The photo-induced change, photodarkening in this case, is clearly observed upon the illumination with a pulsed

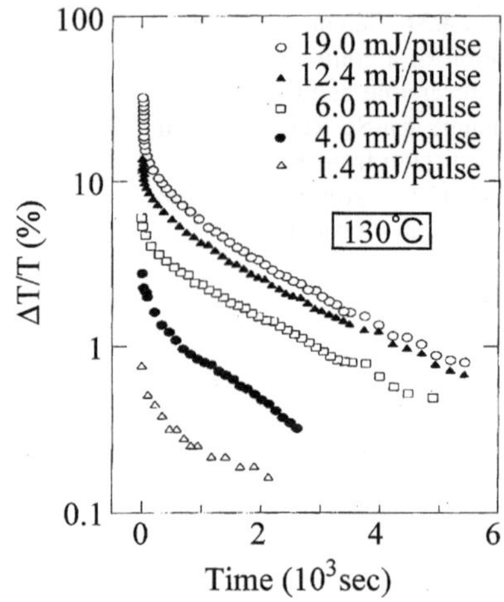

Fig. 4.32 Logarithmic plots of $\Delta T/T$ versus time for liquid sulphur at 130°C after illumination by a single laser pulse with different intensities. Reprinted from Sakaguchi and Tamura [88] with permission from IOP Publishing Ltd.

laser with an intensity of about 1.0 mJ/pulse. With increasing laser power, the value of $\Delta T/T$ increases, while the relaxation time remains unchanged. With further increase of the laser power, higher than 10 mJ/pulse, a new absorption feature with shorter relaxation time appears. With increasing temperature, the relaxation time becomes shorter.

Figure 4.33 shows the plots of $\Delta T/T$ versus time at 130 °C during repeated illumination by a pulsed laser at 10 Hz and after stopping the illumination. Four curves are obtained for different intensities of the laser (0.05, 0.1, 0.2 and 1.0 mJ/pulse). The intensity of the pulsed laser is quite small compared to that used for the experiment depicted in Fig. 4.32. When the pulsed laser with 0.05 mJ/pulse is used, $\Delta T/T$ gradually increases with time and eventually saturates after 15 min. Assuming that the photon energy is totally consumed for heating the sample, the temperature of liquid sulphur is estimated to increase by 1° when a laser pulse (0.1 mJ/pulse) illuminates the sample, i.e. the temperature increase of the sample is negligibly small.

After stopping the illumination, $\Delta T/T$ exponentially decreases. When a stronger laser pulse (0.1 mJ/pulse) illuminates the sample, the saturated value of $\Delta T/T$ becomes larger and the time to reach the saturation value becomes shorter. The fast relaxation process becomes clearer when the intensity of pulsed laser increases. The logarithmic plots of $\Delta T/T$ after stopping the illumination are shown in Fig. 4.34. One can clearly see a crossover between the fast and slow relaxation processes. The values of the relaxation time are 60 s and 10–20 min, respectively.

Possible Atomic-Scale Mechanisms

The faster and slower relaxation processes have been attributed by the authors to the formation of long polymeric chains and charge-transfer complexes. The formation of polymeric chains has been further substantiated by Shimojo et al. [90] who performed

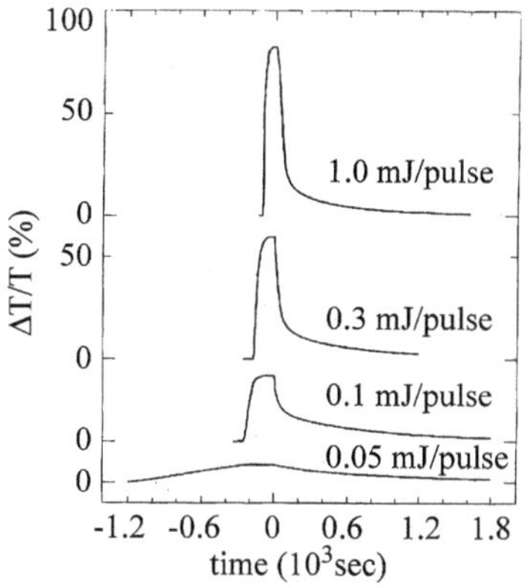

Fig. 4.33 Plots of $\Delta T/T$ versus time for liquid sulphur at 130°C during repeated illumination by a weak pulsed laser at 10 Hz and after stopping the illumination. The origin of the horizontal axis indicates the moment at which the illumination stops. Reprinted from Sakaguchi and Tamura [88] with permission from IOP Publishing Ltd.

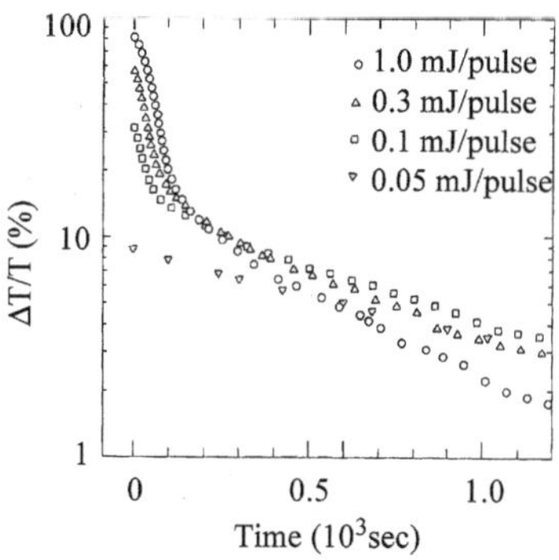

Fig. 4.34 Logarithmic plots of $\Delta T/T$ versus time for liquid sulphur at 130°C after stopping the repeated illumination of the pulsed laser. Reprinted from Sakaguchi and Tamura [88] with permission from IOP Publishing Ltd.

ab initio molecular-dynamic (MD) simulations of an isolated S_8 molecule in order to investigate the instability of an S_8 ring illuminated by a pulsed laser. They have shown that a bond in the S_8 ring is broken immediately after an electron in the highest occupied molecular orbital (HOMO) is excited to the lowest unoccupied molecular orbital (LUMO). This result strongly supports the possibility of photo-induced bond breaking in the S_8 ring followed by polymerisation in liquid sulphur. Figure 4.35 shows the time dependence of the S–S bond lengths in a S_8 cluster. The origin of time ($t = 0$) is the time at which an electron in the HOMO is excited to the LUMO.

The relaxation process was also studied by Munejiri et al. [91]. The most interesting finding is that the S_8 chain produced by the electron excitation does not reconstruct the original S_8 ring during the relaxation process but forms a 'tadpole-shaped' structure (Fig. 4.36). The authors pointed out that the tadpole is very stable and suggested that the tadpole is one of the possible structures for experimentally observed long-living products. The tadpole has the structure that contains a threefold coordinated site neighboured by a singly coordinated atom with dangling bond, which is reminiscent of an intimate VAP defect proposed to explain the photo-induced changes in

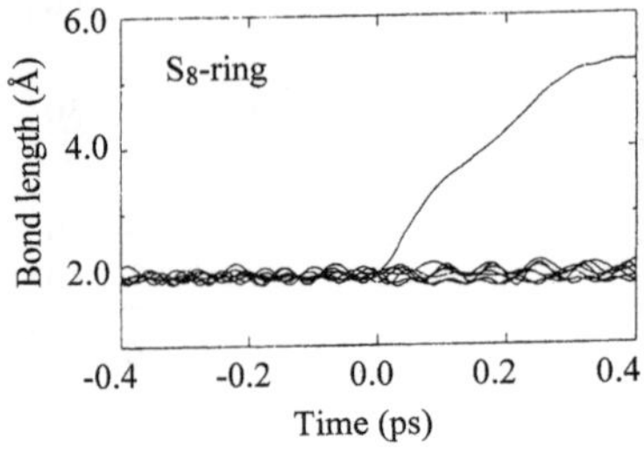

Fig. 4.35 Time dependence of the S–S bond length in an isolated S_8 ring molecule. The bond lengths for eight S–S bonds are displayed. The origin of time is the instance at which an electron in the HOMO is excited to the LUMO. Reprinted from Shimojo et al. [90] with permission from IOP Publishing Ltd.

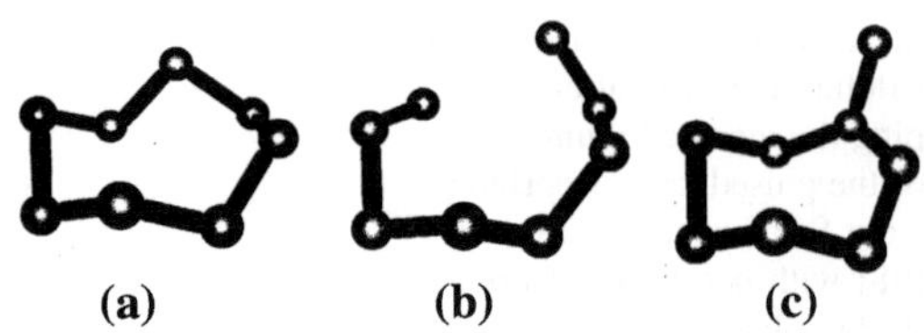

Fig. 4.36 A S_8-ring, a S_8-chain with a broken bond and a tadpole-shaped S_8-molecule appearing in the relaxation process. Reprinted from Munejiri et al. [91] with permission from IOP Publishing Ltd.

chalcogenide glasses. Unfortunately, Munejiri et al. [91] do not give any description about the charge distribution in the tadpole.

4.5.2 Liquid Selenium

Trigonal selenium, the most stable crystalline form under normal condition, consists of helical chains. Lone-pair electrons that are localised around selenium atoms and do not participate in covalent bonding play a crucial role in stabilising the helical structure of trigonal selenium [92, 93]. They occupy the highest filled states forming the valence band, while the empty anti-bonding σ^* states form the conduction band. Between the valence and conduction bands there is an energy gap of about 2 eV, so that trigonal selenium is a semiconductor. The bond length, bond angle and the dihedral angle of long polymeric chains in liquid selenium near the melting point are almost the same as those in trigonal selenium. As mentioned above, the structural disorder in liquid selenium comes from the random variation of the sequence of dihedral angles along the chain in liquid selenium.

The semiconductor–metal transition observed near the critical point in liquid selenium must be related to instability of the chain structure, in which lone-pair electrons play a crucial role.

Sakaguchi and Tamura [94] carried out measurements of transient absorption spectra of liquid selenium after illumination with a pulsed laser which excited lone-pair electrons into the σ^*-orbitals and observed a large change in the transmitted light intensity upon illumination with a pulsed laser. In addition, detailed analysis of the results suggested that the optical gap decreases with time and disappears after 0.5 μs when the pulsed laser illuminates liquid selenium at 350 °C with the intensity of 15 mJ/pulse. They also carried out measurements of the transient direct-current (dc) conductivity and observed a large increase of dc conductivity by a factor 105 on the laser illumination [95].

Hoshino and Shimojo [96] asserted that photo-induced bond breaking can be considered as an elemental process of the photo-induced semiconductor–metal transition; laser illumination may produce numerous short chains that have energy eigenvalues in the middle of the gap and have wave functions with large amplitude near the ends of short chains. The authors suggested that, as a result of these bond breaking acts, the gap of the electronic density of states disappears and the system becomes metallic.

4.6 Photo-Induced Site-Selective Oxidation

While strictly speaking photo-induced oxidation is not a metastable process; it may have serious effect on other processes and for this reason is also mentioned here. The effect of photo-oxidation was first illustrated by a drastic decrease of sticking coefficients of some metals (e.g. Zn and Cd) onto films of chalcogenide glasses illuminated in air. While the growth of a metallic film starts immediately on a non-illuminated surface, the process is greatly delayed for an irradiated area [97, 98]. Subsequent studies [99–101] demonstrated that upon exposure to air As or Ge sites on the glass surface (in As_2S_3 and GeS_2, respectively) were selectively oxidised upon photoexposure in air.

We would like to note here that similar to the above, oxidation of GeTe, the simplest representative of the phase change materials, also starts at Ge sites [102]. Photo-oxidation of $Ge_2Sb_2Te_5$ has also been reported [103].

References

1. R. Chang, Mater. Res. Bull. **2**, 145 (1967)
2. J.S. Berkes, S.W. Ing, W.J. Hillegas, J. Appl. Phys. **42**, 4908 (1971)
3. S. Keneman, Appl. Phys. Lett. **19**, 205 (1971)
4. H. Hamanaka, Ka. Tanaka, S. Iizima, Solid State Commun. **23**, 63 (1977)
5. J.P. de Neufville, S.C. Moss, S.R. Ovshinsky, J. Non-Cryst. Solids **13**, 191 (1974)
6. M.B. Myers, E.J. Felty, Mater. Res. Bull. **2**, 715 (1967)
7. A.V. Kolobov, B.T. Kolomiets, V.M. Lyubin, N. Sebastian, M.A. Taguirdzhanov, J. Hajto, J. Sov. Phys. Solid State **24**, 1062 (1982)

8. P. Nagels, E. Sleeckx, R. Callaerts, E. Marquez, J. Gonzalez, A. Bernal-Oliva, Solid State Commun. **102**, 539 (1997)
9. N.F. Mott, E.A. Davis, Philos. Mag. **17**, 1269 (1968)
10. Ka. Tanaka, J. Non-Cryst. Solids **35/36**, 1023 (1980)
11. K. Tanaka, H. Hisakuni. J. Non-Cryst. Solids **198**, 714 (1996)
12. K. Tanaka, Jpn. J Appl. Phys. **25**, 779 (1986)
13. M. Senda, N. Yoshida, K. Shimakawa, Philos. Mag. Lett. **79**, 375 (1999)
14. V.L. Averyanov, A.V. Kolobov, B.T. Kolomiets, V.M. Lyubin, J. Non-Cryst. Solids **45**, 343 (1981)
15. S.B. Gurevich, N.N. Ilyashenko, B.T. Kolomiets, V.M. Lyubin, V.P. Shilo, Phys. Stat. Sol. (a) **26**, K127 (1974)
16. E. Mytilineou, P.C. Taylor, E.A. Davies, Solid State Commun. **35**, 497 (1980)
17. V.L. Averianov, A.V. Kolobov, B.T. Kolomiets, V.M. Lyubin, Phys. Stat. Sol. (a) **57**, 81 (2006)
18. Z.U. Borisova, *Glassy Semiconductors* (Plenum, New York, 1981)
19. K. Tanaka, J. Non-Cryst. Solids **59/60**, 925 (1983)
20. H. Hamanaka, Ka. Tanaka, K. Tsui, S. Minomura, J. de Phys. Paris **42**, C4-399 (1981)
21. Ka. Tanaka, in *AIP Conference Proceedings: Structure and Excitations of Amorphous Solids*, ed. by G. Lucovsky, F.L. Galeener. vol. 31 (American Institute of Physics, Williamsburg, 1976), p. 148
22. V.L. Averyanov, B.T. Kolomiets, V.M. Lyubin, M.A. Taguirdzhanov, in *Proceedings of 7th International Conference on Amorphous and Liquid Semiconductor*, ed. by W. Spear (CICL, University of Edinburgh, 1977), p. 802
23. K. Shimakawa, S. Inami, S.R. Elliott, Phys. Rev. B **42**, 11857 (1990)
24. K. Shimakawa, S.R. Elliott, Phys. Rev. B **38**, 12479 (1988)
25. H. Naito, T. Teramine, M. Okuda, T. Matsushita, J. Non-Cryst. Solids **97**, 1231 (1987)
26. B.T. Kolomiets, V.M. Lyubin, Mater. Res. Bull. **13**, 1343 (1978)
27. B.T. Kolomiets, S.S. Lantratova, V.M. Lyubin, V.P. Shilo, Sov. Phys. Solid State **21**, 594 (1979)
28. H. Koseki, A. Odajima, Jpn. J. Appl. Phys. **22**, 542 (1983)
29. B.T. Kolomiets, V.M. Lyubin, V.P. Shilo, Fiz. Khim. Stekla (in Russian) **4**, 351 (1978)
30. Ka. Tanaka, Appl. Phys. Lett. **26**, 243 (1975)
31. H. Hisakuni, K. Tanaka. Appl. Phys. Lett. **65**, 2925 (1994)
32. Y. Ikeda, K. Shimakawa, J. Non-Cryst. Solids **338/340**, 539 (2004)
33. K. Tanaka, Phys. Rev. B **57**, 5163 (1998)
34. K. Tanaka, Philos. Mag. Lett. **79**, 25 (1999)
35. R.A. Street, Solid State Commun. **24**, 363 (1977)
36. N. Itoh, M. Stoneham, *Materials Modification by Electronic Excitation* (Cambridge University, Cambridge, 2001)
37. Ka. Tanaka, H. Hamanaka, S. Iizima, in *Proceedings of 7th International Conference on Amorphous and Liquid Semiconductors*, ed. by W. Spear (CICL, University of Edinburgh, 1977), p. 787
38. V.E. Krasnov, V.G. Remesnik, Avtometria **4**, 101 (1980)
39. A.V. Kolobov, B.T. Kolomiets, O.V. Konstantinov, V.M. Lyubin, J. Non-Cryst. Solids **45**, 335 (1981)
40. A.V. Kolobov, V.M. Lyubin, M.A. Taguirdzhanov, Solid State Commun. **41**, 453 (1982)
41. H. Tsutsu, K. Tamura, H. Endo, Solid State Commun. **52**, 877 (1984)
42. K. Tanaka, Phys. Rev. B **30**, 4549 (1984)
43. G. Pfeiffer, M.A. Paesler, J. Non-Cryst. Solids **114**, 130 (1989)
44. M. Frumar, A.P. Firth, A.E. Owen, Philos. Mag. B **50**, 463 (1984)
45. C.Y. Yang, M.A. Paesler, D.E. Sayers, Phys. Rev. B **36**, 9160 (1987)
46. M.A. Paesler, G. Pfeiffer, J. Non-Cryst. Solids **137/138**, 967 (1991)
47. Q. Ma, W. Zhou, D.E. Sayers, M.A. Paesler, Phys. Rev. B **52**, 10025 (1995)
48. S.R. Elliott, J. Non-Cryst. Solids **81**, 71 (1986)

49. K. Shimakawa, N. Yoshida, A. Ganjoo, Y. Kuzukawa, J. Singh, Philos. Mag. Lett. **77**, 153 (1998)
50. Y. Watanabe, H. Kawazoe, M. Yamane, Phys. Rev. B **38**, 5677 (1988)
51. A.V. Kolobov, H. Oyanagi, Ka. Tanaka, K. Tanaka, Phys. Rev. B **55**, 726 (1997)
52. A.V. Kolobov, H. Oyanagi, A. Roy, Ka. Tanaka, J. Non-Cryst. Solids **232/234**, 80 (1998)
53. G. Chen, H. Jain, S. Khalid, J. Li, D.A. Drabold, S.R. Elliott, Solid State Commun. **120**, 149 (2001)
54. A.V. Kolobov et al., J. Photoluminescence **83/84**, 205 (1999)
55. A.V. Kolobov, H. Oyanagi, Ka. Tanaka. MRS Bull. **24**, 32 (1999)
56. J.M. Lee, M.A. Paesler, D.E. Sayers, A. Fontaine, J. Non-Cryst. Solids **123**, 295 (1990)
57. J. Cernogora, F. Mollot, C. Benôit à la Guillaume, M. Jouanne, Solid State Commun. **19**, 465 (1976)
58. J.Z. Liu, P.C. Taylor, Phys. Rev. Lett. **59**, 1938 (1987)
59. J.Z. Liu, P.C. Taylor, Phys. Rev. B **41**, 3163 (1990)
60. S.H. Hunter, A. Bienenstock, T.M. Hayers, in *Proceedings of 7th International Conference on Amorphous and Liquid Semiconductors*, ed. by W. Spear (CICL, University of Edinburgh, 1977), p. 78
61. W. Salanck, J. Bcrkcs, Solid State Commun. **13**, 1721 (1973)
62. K. Shimakawa, T. Kato, T. Hamagishi, J. Non-Cryst. Solids **338**, 548 (2004)
63. H. Fritzsche, Solid State Commun. **99**, 153 (1996)
64. N. Toyosawa, K. Tanaka, Phys. Rev. B **56**, 7416 (1997)
65. S.G. Bishop, U. Strom, P.C. Taylor, Phys. Rev. Lett. **36**, 543 (1976)
66. C. Benôit à la Guillaume, F. Mollot, J. Cernogora, in *Proceedings of 7th International Conference on Amorphous and Liquid Semiconductors*, ed. by W. Spear (CICL, University of Edinburgh, 1977), p. 612
67. D.K. Biegelsen, R.A. Street, Phys. Rev. Lett. **44**, 803 (1980)
68. S. Yamasaki, T. Umeda, J. Isoya, A. Matsuda, Ka. Tanaka, in *Proceedings of the Materials Research Society Meeting* ed. by M. Hack, E.A. Schiff, S. Wagner, R. Schropp, A. Matsuda, Amorphous Silicon Technology, vol. 420 (Pittsburgh, 1996), p. 587
69. J. Hautala, W.D. Ohlsen, P.C. Taylor, Phys. Rev. B **38**, 11048 (1988)
70. K. Shimakawa, S. Inami, T. Kato, S.R. Elliott, Phys. Rev. B **46**, 10062 (1992)
71. D. Vanderbilt, J.D. Joannopoulos, Phys. Rev. B **23**, 2596 (1981)
72. D. Vanderbilt, J.D. Joannopoulos, Phys. Rev. Lett. **49**, 823 (1982)
73. A.V. Kolobov, M. Kondo, H. Oyanagi, R. Durny, A. Matsuda, Ka. Tanaka, Phys. Rev. B **56**, 485 (1997)
74. M. Abkowitz, J. Chem. Phys. **46**, 4537 (1967)
75. S.G. Bishop, U. Strom, P.C. Taylor, Phys. Rev. Lett. **34**, 1346 (1975)
76. F. Mollot, J. Cernogora, C. Benôit à la Guillaume, Phys. Stat. Sol. (a) **21**, 281 (1974)
77. R.A. Street, Adv. Phys. **25**, 397 (1976)
78. S.R. Elliott, Solid State Commun. **27**, 749 (1978)
79. A. Ganjoo, K. Shimakawa, N. Yoshida, T. Ohno, A.V. Kolobov, Y. Ikeda, Phys. Rev. B **59**, 14856 (1999)
80. C.K. Wong, G. Lucovsky, J. Bernholc, J. Non-Cryst. Solids **97/98**, 1171 (1987)
81. A.V. Kolobov, M. Kondo, H. Oyanagi, A. Matsuda, Ka. Tanaka, Phys. Rev. B **58**, 12004 (1998)
82. X. Zhang, D.A. Drabold, Phys. Rev. Lett. **83**, 5042 (1999)
83. J. Li, D.A. Drabold, Phys. Rev. Lett. **85**, 2785 (2000)
84. T. Uchino, D.D. Clary, S.R. Elliott, Phys. Rev. Lett. **85**, 3305 (2000)
85. O.I. Shpotyuk, A.O. Matkovskii, J. Non-Cryst, Solids **176**, 45 (1994)
86. O.I. Shpotyuk, V.O. Balitska, Phys. Stat. Sol. (a) **165**, 295 (1998)
87. Y. Sakaguchi, K. Tamura, in *Photo-Induced Metastability in Amorphous Semiconductors*, ed by A.V. Kolobov (Wiley-VCH, Weinheim, 2003), p. 199
88. Y. Sakaguchi, K. Tamura, J. Phys. Condens. Matter **7**, 4787 (1995)
89. S. Hosokawa, T. Matsuoka, K. Tamura, J. Phys. Condens. Matter **6**, 5273 (1994)

90. F. Shimojo, K. Hoshino, Y. Zempo, J. Phys. Condens. Matter **10**, L177 (1998)
91. S. Munejiri, F. Shimojo, K. Hoshino, J. Phys. Condens. Matter **12**, 7999 (2000)
92. H. Fukutome, Prog. Theor. Phys. **71**, (1984)
93. W.A. Harrison, *Electronic Structure and the Properties of Solids: the Physics of the Chemical Bond* (Dover Publications, New York, 1989)
94. Y. Sakaguchi, K. Tamura, J. Phys. Condens. Matter **10**, 2209 (1998)
95. Y. Sakaguchi, K. Tamura, J. Phys. Condens. Matter **11**, 659 (1999)
96. K. Hoshino, F. Shimojo, J. Phys. Condens. Matter **10**, 11429 (1998)
97. A.V. Kolobov, V.M. Lyubin, Fiz. Tverd. Tela **28**, 3567 (1986)
98. G.E. Bedelbaeva, A.V. Kolobov, V.M. Lyubin, Philos. Mag. B **60**, 689 (1989)
99. A.V. Kolobov, J.P.S. Badyal, R.M. Lambert, Surface Sci. **222**, L819 (1989)
100. J.H. Horton, C. Hardacre, C.J. Baddeley, G.D. Moggridge, R.M. Ormerod, R.M. Lambert, J. Phys. Condens. Matter **8**, 707 (1996)
101. J.H. Horton, K.L. Peat, R.M. Lambert, J. Phys. Condens. Matter **5**, 9037 (1993)
102. L.V. Yashina, R. Puettner, V.S. Neudachina, T.S. Zyubina, V.I. Shtanov, M.V. Poygin, J. Appl. Phys. **103**, 094909 (2008)
103. B.S. Lee, Y. Xiao, S.G. Bishop, J.R. Abelson, S. Raoux, V.R. Deline, M.H. Kwon, K.B. Kim, B.K. Cheong, H. Li, P.C. Taylor, in *Chalcogenide Alloys for Reconfigurable Electronics. MRS Symposium Proceedings*, ed. by P.C. Taylor, A.V. Kolobov, A.H. Edwards, J. Maimom, vol. 918 (MRS, Warrendale, Pennsylvania, 2006), p. 113

Chapter 5
Photo-Induced Anisotropy

5.1 Optical Linear Dichroism and Birefringence

5.1.1 Photo-Induced Optical Anisotropy

The results presented in the previous chapter do not take into account the polarisation of the incident light. It was in the late 1970s that the first publications about photo-induced anisotropy in initially isotropic films of glassy As_2S_3 appeared [1–3]. Both photo-induced dichroism ($|\alpha_{\parallel} - \alpha_{\perp}|$) and photo-induced birefringence ($|n_{\parallel} - n_{\perp}|$) were reported. Depending on the sign of the difference, the dichroism and birefringence can be either positive and negative.

To reflect the fact that not only the value of the absorption changed upon exposure to polarised light but also the orientation of the polarisation plane of the light played an important role, photo-induced anisotropic effects are called vectoral, as opposed to photodarkening that is a scalar effect.

5.1.2 Optical Reversibility of Induced Anisotropy

An important finding was that photo-induced dichroism can be reoriented by changing the polarisation of the incident light. An identical dichroism is created afterwards in an orthogonal direction, i.e. reversal of photodichroism is possible. Should circularly polarised or unpolarised light be used in the second irradiation step, the induced dichroism is destroyed [4]. This behaviour is demonstrated in Fig. 5.1. Such reorientation can be performed repeatedly without any sign of a decrease in the effect. Interestingly, this light-induced reorientation or erasure of photo-induced dichroism is not associated with any change in the scalar photodarkening that continues monotonically with exposure time (Fig. 5.1).

A. V. Kolobov and J. Tominaga, *Chalcogenides*, Springer Series in Materials Science 164, DOI: 10.1007/978-3-642-28705-3_5,

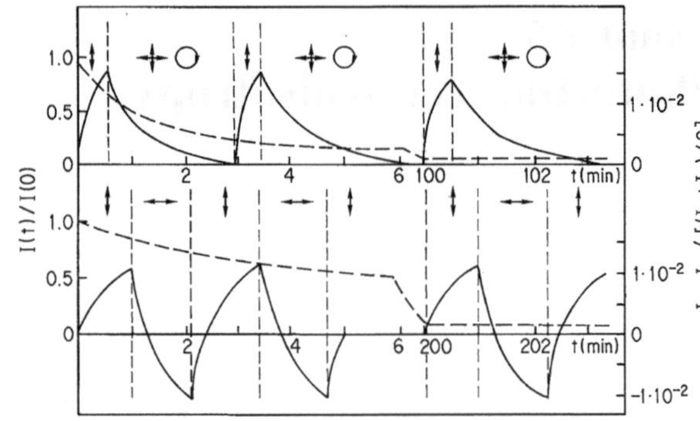

Fig. 5.1 Kinetics of photodarkening (*dashed line*) and photo-induced dichroism (*solid line*) in an $As_{50}Se_{50}$ film. *Arrows* represent linear polarisation and *circles* refer to circular polarisation. Reprinted from Lyubin and Tikhomirov [15] with permission from Elsevier

It was subsequently found that the existing dichroism is destroyed within a time span much shorter than that necessary for its initial creation [5, 6], and that preliminary exposure of a chalcogenide film by unpolarised light also results in faster kinetics of the anisotropic change when the film is subsequently exposed to linearly polarised light [7, 8] as illustrated by Fig. 5.2.

5.1.3 Dependence of Photo-Induced Anisotropy on Exposure Parameters

Light-Intensity Dependence

The rate with which anisotropy is induced increases for higher light intensities [9]. At the same time, there is no general consensus whether the saturated state of photo-induced anisotropy depends on the intensity of the inducing light. While most groups agree that the magnitude of the saturated dichroism remain unchanged for different light intensities [10–12], there are also reports that the magnitude of saturated anisotropy increases for higher intensities of the light [13]. In the latter case, however, the intensities used were rather low and it may be informative to remind the reader that for scalar photodarkening the reciprocity between the light intensity and exposure time holds for a wide range of intensities but deviation from the constant saturated value was observed at elevated temperatures and/or low light intensities [14]. It could be that the same is true for the photo-induced anisotropy but detailed studies have yet to be performed.

Spectral Dependence

Figure 5.3 shows the excitation energy dependence of photo-induced dichroism in As_2S_3 [15]. While the photodarkening efficiency (cf. Fig. 4.2) increases with photon

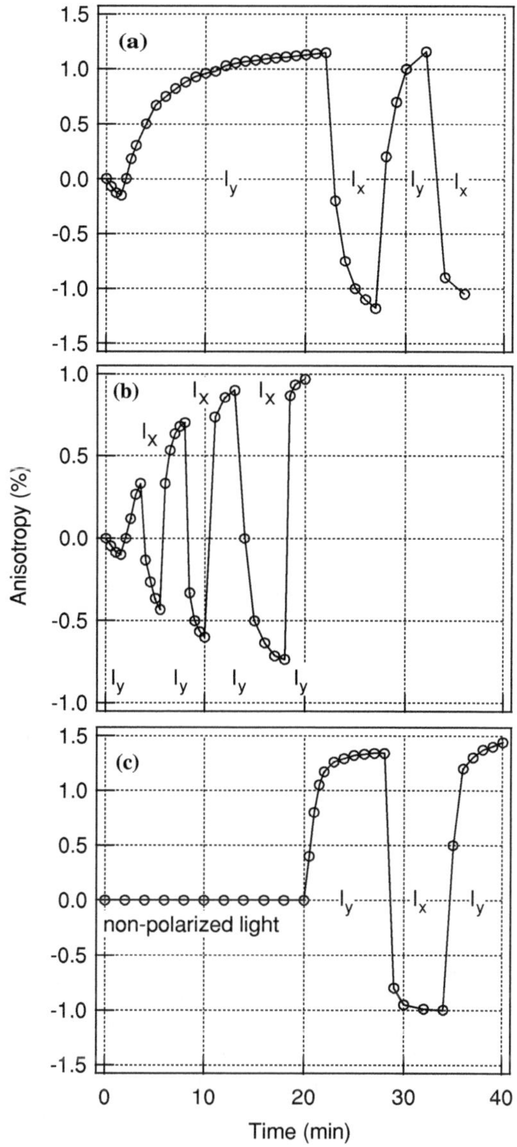

Fig. 5.2 Kinetics of dichroism generation and reorientation in an $As_{50}Se_{50}$ film under the action of linearly polarised laser light with two orthogonal directions of electrical vector (x and y) when the reorientation starts after the dichroism saturation (**a**), when the reorientation takes place during the initial stages of the dichroism growth (**b**), and when the polarised light irradiation starts after prolonged illumination with unpolarised light (**c**). Reprinted from Lyubin et al. [7] with permission from Elsevier

energy and levels off at energies slightly above the bandgap, the efficiency of photo-induced anisotropic effects peaks at photon energies around the bandgap (Fig. 5.3). This results strongly suggest that anisotropic effects are temperature activated.

Reflectance difference study of photo-induced anisotropy [16] demonstrated that the sign of the effects changes as the photon energy of the (probing) light is varied (Fig. 5.4)

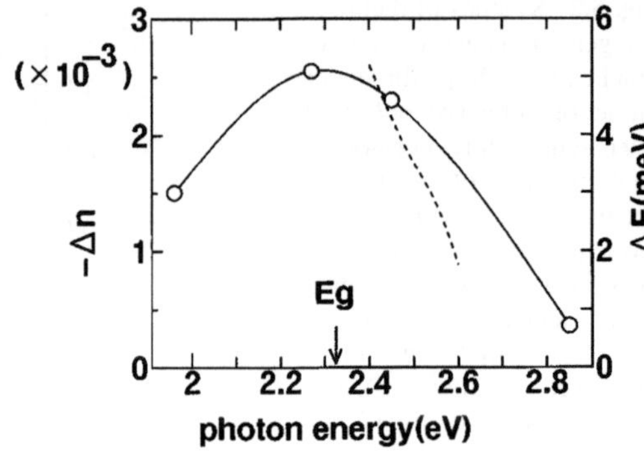

Fig. 5.3 Excitation energy dependence of photo-induced anisotropy in As_2S_3 glass. Reprinted with permission from Tanaka et al. [11]. Copyright 1996 by the American Physical Society

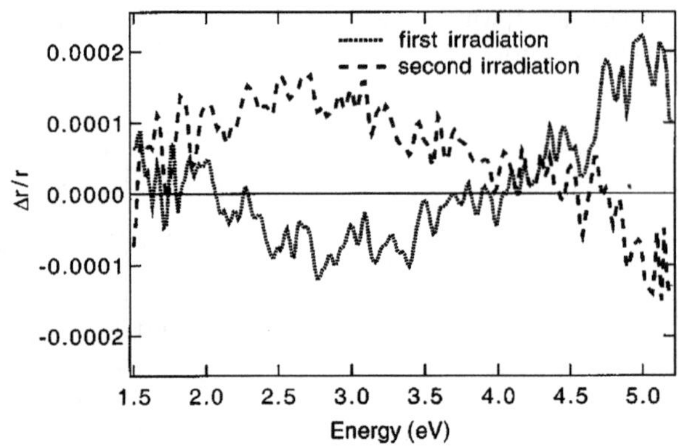

Fig. 5.4 Photo-induced reflectance anisotropy ($\Delta r/r$) induced by x-polarised light of an Ar^+-ion laser in As_2S_3 (*solid line*) and its subsequent reorientation upon exposure to y-polarised light (*dashed line*). Reprinted with permission from Kolobov et al. [16]. Copyright 1997 by the American Physical Society

Temperature Dependence

Figure 5.5 shows the temperature dependence of photo-induced anisotropy. A maximum is clearly seen for Se and As_2S_3 [11] suggesting that thermal excitation is indispensable for photo-induced anisotropy. A similar result was obtained for As_2Se_3 and AsSe [11, 12, 17]. At the same time, photodarkening typically increases as the temperature is decreased (cf. Fig. 4.5). An interesting observation is that the location of the maximum shifts with the excitation energy, namely, it moves to higher temperatures for lower photon energies of the inducing light which led to a speculation [18] that the photo-induced anisotropy is a cooperative photo-thermal process.

External Electric Field

The effect of external electric field on the photo-induced anisotropy could not be detected irrespective of the orientation of electric field [9].

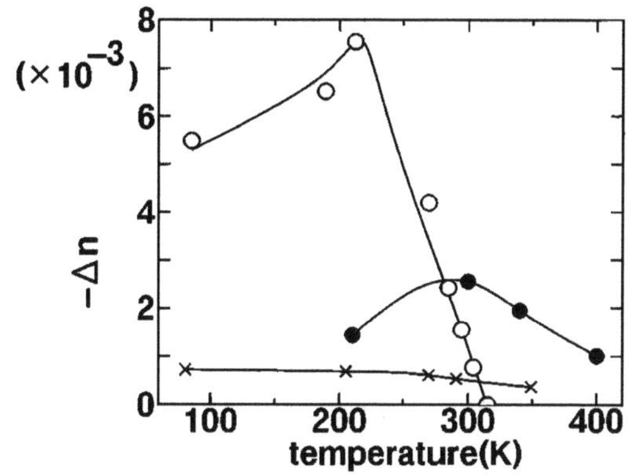

Fig. 5.5 The photo-induced birefringence Δn induced by bandgap illumination in Se (*open circles*), As_2S_3 (*closed circles*) and As_2Se_3 (*crosses*) as a function of temperature at which excitation and probe are performed. Reprinted with permission from Tanaka et al. [11]. Copyright 1996 by the American Physical Society

Thermal Erasure of Photo-Induced Anisotropy

Above we have already described the destruction of photo-induced anisotropy through subsequent exposure to light with a different polarisation. At the same time, similar to scalar photodarkening, induced photoanisotropy can be annealed out thermally. Thermal destruction of the photodarkening and photodichroism has been studied by various authors and the results are rather contradictory. Thus, while several groups reported that dichroism disappears faster and at lower temperatures than photodarkening [3, 15, 19, 20], there are also reports that contradict these observations [11]. In a comprehensive review on photo-induced anisotropy [18] Tanaka argues that this inconsistency may be caused by the fact that the vectoral effects are rather small and accurate detection of small changes may be complicated. He also points out that the anisotropy annealing may proceed differently in different materials as is also the case for photodarkening that is annealed out gradually in stoichiometric As_2Se_3 but has a pronounced threshold nature in As-rich glasses [21].

The kinetics of the dark relaxation of photo-induced anisotropy were found to obey a fractional-exponential Kohlrausch law $\exp(-t/\tau)^{\gamma}$ over a wide time interval except in the initial stage when it is better described by the Debye law $\exp(-t/\tau_D)$ [22].

5.1.4 Anisotropy Versus Photodarkening

In early papers it was suggested that photo-induced anisotropy and photodarkening are different facets of the same phenomenon. Subsequent studies have revealed substantial differences between the two processes, such as the different spectral and temperature dependences mentioned above. Perhaps the strongest argument in favour of the different natures of photodarkening and photoanisotropy, however, is the fact that the largest value of photo-induced dichroism is observed in Sb_2S_3 and

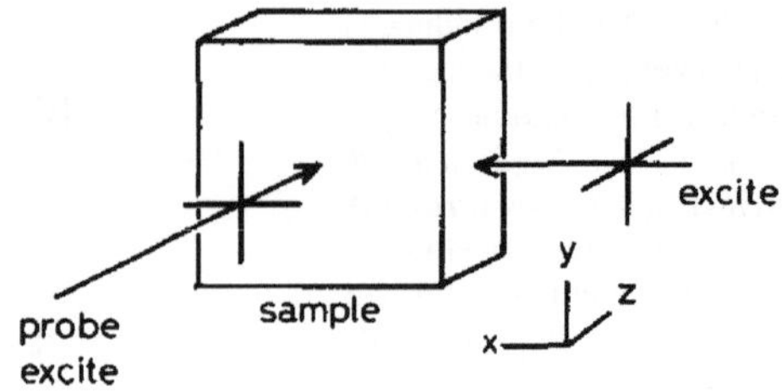

Fig. 5.6 Illumination scheme for inducing linear anisotropy by unpolarised (circularly polarised) light. Reprinted from Tanaka et al. [25] with permission from Elsevier

$Ge_{28.5}Pb_{15}S_{56.5}$ [20], that is, in materials that do not exhibit reversible photodarkening.

5.1.5 Anisotropy Induced by Non-Polarised and Circularly Polarised Light

Fritzsche's Prediction

While in the results described above chalcogenide glasses become anisotropic under exposure to linearly polarised light, it should be mentioned that unpolarised light may under certain circumstances also generate anisotropy. This effect was first predicted by Fritzsche [23] based on the fact that the polarisation vector is oriented perpendicular to the Poynting vector of the light, i.e. even upon exposure to unpolarised light the directions along the propagation of the light and perpendicular to it should exhibit anisotropic properties.

Experimental Observations

This prediction was confirmed by Tihkomirov and Elliott who demonstrated that unpolarised light can induce anisotropy in a chalcogenide glass (bulk $As_2S_3I_{0.8}$ and thin AsSe films were used) if the Poynting vector of the incident light is perpendicular to the probing beam (for a bulk sample) or at glancing angles (for a thin film sample) [24]. The authors concluded that the effect was not caused by dichroism of absorption but rather was due to dichroism of scattering.

More quantitative studies of photoanisotropy induced by unpolarised light were performed by Tanaka et al. [25]. These authors performed birefringence measurements using a thick (50 μm) annealed As_2S_3 film and a He-Ne laser. The probe light was incident upon the front surface while the excitation beam was incident upon a side surface and its polarisation was varied (Fig. 5.6). As the sample was excited by the vertically polarised light, negative anisotropy was generated with a magnitude comparable to the case when the exciting light was incident upon the front surface.

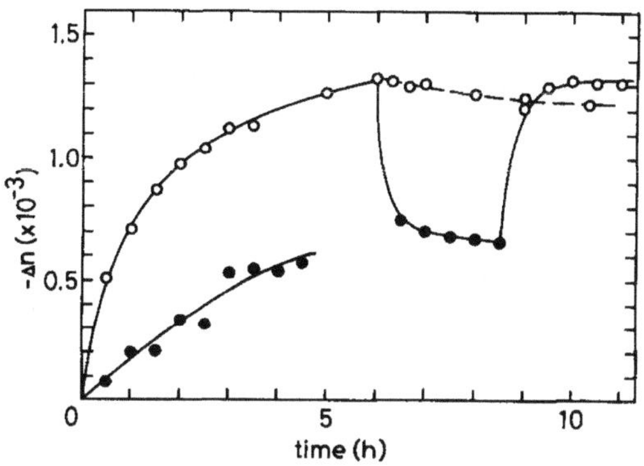

Fig. 5.7 Photo-induced birefringence Δn in As_2S_3 induced by linearly (*open circles*) and circularly (*closed circles*) polarised light incident from a side surface as shown in Fig. 5.6. Reprinted from Tanaka et al. [25] with permission from Elsevier

Subsequent exposure to unpolarised (or circularly polarised light) resulted in a decrease in the induced birefringence. The same saturated value of birefringence was achieved when the sample was directly exposed to the unpolarised light (Fig. 5.7).

Photo-Induced Gyrotropy

Application of circularly polarised light leads to the creation of circular anisotropy (gyrotropy) [26]. Unexpectedly, photo-induced gyrotropy could also be induced by linearly polarised light, although its magnitude in this case was lower than in the case of illumination with circularly polarised light [26]. The observed gyrotropy provides evidence for existence in As-based chalcogenides of helical structures proposed from structural modelling [27].

A somewhat related effect, the photo-induced Faraday effect, was also observed in various chalcogenide glasses [28]. The authors concluded that this phenomenon was related to photodarkening.

Photo-Induced Scattering Anisotropy

In early works, photo-induced anisotropy was studied through transmission measurements. It should be remembered, however, that not only absorption but also scattering can effect the transmission. Photo-induced scattering has indeed been detected (Fig. 5.8) and studied [26, 29–31]. Through angle-dependent transmission measurements using carefully polished bulk As_2S_3 glass and a He-Ne laser it was shown that the transmitted light intensity dropped faster for smaller angles (Fig. 5.9). The change at higher angles was attributed to induced scattering. It was shown that the induced scattering is anisotropic and could be reversed by changing the polarisation of the light.

It was also found that the scattering anisotropy was anticorrelated with the transmission anisotropy (Fig. 5.10) which led the authors to conclude that the whole family

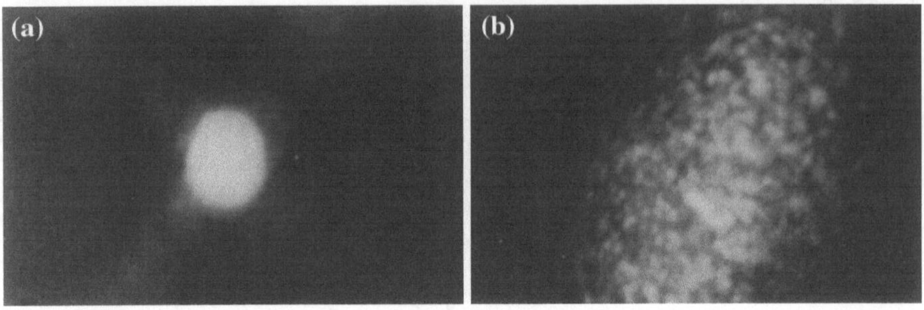

Fig. 5.8 Images of a laser beam transmitted through a 2.5 cm thick As_2S_3 observed on the screen during the initial (**a**) and final (**b**) stages of illumination. Reprinted from Lyubin and Tikhomirov [26] with permission from Elsevier

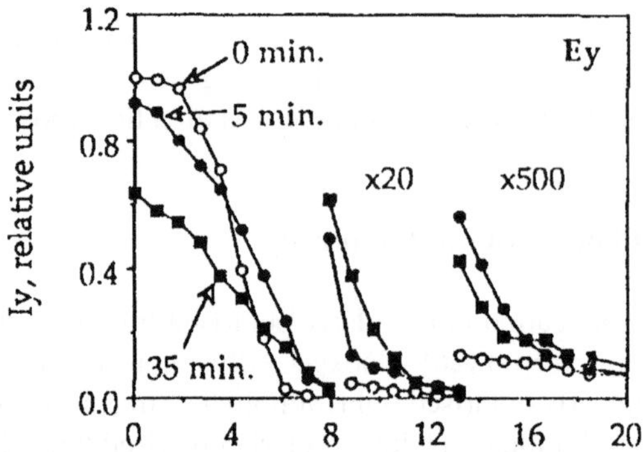

Fig. 5.9 Angular distribution of intensity of light transmitted through As_2S_3 sample as a function of irradiation time. Reprinted from Lyubin et al. [29] with permission from Elsevier

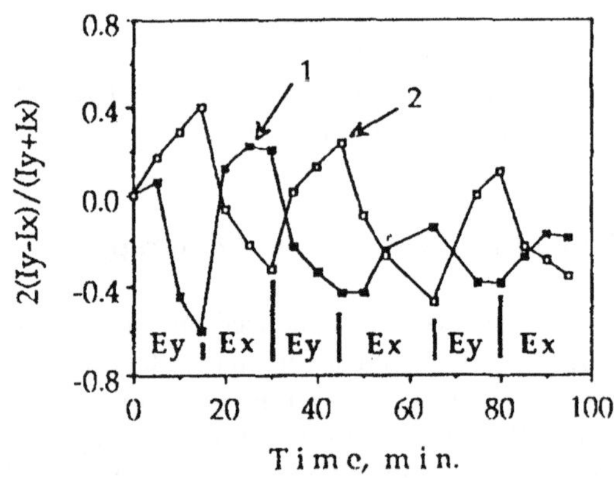

Fig. 5.10 Kinetics of transmittance anisotropy (*1*) and scattering anisotropy (*2*) changes induced by two orthogonal polarisations of the light. Reprinted from Lyubin et al. [29] with permission from Elsevier

of photo-induced anisotropy has scattering anisotropy as its basis [29, 30]. This point of view, however, is not generally accepted and it was argued that the scattering is not genuine but results from the formation of self-focusing structures [32] formed through photo-induced refractive index change.

5.2 Photo-Induced Anisotropy of Other Properties

While most studies concentrate on photo-induced anisotropy of optical properties, exposure to polarised light also causes anisotropic changes in other properties of chancogenide glasses as will be reviewed below.

5.2.1 Anisotropic Photoluminescence

The photoluminescence of chalcogenide glasses depends on the polarisation of the light [10, 33]. The effect was called the "polarisation memory". It was subsequently reported that in Ge-S glasses optical anisotropy of the luminescence and photo-induced anisotropy in absorption had different composition dependencies [34]. From this result it was suggested that the polarisation memory of the photoluminescence should not be considered as a manifestation of the structural change that results in anisotropic optical properties.

5.2.2 Anisotropy of Photoconductivity

Photo-induced optical anisotropy is accompanied by photo-induced anisotropy of the photoconductivity. As the former, the latter is also optically reversible [35, 36]. The authors relate the photo-induced anisotoropy of the conductivity to the formation and reorientation of microanisotropic fragments in the glass, similar to the case of optically induced anisotropy.

5.2.3 Polarised Photodoping

Provided photodoping (see Chap. 7) is performed using linearly polarised light, a large photodichroism is induced [37]. An essentially similar value of photodichroism albeit with a faster generation rate can be induced by the same light on a sample previously photodoped with unpolarised light. Interestingly, when the authors tried to induce absorption anisotropy in a sample that was pre-doped with Ag thermally, the magnitude of the effect was much smaller and the sign of the induced dichroism changed.

5.2.4 Anisotropic Optomechanical Effect

Using AsSe deposited atop of a cantilever Krecmer et al. [38] have found that exposure to linearly polarised light causes reversible polarisation-dependent nanocontraction and dilatation [38, 39]. The observed optomechanical effect correlates with the photo-induced dichroism. The authors have demonstrated that contraction occurs along the direction of the electric field of the inducing light while the material expands in the orthogonal direction and proposed that this effect may be used for optically controlled nanoindentors.

5.2.5 *Anisotropic Mass Flow*

The exposure of chalcogenide glasses to linearly polarised light results in an anisotropic volume change (Fig. 5.11) suggesting anisotropic mass flow under photoexcitation with linearly polarised light [40]. It is not clear whether the created relief can be erased by changing the polarisation of the acting light.

5.2.6 *Macroscopic Anisotropic Deformations*

Exposure to rather low intensity light of As_2S_3 flakes (lying on a frosted glass plate) with thicknesses of 1–10 μm and surface areas such that the complete area was irradiated resulted in a rather unusual response [41, 42]. Upon illumination by 2.3 eV light, the flake first curls cylindrically in U-shape with the cylinder axis perpendicular (horizontal in Fig. 5.12) to the electric field (vertical) of light (panel *b*). In some cases, the flake tends to stand up on the substrate. Then, the cylinder elongates in the perpendicular direction reducing the projected sample size parallel to the electric field (panel *c*), and finally, becomes spirally shaped (panels *d*).

The explanation of this unusual behaviour has been offered in terms of radiation force (torque) and photo-induced fluidity [41, 42].

5.2.7 *Anisotropic Photocrystallisation*

In 1997 three groups independently reported anisotropic crystallisation of selenium and Se-based films [43–45]. Once crystallised, the induced dichroism could not be reversed.

Figure 5.13 shows the maximum induced birefringence in Se as a function of temperature [44]. One can see a rather small change at temperatures below T_g that is related to the reversible photo-induced anisotropy within the amorphous phase described earlier. Photo-induced birefringence disappears around T_g and then exhibits a very strong peak around 350 K indicative of crystallisation [44]. The sharp drop at higher temperatures was attributed to the onset of isotropic thermal crystallisation. From X-ray diffraction studies the same authors concluded that in the crystalline phase Se chains align perpendicular to the polarisation vector of the light.

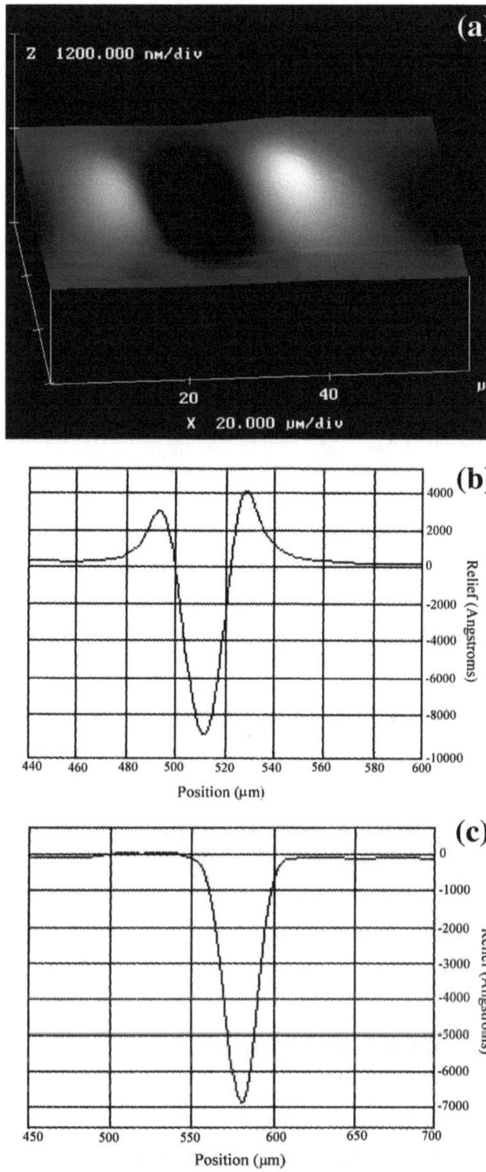

Fig. 5.11 **a** An AFM image of an area exposed by linearly polarised light (power ca. 10 mW, exposure time ca. 480 s, illumination spot size ca. 40–50 µm) (*vertical axis* is in nm and the *horizontal axis* is in µm). **b** Surface profile probed along the polarisation direction. **c** Surface profile probed perpendicular to the polarisation direction. Reprinted with permission from Saliminia et al. [40]. Copyright 2000 by the American Physical Society

It is of interest to note that simultaneous exposure of an amorphous Se film to two linearly polarised laser beams with different wavelengths ($\lambda = 676$ nm and $\lambda = 488$ nm) may result in both an increase and a decrease in the crystallisation rate depending on the mutual orientation of the polarisation planes of the two beams [46]. Thus if the two beams are polarised in the same direction the crystallisation is suppressed, while for the orthogonal polarisations the crystallisation speed is increased (Fig. 5.14). The result was explained in terms of different response of Se electrons to sub-bandgap

Fig. 5.12 Photo-induced deformations of As_2S_3 glass flakes induced by 2.3 eV (532 nm) light with an intensity of 40 mW and a vertical polarisation axis. The green bars represent the polarisation direction, and indicate the length of 0.1 mm. **a** a 4-μm-thick flake placed on frosted glass exposed for **b** 0.25, **c** 4 and **d** 17 h. Reprinted with permission by Japan Society of Applied Physics from Tanaka [41]

and super-bandgap photons. Obviously, such an effect is only possible if the process is due to photoexcitation. Should the crystallisation have been simply due to laser-induced heating effect, the crystallisation rate would not have changed depending on the polarisation of the light.

5.3 Structural Studies

Structural studies of photo-induced anisotropy were first conducted by Tanaka et al. [11] who used an X-ray diffraction technique to investigate photo-induced anisotropy in As_2S_3. The observed results were interpreted within a layer-cluster model (see below).

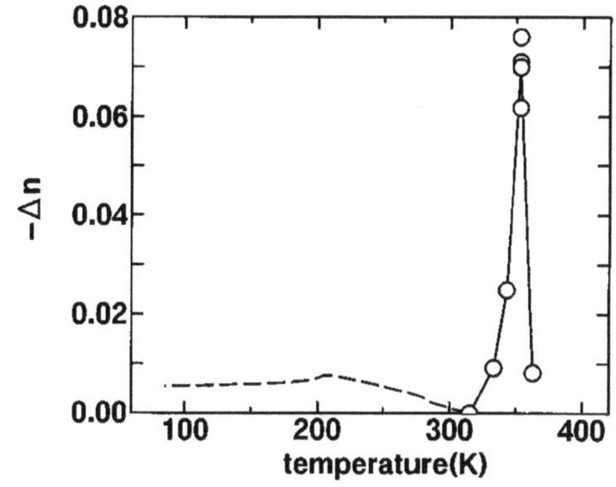

Fig. 5.13 Maximal photo-induced birefringence in Se as a function of the temperature at which the sample is exposed to linearly polarised light ($\lambda = 633\,\text{nm}$ and $W=300\,\text{mW/cm}^2$). Reprinted with permission from Ishida and Tanaka [44]. Copyright 1997 by the American Physical Society

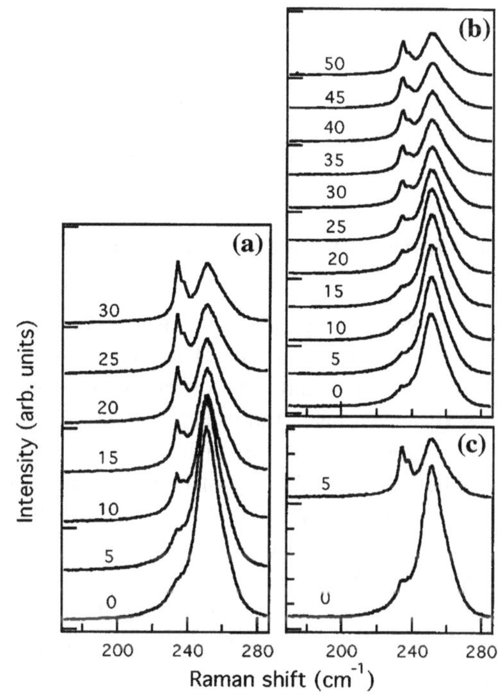

Fig. 5.14 Comparison of modification of Raman spectra of an a-Se film photocrystallised by irradiation with the 676 nm light of a Kr^+-laser ($400\,\text{W/cm}^2$) solely (**a**) and simultaneously with irradiation by the 488 nm light from an Ar^+-laser ($200\,\text{W/cm}^2$) whose polarisation plane is oriented parallel (**b**) and perpendicular (**c**) to that of the Kr^+-laser light. The peak located at $255\,\text{cm}^{-1}$ is characteristic of the amorphous phase and the emerging peak located at $237\,\text{cm}^{-1}$ arises from the crystalline phase. Irradiation time (in min) is indicated next to each curve. Reprinted with permission from Roy et al. [46]. Copyright 1998 by the American Institute of Physics

Subsequent X-ray absorption studies performed on elemental selenium have found that the Se-Se bond lengths were the same for the two orthogonal directions. At the same time, a reversible shift in the K-edge absorption as a function of the polarisation of the light with respect to the probing X-rays has been detected (Fig. 5.15). The authors concluded that the photo-induced anisotropy results not from the bond elongation but from bond reorientation in the direction selected by the light polarisation. Namely, the irradiation with linearly polarised light results in the formation of

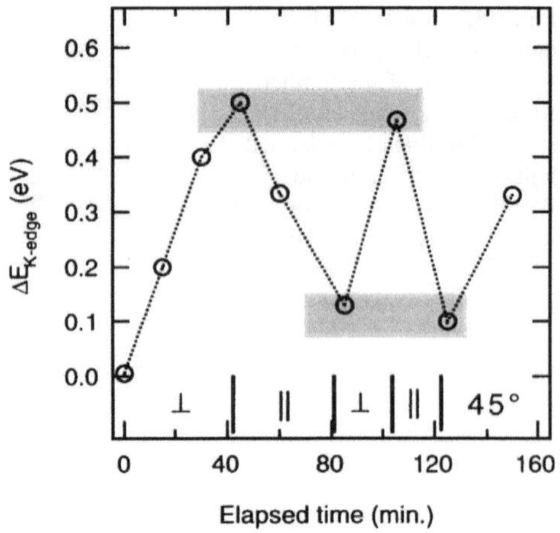

Fig. 5.15 Kinetics of the light-induced K-edge shift upon exposure of amorphous Se to linearly polarised light whose orientation is varied. The mutual orientation of the polarisation planes of the light and X-rays is shown at the bottom. *Dotted lines* are guides for the eye and *grey bars* indicate the reproducibility in repeated measurements (uncertainties). Reprinted with permission from Kolobov et al. [47]. Copyright 2001 by the American Physical Society

quasiparallel chains with the c-axis oriented perpendicular to the polarisation plane of the light [47].

5.4 Models of Photoanisotropy

In early works (e.g. [3]), it was suggested that the mechanism responsible for photodichroism was identical to that of scalar photostructural changes assuming the existence of randomly oriented microanisotropic fragments in chalcogenide films. However, later studies [15, 20] have clearly demonstrated differences between the two processes such as different inducing kinetics, different annealing kinetics and, finally, the possibility of inducing photodichroism in films that do not exhibit reversible photodarkening. It seems more likely that photodarkening and photo-induced anisotropic effects are independent and different mechanisms are responsible for these two effects.

The first comprehensive phenomenological model has been suggested by Fritzsche in [23, 48, 49] who suggested that chalcogenide glasses are locally strongly anisotropic and these anisotropic microvolumes can be grouped into three categories depending on the direction of the projection of their dielectric tensor (along axes x, y or z). In a macroscopically isotropic glass the numbers of the microvolumes are equal; that is, $N_x = N_y = N_z = N/3$, where N is the total number of anisotropic microvolumes. On irradiation with linearly polarised light (in the z-direction), the absorption predominantly takes place in microvolumes oriented along the z-axis. The local structural change in the microvolume can either change the anisotropy of the microvolume or leave it intact. In the former case, the microelement is lost for the z-polarisation and a redistribution takes place (Fig. 5.16):

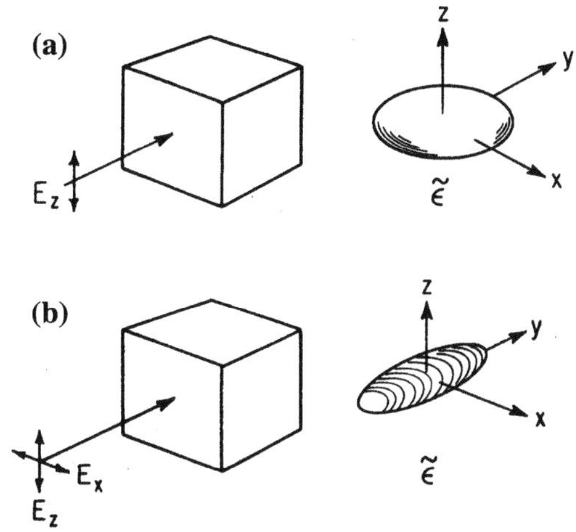

Fig. 5.16 Dielectric tensor $\tilde{\epsilon}$ after exposure to linearly polarised light (**a**) and to unpolarised light (**b**). Reprinted with permission from Fritzsche [23]. Copyright 1995 by the American Physical Society

$$-\Delta N_z = \Delta N_x + \Delta N_y \tag{5.1}$$

This process is more important in the case when the excitation photon energy is close to that of the forbidden gap. At higher photon energies, photocreated electron-hole pairs can drift apart. This reduces their chance to recombine geminately and leads to a loss of the correlation between the polarisation of the incident light and the resultant changes in the local anisotropies in agreement with the experimentally observed spectral dependence of photo-induced anisotropy.

An important prediction of this model is the possibility of inducing optical anisotropy by unpolarised light. In this case, the bonds oriented perpendicular to the k vector and parallel to the E vector of the light will be most absorbing and anisotropy between the direction of the light propagation and the plane of the E and H vectors of the light should be observed. This prediction has been subsequently confirmed experimentally [24, 25].

Lyubin and Tikhomirov were the first to suggest that photo-induced anisotropic effects arise from the interaction of light with native defects or weak bonds in the glasses whose excitation energy should be below E_g. It was argued that light can reorient such native defects without changing their dipole moments (involved in the optical absorption matrix element). In this case photo-induced anisotropy is not accompanied by a change in the absorption [15].

An extension of this model was the suggestion that photo-induced defects may also play a role. Within this approach it is natural that the initial creation cycle is slower than subsequent reorientation cycles or that the preliminary exposure results in a higher rate of anisotropy generation. Analogous to photo-induced ESR it was speculated that initial exposure results in the generation and re-orientation of defects while subsequent cycles only act to reorient the already existing defects, which accounts for the faster kinetics during the subsequent cycles.

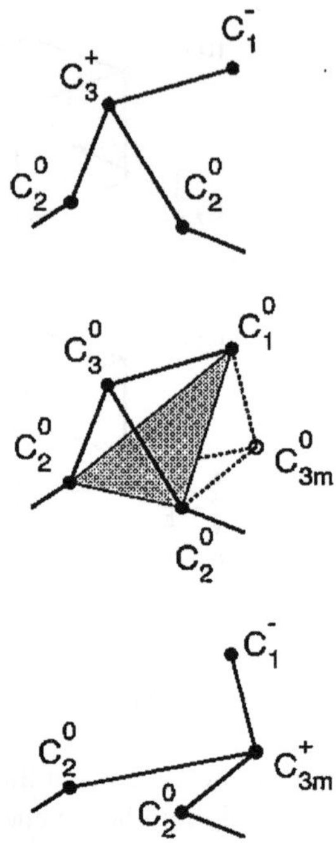

Fig. 5.17 The flip of the pyramid involving an IVAP centred at a C_3^+ atom proposed to be responsible for photo-induced anisotropy in a-Se (*top*). The IVAP configuration before optical excitation (*middle*). The transient configuration after optical excitation, in which the electron excited from a lone-pair orbital of the C_1^- atom is trapped in an anti-bonding orbital localised at the C_3^0 atom. The C_3^0 atom has the possibility to tunnel through the plane, shown hatched, to the mirror-image position represented by the dotted bonds (*bottom*). The final metastable IVAP conguration, after the pyramid flip, proposed to be responsible for photo-induced anisotropy. Reprinted with permission from Tikhomirov et al. [17]. Copyright 1997 by the Americal Physical Society

Regarding the microscopic models of the anisotropic centres, various possibilities have been suggested. One group of models considers anisotropy to be a result of bond flipping at intimate valence alternation pair sites [17] and is schematically illustrated in Fig. 5.17.

The authors of this model argue that the concentration of native defects present in amorphous chalcogenides is sufficient to explain the experimentally observed values of the anisotropy [50]. A related explanation was offered to account for the reversible anisotropic dilatation and contraction [38] which did not consider the presence of VAPs and also assumed the flipping of the bonding plane.

Because the sign of anisotropy depends on the energy of probing photons it was argued that anisotropic formation of dynamic interchain bonds was followed by the redistribution of spatial orientation of bonding and non-bonding electrons [8, 16]. Reorientation of the bonds provides a natural explanation of the photo-induced anisotropy (Fig. 5.18).

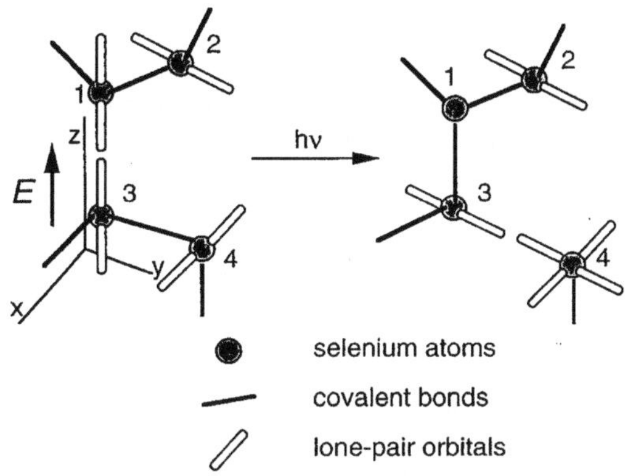

Fig. 5.18 Excitation of lone-pair electrons with linearly polarised light results in the formation of additional (interchain) bonds. Subsequent recombination leads to spatial redistribution of bonding and non-bonding electrons. Reprinted with permission from Kolobov et al. [8]. Copyright 1997 by the American Physical Society

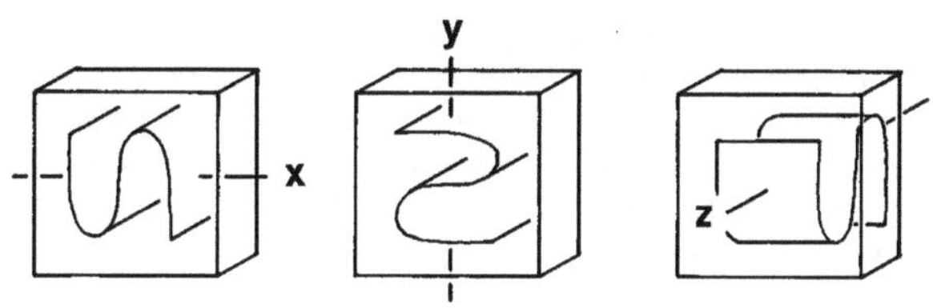

Fig. 5.19 Schematic illustrations of an oriented layer cluster which may be produced with illumination by light linearly polarised along the x, y and z directions. Only one oriented cluster is illustrated for simplicity. Reprinted with permission from Tanaka et al. [11]. Copyright 1996 by the American Physical Society

Although most researchers agree on the importance of charged VAPs, there are alternative explanations. For example, an explanation involving quasi-molecular defects, as argued by its authors, provides a better explanation for photo-induced gyrotropy [51]. Another approach (Fig. 5.19) is based on the assumption that remnants of crystalline layers are present in the glass and their presence and orientation can account for the observed anisotropy [11]. In a sense the latter model is close to the approach to photodarkening suggested in [52], which is also based on the assumption that small layer-like fragments exist in the glass.

While the majority of models consider photo-induced anisotropy in terms of photon-assisted rearrangement or rotation of some intrinsic structural units, in [53, 54] a model considering photo-induced anisotropy as a purely electronic phenomenon was also proposed. This model relates photoanisotropy to photogeneration of oriented electron-hole pairs. Within this framework, reversible reorientations of the anisotropy axis are interpreted as photostimulated generation of new geminate pairs in parallel to photon-assisted relaxation of the old pairs rather than photostimulated rotation of existing dipoles.

References

1. V.G. Zhdanov, V.K. Malinovskii, Sov. Tech. Phys. Lett. **3**, 943 (1977)
2. V.E. Karnatovskii, V.G. Tsukerman, Sov. J. Quantum Electron. **8**, 803 (1978)
3. V.G. Zhdanov, B.T. Kolomiets, V.M. Lyubin, V.K. Malinovskii, Phys. Stat. Sol. (a) **52**, 621 (1979)
4. T. Kosa, I. Janossy, Philos. Mag. B **64**, 355 (1991)
5. V. Lyubin, M. Klebanov, Phys. Rev. B **53**, 11924 (1996)
6. H. Lee, S. Jang, E. Kim, H. Chung, Jpn. J. Appl. Phys. **40**, 3965 (2001)
7. V. Lyubin, M. Klebanov, V.K. Tikhomirov, G.J. Adriaenssens, J. Non-Cryst. Solids **198/200**, 719 (1996)
8. A.V. Kolobov, V. Lyubin, T. Yasuda, M. Klebanov, Ka. Tanaka, Phys. Rev. B **55**, 8788 (1997)
9. P. Hertogen, G.J. Adriaenssens, J. Non-Cryst. Solids **266**, 948 (2000)
10. K. Kimura, K. Murayama, T. Ninomiya, J. Non-Cryst. Solids **77**, 1203 (1985)
11. K. Tanaka, K. Ishida, N. Yoshida, Phys. Rev. B **54**, 9190 (1996)
12. G.J. Adriaenssens, V.K. Tikhomirov, S.R. Elliott, J. Non-Cryst. Solids **227**, 688 (1998)
13. J.M. Lee, G. Pfeiffer, M.A. Paesler, D.E. Sayers, A. Fontaine, J. Non-Cryst. Solids **114**, 52 (1989)
14. V.L. Averyanov, A.V. Kolobov, B.T. Kolomiets, V.M. Lyubin, J. Non-Cryst. Solids **45**, 343 (1981)
15. V.M. Lyubin, V.K. Tikhomirov, J. Non-Cryst. Solids **114**, 133 (1989)
16. A.V. Kolobov, V. Lyubin, T. Yasuda, Ka. Tanaka, Phys. Rev. B **55**, 23 (1997)
17. V.K. Tikhomirov, G.J. Adriaenssens, S.R. Elliott, Phys. Rev. B **55**, R660 (1997)
18. K. Tanaka, in *Handbook of Advanced Electronic and Photonic Materials*, vol. 5, ed. by H.S. Nalwa (Academic Press, San Diego, 2001), p. 120
19. R. Grigorovici, A. Vancu, L. Ghita, J. Non-Cryst. Solids **59**, 909 (1983)
20. V.M. Lyubin, V.K. Tikhomirov, Sov. Phys. Solid State **32**, 1069 (1990)
21. V.L. Averianov, A.V. Kolobov, B.T. Kolomiets, V.M. Lyubin, Phys. Stat. Sol. (a) **57**, 81 (2006)
22. V.K. Tikhomirov, JETP Lett. **57**, 821 (1993)
23. H. Fritzsche, Phys. Rev. B **52**, 15854 (1995)
24. V.K. Tikhomirov, S.R. Elliott, Phys. Rev. B **49**, 17476 (1994)
25. K. Tanaka, M. Notani, H. Hisakuni, Solid State Commun. **95**, 461 (1995)
26. V.M. Lyubin, V.K. Tikhomirov, J. Non-Cryst. Solids **135**, 37 (1991)
27. G. Pfeiffer, C.J. Brabec, S.R. Jefferys, M.A. Paesler, Phys. Rev. B **39**, 12861 (1989)
28. V.K. Tikhomirov, M.M. Chervinskii, Sov. Phys. Solid State **35**, 566 (1993)
29. V.M. Lyubin, M. Klebanov, S. Rosenwaks, V. Volterra, J. Non-Cryst. Solids **164/166**, 1165 (1993)
30. M. Klebanov, V. Lyubin, V.K. Tikhomirov, Solid State Commun. **89**, 539 (1994)
31. V. Lyubin, M. Klebanov, Semiconductors **32**, 817 (1998)
32. H. Hisakuni, K. Tanaka, Solid State Commun. **90**, 483 (1994)
33. K. Kimura, K. Murayama, T. Ninomiya, J. Non-Cryst. Solids **59**, 913 (1983)
34. K. Iakoubovskii, P.W. Hertogen, G.J. Adriaenssens, J. Non-Cryst. Solids **240**, 237 (1998)
35. V. Lyubin, M. Klebanov, V.K. Tikhomirov, Phys. Rev. Lett. **87**, 216806 (2001)
36. V. Lyubin, M. Klebanov, V.K. Tikhomirov, J. Non-Cryst. Solids **299**, 945 (2002)
37. A.V. Kolobov, V.M. Lyubin, V.K. Tikhomirov, Philos. Mag. Lett. **65**, 67 (1992)
38. P. Krecmer, A.M. Moulin, R.J. Stephenson, T. Rayment, M.E. Welland, S.R. Elliott, Science **277**, 1799 (1997)
39. M. Stuchlik, P. Krecmer, S.R. Elliott, J. Optoelectron. Adv. Mater. **3**, 361 (2001)
40. A. Saliminia, T.V. Galstian, A. Villeneuve, Phys. Rev. Lett. **85**, 4112 (2000)
41. K. Tanaka, Appl. Phys. Express **1**, 012006 (2008)
42. K. Tanaka, A. Saitoh, N. Terakado. J. Non-Cryst. Solids **355**, 1828 (2009)
43. V. Lyubin, M. Klebanov, M. Mitkova, T. Petkova, Appl. Phys. Lett. **71**, 2118 (1997)
44. K. Ishida, K. Tanaka, Phys. Rev. B **56**, 206 (1997)

45. V.K. Tikhomirov, P. Hertogen, C. Glorieux, G.J. Adriaenssens, Phys. Stat. Sol. (a) **162**, R1 (1997)
46. A. Roy, A.V. Kolobov, Ka. Tanaka, J. Appl. Phys. **83**, 4951 (1998)
47. A.V. Kolobov, H. Oyanagi, Ka. Tanaka, Phys. Rev. Lett. **87**, 145502 (2001)
48. H. Fritzsche, Philos. Mag. B **68**, 561 (1993)
49. H. Fritzsche, J. Non-Cryst. Solids **164/166**, 1169 (1993)
50. V.K. Tikhomirov, S.R. Elliott, Phys. Rev. B **51**, 5538 (1995)
51. S.A. Dembovsky, E.A. Chechetkina, J. Non-Cryst. Solids **85**, 346 (1986)
52. K. Shimakawa, N. Yoshida, A. Ganjoo, Y. Kuzukawa, J. Singh, Philos. Mag. Lett. **77**, 153 (1998)
53. V.I. Arkhipov, E.V. Emelianova, P. Hertogen, G.J. Adriaenssens, Philos. Mag. Lett. **79**, 463 (1999)
54. E.V. Emelianova, P. Hertogen, V.I. Arkhipov, G.J. Adriaenssens, J. Non-Cryst. Solids **266**, 954 (2000)

Chapter 6
Photo-Induced Change of the Phase State

6.1 Photocrystallisation

6.1.1 Photocrystallisation of Selenium

Amorphous selenium is known to crystallise readily. Several works have been reported on the photocrystallisation of a-Se. Dressner and Stringfellow [1] reported that the radial growth rates of spherulitic crystals in a-Se films are sharply enhanced, by an order of magnitude, upon irradiation with light that has a wavelength shorter than a critical value. Based on the observation that the crystal growth rate was controlled by the production of electron–hole pairs and not by the density of the absorbed flux, they have concluded that this behaviour reflects primarily the photo-, rather than the thermal, effect of the irradiation. The transformation of Se from the amorphous to crystalline state depends strongly on the power of laser irradiation with pronounced threshold behaviour [2]. Anisotropic crystallisation of a-Se under the action of linearly polarised light [3–7] is another strong argument in favour of the electronic nature of this process.

Photocrystallisation has been shown to have a complicated behaviour if induced by simultaneous irradiation of the two lasers (Kr^+, $\lambda = 676$ nm, and Ar^+, $\lambda = 488$ nm) and can be either enhanced or slowed down depending on polarisation geometry of the two beams (cf. Fig. 5.14) [8]. Thus while for the case in which the polarisation planes of the two light sources are oriented perpendicular to each other the photocrystallisation rate is enhanced, for the parallel orientation suppression of the photocrystallisation rate is observed. It has been suggested that crystallisation induced by sub-bandgap light (Kr^+ laser) and cross-bandgap light (Ar^+ laser) have different mechanisms and opposite polarisation dependencies. Crystallites induced by one light source can be either enhanced or destroyed by the other light source. The observed photoamorphisation of a partly crystallised a-Se film (Fig. 6.1) [8] supports this suggestion.

A. V. Kolobov and J. Tominaga, *Chalcogenides*, Springer Series in Materials Science 164, DOI: 10.1007/978-3-642-28705-3_6,

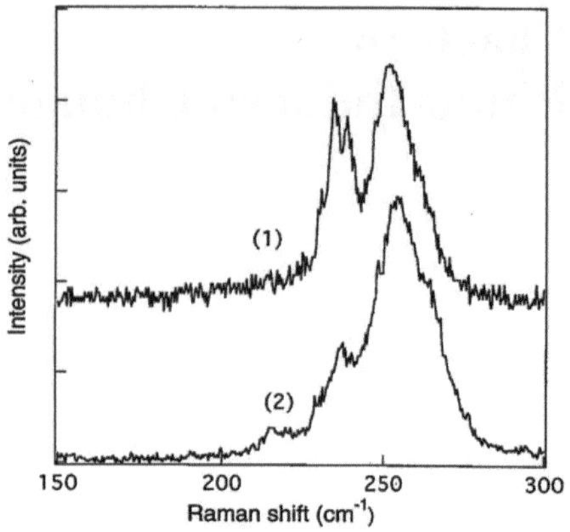

Fig. 6.1 Photoamorphisation at 77 K of a partially crystallised a-Se film. The "crystalline" peak located at 237 cm^{-1} (*1*) disappears upon photoillumination (*2*). Reprinted from Kolobov and Ka. Tanaka, [9] with permission by Elsevier

6.1.2 Crystallisation of Phase-Change Materials Through the Use of Coherent Phonons

While the crystallisation process in phase-change materials is generally believed to arise from laser- or current pulse-induced heating, it has been recently demonstrated that using coherent phonons excitation the layered $GeTe/Sb_2Te_3$ structures (cf. Sect. 8.3.1) can be switched between two structural states using energy that is not sufficient to heat the material. Figure 6.2 shows the effect of the time interval between two change-inducing pulses. As the delay time gets to 276 fs, i.e. the value corresponding to the frequency of the crystal-like mode, harmonic oscillations are resonantly enhanced (panel (a)). At the same time, the peak position of the local A_1 mode shifts from 3.84 THz corresponding to the amorphous-like phase to 3.68 THz corresponding to the crystal-like phase [10].

It is interesting to note that the exposure of the structure to intense fs pulses results in the transient softening of the network (Fig. 6.3) during ca. 600 fs following the pulse as is evidenced by the frequency kinetics of the local A_1 mode. After this the frequency either reverts to that of the initial state (in case of a single pulse) or levels off at the level corresponding to the crystal-like phase (in case of coherent excitation) [10].

The authors of Ref. [10] conclude that the pump pulse pair selectively modulates the bonding of Ge-Te in the covalently bonded fragments of the amorphous-like phase (referred to as $GeTe_4$ in the cited work) making Ge atoms migrate into octahedral symmetry positions charascteristic of the crystal-like phase along the [111] direction. The process supposedly requires little energy because chemical bonds in the excited state are more easily reconfigured than in the ground state [11].

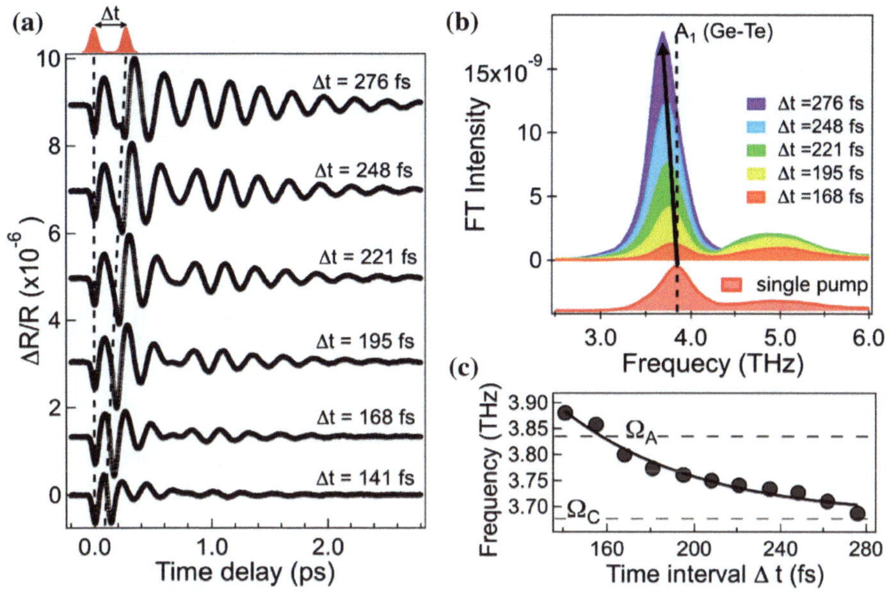

Fig. 6.2 Coherent vibrations excited in the "amorphous" phase by the pump pulse pair as the time interval Δt between the pulses is varied (**a**). Fourier-transformed spectra from the time-domain date demonstrate that the frequency of the local mode A_1 shifts to lower frequencies (**b**) as is also shown in (**c**). Reprinted from Makino et al. [10] with permission from Optical Society

Fig. 6.3 Transient frequency of the coherent A_1 mode obtained by a wavelet analysis from the time-domain signal with single and pump-pulse pair excitations. Reprinted from Makino et al. [10] with permission from Optical Society

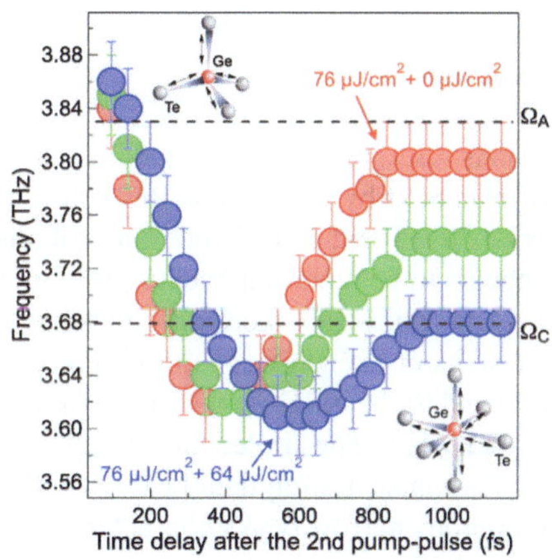

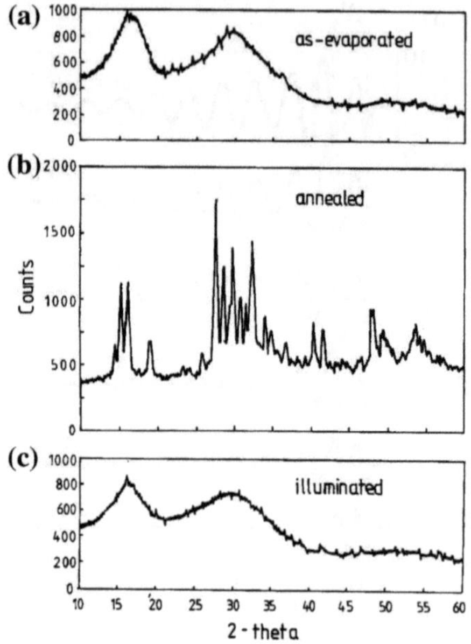

Fig. 6.4 X-ray diffraction patterns (CuK_α radiation) of (**a**) as-evaporated, (**b**) crystallised, and (**c**) photo-vitrified $As_{50}Se_{50}$ film on a silica substrate. Reprinted from Elliott and Kolobov [12] with permission from Elsevier

6.2 Photo-Induced Loss of Long-Range Order

6.2.1 Athermal Photoamorphisation

$As_{50}Se_{50}$

An $As_{50}Se_{50}$ film crystallised on a silica glass substrate becomes amorphous again if irradiated by continuous low-intensity light [12–16]. XRD patterns of thermally crystallised and illuminated amorphous films are shown in Fig. 6.4. This change appears to be athermal because the light-induced heating is negligible.

Amorphisation, therefore, does not take place by conventional local melting and subsequent quenching to the vitreous phase. The positions of crystallised films correspond to the XRD powder pattern of As_4Se_4. Realgar-like As_4Se_4 is a molecular crystal [17]. The molecular structure of the crystallised film is further confirmed by Raman scattering data [15] (Fig. 6.5) where the spectrum of the crystal exhibits numerous narrow peaks characteristic of a molecular solid. Interestingly, after photocrystallisation no trace of molecular modes is left: the Raman spectrum of the photoamorphised material exhibits a single broad peak characteristic of a cross-linked amorphous solid.

Each As_4Se_4 molecule (Fig. 1.16) has two intramolecular homopolar As–As bonds; the shortest intermolecular separation (3.42 Å) is between an As atom in one molecule and a Se atom in the other. The As–As bond within the As_4Se_4 molecule (2.57 Å) is significantly longer than in, say, a-As (2.49 Å) and the As–As-Se bond

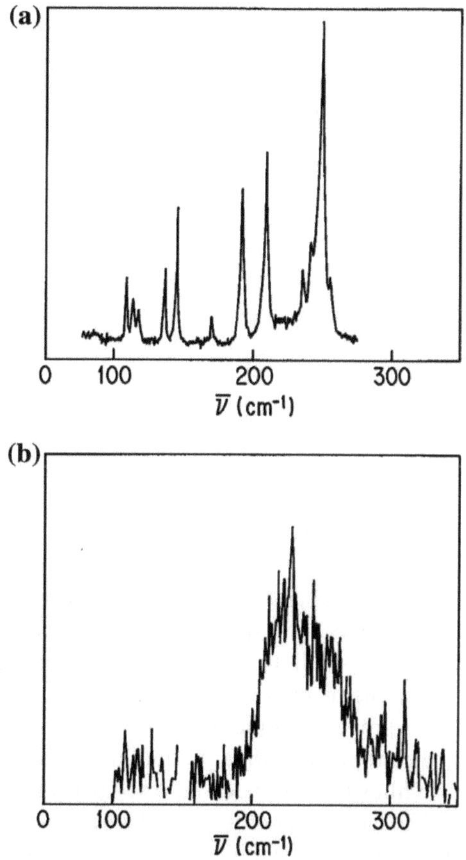

Fig. 6.5 Raman spectra of $As_{50}Se_{50}$ film in the crystallised (**a**) and photo-amorphised (**b**) states. Reprinted from Kolobov and Elliott [15] with permission from Elsevier

angle subtended at either one of the As atoms comprising the As–As bonds (101.2°) is also larger than the average bond angle in a-As (98°). The authors conclude that such intramolecular bonds are rather strained and, as a result, the electronic states associated with these bonds are likely to be at, or just above, the top of the valence band [12, 15]. Optical illumination might be expected preferentially to involve the excitation of such states, leading to bond scission; homopolar As–As bond formation could then be formed between closest atoms of neighbouring molecules, leading to a more cross-linked, and amorphous, structure. In addition, intramolecular As-Se bonds may also be broken upon illumination, giving rise to intermolecular As-Se bonds formation.

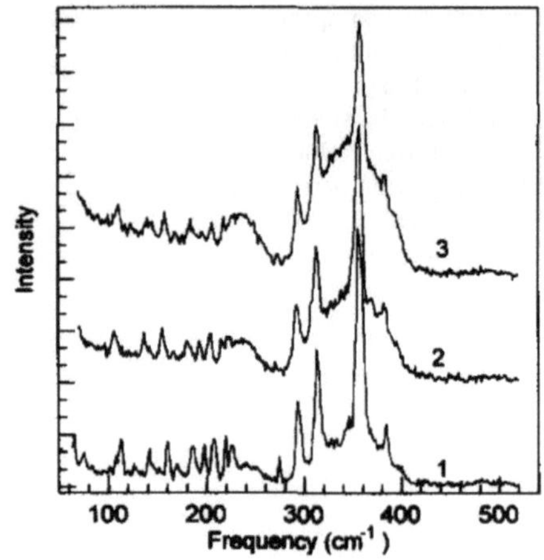

Fig. 6.6 Raman spectra for As_2S_3 (orpiment) at $T = 36$ K, after excitation by a Kr^+ laser ($\lambda = 568.2$ nm): 1, $I = 1$ W/cm^2); 2, $I = 3$ W/cm^2); 3, $I = 8$ W/cm^2). Reprinted from Frumar et al. [18] with permission from Elsevier

As_2S_3 and a-Se

Photo-induced amorphisaton has also been observed in As_2S_3 [18] (Fig. 6.6). In the case of As_2S_3, the starting material was natural orpiment crystals. When irradiated by Kr^+ laser light ($\lambda = 568$ nm) the spectrum gradually transforms and a broad feature characteristic of a glass emerges. Investigation of the temperature dependence of the process led the authors to the conclusion that amorphisation has essentially an electronic and not a thermal origin. Photoamorphisation of partially crystallised a-Se films has also been reported [8] (Fig. 6.1).

The fact that photo-induced athermal amorphisation has been observed in several different chalcogenide alloys indicates that this phenomenon may be rather general. Later in this volume, photo-assisted amorphisation of $Ge_2Sb_2Te_5$ is also described (cf. Sect. 11.1.4).

6.2.2 Photo-Induced Viscosity/Fluidity

In a series of experiments Nemilov and Tagantsev [19–22] demonstrated that the equilibrium viscosity of chalcogenide glasses decreased under the action of light. Among the most interesting findings was an observation that the *influence of light on viscosity decreases with increasing temperature*. The term 'photoviscous' was suggested to account for the athermal nature of this effect.

Subsequently, deformation of flakes of As_2S_3 under a combined action of light and mechanical stress was demonstrated and the term 'photofluidity' was suggested [23]. It was also found that while in the dark sample fluidity increases with an increase in temperature, in the case of photo-induced fluidity the trend is the opposite: the lower the temperature, the more "fluid" the glass becomes under photoexcitation

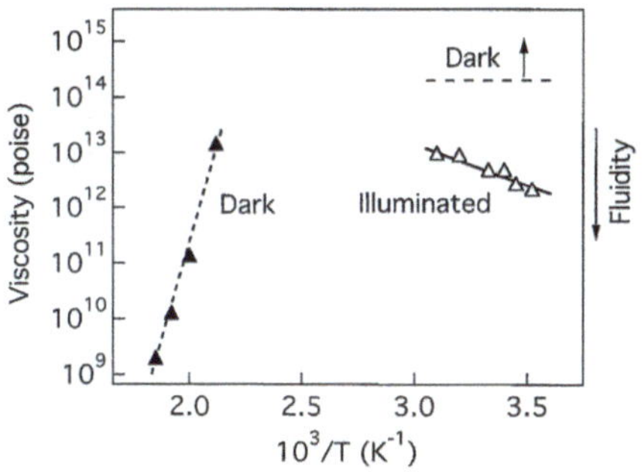

Fig. 6.7 Temperature dependencies of *dark* and photo-induced viscosity. Opposite behaviours demonstrate that the mechanism of photofluidity is not of thermal origin. From Hisakini and Tanaka [23]. Reprinted with permission from AAAS

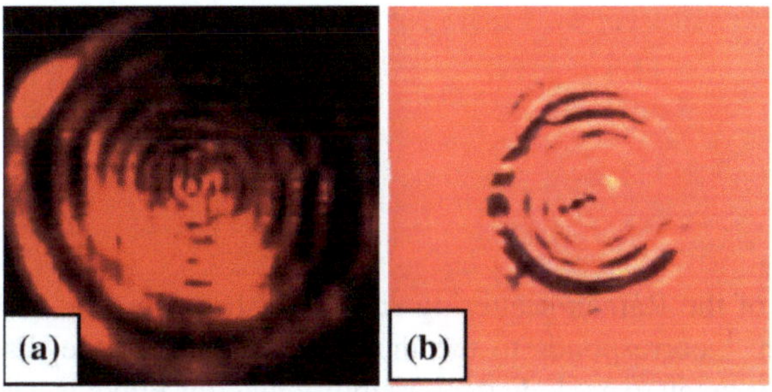

Fig. 6.8 Reflected light microphotographs. **a** Laser light focused into the spot with a diameter of 30 μm. An interference structure of the spot is due to an interference filter installed into the laser path. **b** a-Se film illuminated with the laser spot shown in (**a**) for 5 min at 77 K. Reprinted with permission from Poborchii et al. [25]. Copyright 1999 by the American Institute of Physics

(Fig. 6.7). The opposite tendencies for photo-induced and dark viscosities are a clear proof that the effect originates from the electronic excitation of the system and not from heating due to light absorption. Two possible mechanisms were suggested, namely, (i) an interchange of covalent bonds through bond rupture and rebonding and (ii) weakening of the intermolecular bonds.

For a recent review of photoplastic effects in chalcogenide glasses the interested reader is referred to [24].

6.2.3 Photomelting of Selenium

It has also been found that selenium, in both amorphous and crystalline forms, can be melted by light at temperatures as low as 77 K [25]. The photomelting was studied visually using an optical microscope (Fig. 6.8) and by Raman scattering.

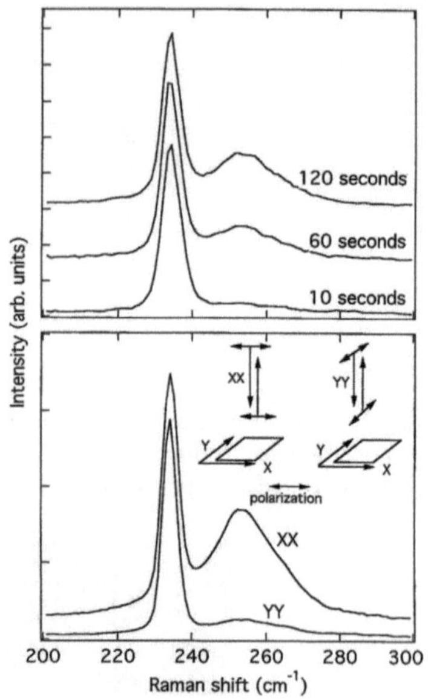

Fig. 6.9 (*top*) Evolution of Raman spectra of bulk polycrystalline t-Se at 77 K during illumination; (*bottom*) XX and YY configurations at 77 K after illumination with a light polarised parallel to the *X* axis. Reprinted with permission from Poborchii et al. [25]. Copyright 1999 by the American Institute of Physics

The results of the Raman scattering study are shown in Fig. 6.9. The growing peak at 252 cm^{-1} corresponds to the increased fraction of the disordered phase. Special care was taken to exclude the possibility that this effect is caused by heating due to light absorption. In particular, comparison of the Stokes and anti-Stokes peak intensities demonstrated that the temperature rise in this process did not exceed 10°.

It was further found that the "amorphous" peak at 252 cm^{-1} was strongly polarised (Fig. 6.9, bottom panel). This result indicates that quasi-free chains produced as a result of photo illumination are strongly oriented with respect to the polarisation of the excitation light. The authors concluded that photomelting of selenium is associated with breaking of interchain bonds between Se chains oriented parallel to the polarisation plane of the inducing light [25].

Athermal photomelting has also been studied by computer simulations and the obtained results suggested that "the occupation of low-lying conduction states gives already weakly bonded atoms additional freedom to diffuse in the network and in extreme conditions to photomelting, while conventional thermal diffusion depends on the lattice dynamics of atoms" [26].

References

1. J. Dresner, G. Stringfellow, J. Phys. Chem. Solids **29**, 303 (1968)
2. A.A. Baganich, V.I. Mikla, D.G. Semak, A.P. Sokolov, A.P. Shebanin, Phys. Stat. Sol. (b) **166**, 297 (1991)

3. V.K. Tikhomirov, P. Hertogen, C. Glorieux, G.J. Adriaenssens, Phys. Stat. Sol. (a) **162**, R1 (1997)
4. K. Ishida, K. Tanaka, Phys. Rev. B **56**, 206 (1997)
5. V. Lyubin, M. Klebanov, M. Mitkova, T. Petkova, J. Non-Cryst. Solids **227**, 739 (1998)
6. V. Lyubin, M. Klebanov, M. Mitkova, T. Petkova, Appl. Phys. Lett. **71**, 2118 (1997)
7. V. Lyubin, M. Klebanov, M. Mitkova, Appl. Surf. Sci. **154**, 135 (2000)
8. A. Roy, A.V. Kolobov, Ka. Tanaka, J. Appl. Phys. **83**, 4951 (1998)
9. A.V. Kolobov, Ka. Tanaka, in *Handbook of Advanced Electronic and Photonic Materials*, vol. 5, ed. by H.S. Nalwa (Academic Press)
10. K. Makino, J. Tominaga, M. Hase, Opt. Express **19**, 1260 (2011)
11. A. Zewail, J. Phys. Chem. A **104**, 5660 (2000)
12. S.R. Elliott, A.V. Kolobov, J. Non-Cryst. Solids **128**, 216 (1991)
13. A.V. Kolobov, S.R. Elliott, Philos. Mag. **71**, 1 (1995)
14. A.V. Kolobov, V.A. Bershtein, S.R. Elliott, J. Non-Cryst. Solids **150**, 116 (1992)
15. A.V. Kolobov, S.R. Elliott, J. Non-Cryst. Solids **189**, 297 (1995)
16. R. Prieto-Alcon, E. Marquez, J.M. Gonzalez-Leal, R. Jimenez-Garay, A.V. Kolobov, M. Frumar, Appl. Phys. A **68**, 653 (1999)
17. P. Goldstein, A. Paton, Acta Cryst. B **30**, 915 (1974)
18. M. Frumar, A.P. Firth, A.E. Owen, J. Non-Cryst. Solids **192/193**, 447 (1995)
19. S.V. Nemilov, D.K. Tagantsev, Fizika i Khimiya Stekla **7**, 195 (1981)
20. D.K. Tagantsev, S.V. Nemilov, Fizika i Khimiya Stekla **15**, 397 (1989)
21. D.K. Tagantsev, S.V. Nemilov, Fizika i Khimiya Stekla **13**, 677 (1987)
22. D.K. Tagantsev, S.V. Nemilov, Fizika i Khimiya Stekla **13**, 132 (1987)
23. H. Hisakuni, K. Tanaka, Science **270**, 974 (1995)
24. S.N. Yannopoulos, M.L. Trunov, Phys. Stat. Sol. (b) **246**, 1773 (2009)
25. V.V. Poborchii, A.V. Kolobov, Ka. Tanaka, Appl. Phys. Lett. **74**, 215 (1999)
26. J. Li, D.A. Drabold, Phys. Rev. Lett. **85**, 2785 (2000)

Chapter 7
Photo-Induced Phenomena in Chalcogenide-Metal Structures

The processes described in the previous chapters are observed in "pure" chalcogenides. The presence of an additional layer of metal such as silver allows for additional processes that involve structural modification. This chapter describes the general features of such processes. For more information, the interested reader is referred to reviews [1, 2].

7.1 Photodoping

7.1.1 Photodissolution

In the mid-1960s Kostyshin et al. [3] reported that illumination of a double-layer structure consisting of a chalcogenide glass and silver by light results in the rapid dissolution of the metal. The process is usually referred to as photodoping although the amount of metal dissolved in the chalcogenide glass can reach 30 at.%. Alternative terms to refer to this process are photodiffusion and photodissolution. Schematically the process is illustrated in Fig. 7.1.

Dissolution kinetics is usually studied by monitoring either the resistivity or the optical transmission of the metallic layer [2, 4]. Alternatively, the use of the X-ray diffraction intensity to monitor the thickness of the layer has been proposed [5]. Usually, three stages are observed, namely, (i) an induction period, (ii) effective dissolution stage and (iii) the completion slowdown stage (Fig. 7.2).

While an induction period has been reported to exist prior to beginning of efficient dissolution, it was argued that the latter was due to the formation of a thin oxide layer between the chalcogenide and the metal [6–9]. Alternative explanations for the existence of the induction period have been suggested in terms of the formation of dendrites located at metallic grain boundaries and the induction period was associated with the formation of a continuous doped layer through a process of lateral diffusion [5, 10].

A. V. Kolobov and J. Tominaga, *Chalcogenides*, Springer Series in Materials Science 164, DOI: 10.1007/978-3-642-28705-3_7,

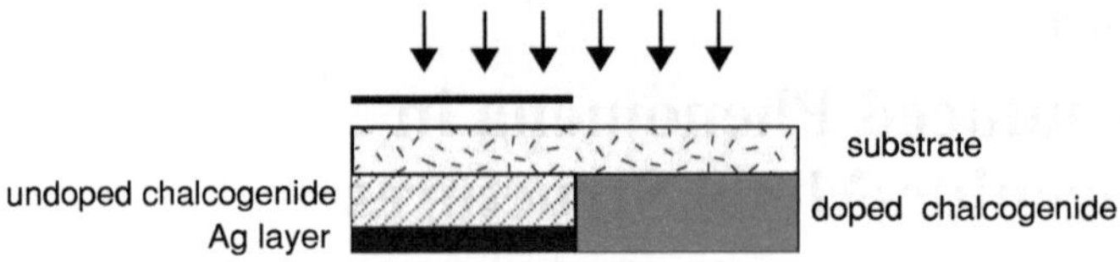

Fig. 7.1 Schematics of the photodissolution process

The dissolution process proper has been reported to obey both linear [4, 11] and square-root [12, 13] time dependencies. This difference has been interpreted by a competition of two limiting factors. When the penetration of silver through the boundary limits the process rate one would expect a linear time dependence. On the other hand, for larger amounts of dissolved silver when diffusion through the doped region becomes the rate-limiting step, the square-root time dependence is anticipated [14].

When the overall process rate becomes diffusion limited, the diffusion coefficient calculated from kinetic curves [15, 16] is $D_{\mathrm{Ag}}^{\mathrm{opt}} = 1.16 \cdot 10^{-11}\,\mathrm{cm^2\,s^{-1}}$ (the light intensity used in the experiment was 200 mW/cm^2) compared to the value of thermal diffusion coefficient $D_{\mathrm{Ag}}^{\mathrm{therm}} = 4 \cdot 10^{-14}\,\mathrm{cm^2\,s^{-1}}$ for Ag^+ ion self-diffusion in darkness has been reported [17]. A similar result, namely that $D_{\mathrm{Ag}}^{\mathrm{opt}} > D_{\mathrm{Ag}}^{\mathrm{therm}}$ by three orders of magnitude was also obtained in [18].

The activation energy of the photodissolution ($E_a \sim 0.1$–0.3 eV, temperature range of 20–120 °C) decreases significantly compared to thermally activated silver diffusion and dissolution ($E_a \sim 1.3$ eV, in a temperature range of 20–120 °C) in the dark [15].

While photodissolution of silver is most studied, group II metals such as Zn or Cd in contact with chalcogenides can also dissolve in the latter and this process changes upon exposure to light [19–21].

7.1.2 Lateral Diffusion

Lateral Diffusion on Insulating Substrates

Of special interest is the so-called lateral diffusion when silver diffuses in the direction *parallel* to the chalcogenide-metal interface as illustrated in Fig. 7.3 where lateral diffusion of silver in As_2Se_3 has been studied using an optical transmission method [22]. Several important conclusions have been drawn for this experiment. Firstly, the constant transmission of the doped region clearly indicates that *the concentration of silver is constant in the doped region*. Secondly, the diffusion process essentially stops after the metallic layer is exhausted suggesting that the process is only possible in the presence of the metallic source.

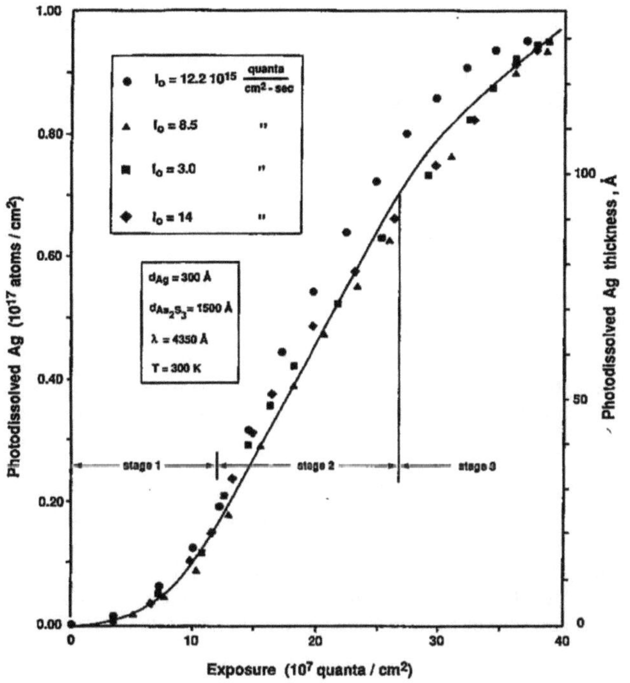

Fig. 7.2 Kinetics of Ag photodissolution process in As_2S_3 showing the induction period (*stage1*), photodissolution proper (*stage 2*), and the exhaustion regime (*stage 3*). Reprinted from Goldschmidt and Rudman [4] with permission from Elsevier

The lateral diffusion process has a strong compositional dependence and is much slower in sulphide glasses despite very efficient photodissolusion in them [22]. Of interest is the result of Ag diffusion in Ni- and Bi-doped chalcogenides [23]. In the former case when the conductivity stays *p*-type with doping, a monotonic increase in the lateral (thermal) diffusion rate has been observed; in contrast, in the Bi-doped samples an initial increase in the lateral diffusion was followed by a sharp drop with an increasing Bi content at the concentration when the conductivity type changed from *p*-type to *n*-type clearly suggesting the role of mobile electrons in the diffusion process.

Lateral Diffusion on Conducting Substrates

A number of investigators found that the lateral diffusion is enhanced if the chalcogenide layer is evaporated onto conductive substrates [24–26]. This result has lead to the conclusion that transport of electrons, that is necessary to preserve the overall neutrality, is the diffusion-limiting step [25, 27]. Since the electronic conductivity of chalcogenides is rather low, the presence of a conductive substrate, that can be used

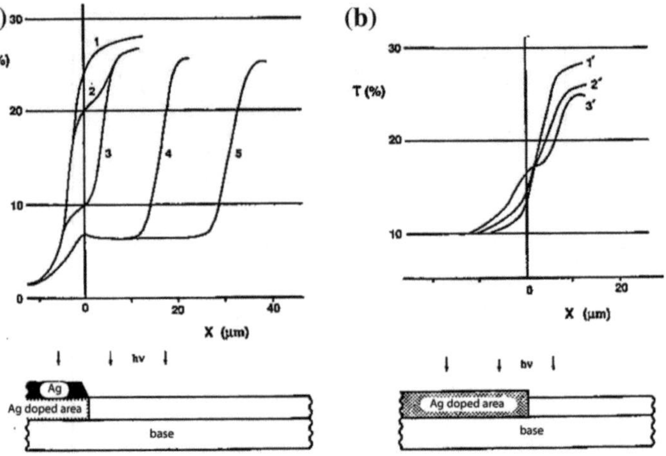

Fig. 7.3 Lateral diffusion of silver in As_2Se_3 films monitored by optical transmittance. The experimental arrangement used is illustrated at the bottom of the figures. **a** Results for a layer of As_2Se_3 (500 Å thick) with an overlayer of 650 Å Ag film, the sharp edge of which is illuminated with light (380 nm, 4 mW cm^{-2}) for (1) 0 min, (2) 5 min, (3) 30 min, (4) 60 min and (5) 80 min. **b** Results for the boundary between Ag-doped and undoped As_2Se_3 when no metallic silver is present: (1′) 0 min, (2′) 20 min, (3′) 150 min. Reprinted from Yamaguchi et al. [22] with permission from Elsevier

as a source of electrons to ensure the charge neutrality of the doping front, enhances the diffusion rate because the electrons can pass laterally through the metal without the need to travel through the doped region.

It was subsequently demonstrated that the enhancement of the diffusion rate also depends on the chemical nature of the metal: the diffusion increase was the strongest for gold and nickel and hardly enhanced at all for samples containing an Al conducing layer [23, 28] although it cannot be excluded that the Al layer was oxidised and the oxide layer prevented electrons from effectively contributing to the Ag diffusion.

Lateral Migration

When lateral processes are considered, one should also take into account a different process, namely, lateral migration. This process was first reported in [29, 30] and consists of the transfer of silver from the non-illuminated region to illuminated regions when a double Ag-chalcogenide layer is illuminated through a mask with a sharp edge.

7.1.3 Diffusion Mechanism

Understanding the photodiffusion mechanism requires answers to the following questions: (i) the charge state of propagating Ag particles, (ii) the nature of silver–chalcogenide interaction, and (iii) the location of actinic light absorption.

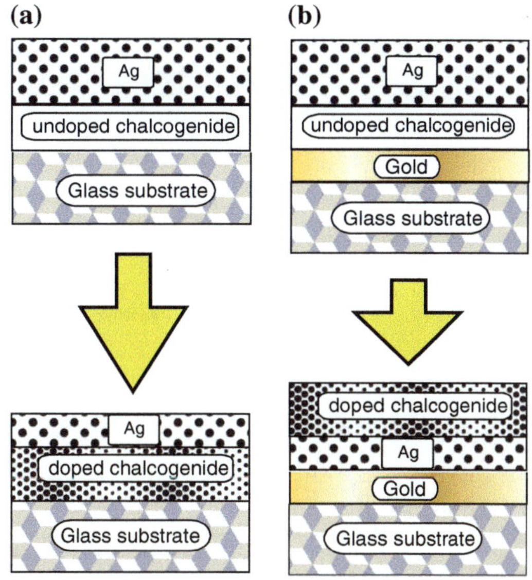

Fig. 7.4 The effect of an electronegative substrate on the final stage of Ag dissolution. When the chalcogenide film is deposited directly onto a glass substrate, the excess silver remains on top (**a**); at the same time if an additional electronegative metal layer is present, the excess silver is transferred to the bottom (**b**)

Charge State of Silver Particles

It is generally believed that Ag-doped chalcogenide glasses are solid electrolites and silver diffuses as positively charged Ag^{+} ions with a flux of electrons in the same direction or a flux of holes in the opposite direction ensuring the overall charge neutrality.

This conclusion was initially drawn based on the effect of an electrical field on the diffusion rate [24, 31, 32]. This idea was subsequently developed in Ref. [23] where the authors used structures containing an additional metallic layer (Fig. 7.4). The authors have demonstrated that dissolution of Ag strongly depends on the electronegativity of the lower metal layer. If the electronegativity of the metal (substrate) is higher than that of the silver, the dissolution of silver continues and the excess silver (compared to Ag solubility limit in the chalcogenide layer) is deposited on the electronegative metal in form of metallic silver. This process, analogous to the dissolution of the electropositive electrode in a solid galvanic cell, strongly supports the suggestion that silver dissolves as positively charged ions.

Step-Like Concentration Profile

The fact that the silver concentration in the doped region is constant with a step-like concentration profile [22] has been accounted for by the observation that there are two separated glass-formation regions in the Ag-As-S system [33–35] (and possibly others) as illustrated by Fig. 7.5. The existing immiscibility gap [27] in the glass-formation diagram naturally accounts for the observed experimental result.

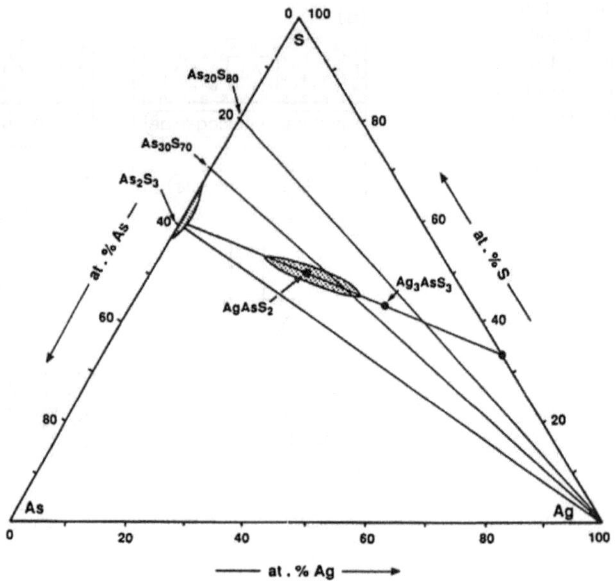

Fig. 7.5 Glass-formation regions in the Ag-As-S system, redrawn after Kawamoto et al. [33]

The current consensus is that photodissolution of silver is a photo-enhanced (percolation) chemical reaction between silver and chalcogenides [27, 36, 37]. Exposure to light modifies the electrochemical potential of the system and thus changes the dissolution process.

Location of Actinic Light Absorption

The most convincing results considering the location of actinic light absorption were obtained using lateral diffusion approach that allowed the authors to illuminate selectively the metal, chalcogenide, metal/doped boundary, and doped/undoped boundary [22]. The experimental set-up and the results are shown in Fig. 7.6. This experiment has allowed the authors to draw the unambiguous conclusion that the most effective driving force for the diffusion is light absorption at the doped/undoped boundary.

7.1.4 Reaction Products and Their Properties

The silver diffusion and dissolution process leads to the formation of homogeneous products in a wide span of compositions based on the dissolution limits of silver in different composition of chalcogenides [38, 4, 15, 39, 40]. The silver dissolution

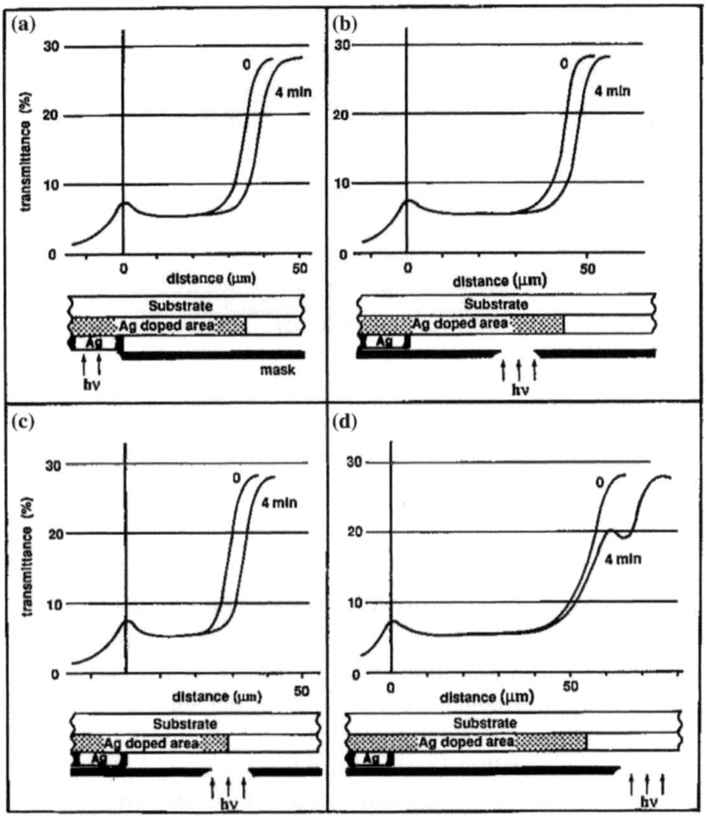

Fig. 7.6 Determination of the site for effective absorption causing lateral diffusion. Illumination was performed using a tungsten lamp (4 mW cm^{-2}) spatially defined by a mask. Reprinted from Yamaguchi et al. [22] with permission from Elsevier

limits were found to be in ca. 10–30 at.% range depending on the composition of the glass.

Doping modifies both the electrical and optical properties of chalcogenides. The electrical conductivity significantly increases and the conductivity type changes from the *p*-type electronic conductivity in undoped chalcogenides to mixed electronic and ionic conductivity in the doped samples. The optical bandgap decreases with doping [2].

7.2 Photodeposition of Silver

Ag-rich chalcogenide glasses containing more than 30 at.% Ag have long been studied for applications as solid-state electrolytes for batteries, since they exhibit

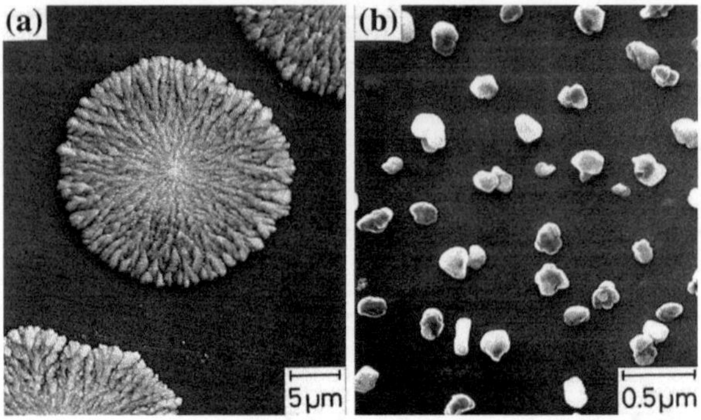

Fig. 7.7 SEM photographs of photodeposited Ag particles on **a** bulk glass and **b** amorphous film. Light from an ultrahigh-pressure Hg lamp with an infrared-cut filter illuminated the samples: (**a**) $Ag_{45}As_{15}S_{40}$ bulk glass ($W = 200\,mW/cm^2$, 15 min); (**b**) $Ag_{65}(Ge_{0.3}S_{0.7})_{35}$ film ($W = 80\,mW/cm^2$, 2 min). Reproduced from Kawaguchi [42] with permission by Wiley–VCH Verlag GmbH

a marked ionic conduction [41]. In 1975, Maruno and Kawaguchi reported that Ag-rich Ag-As-S glasses exhibit a phenomenon opposite to photodoping. Subsequently, these authors conducted systematic studies of this effect [42–47] and a review is given in [48]. When the bandgap light illuminates the surface of a Ag-rich glass, a large number of fine particles are deposited in the illuminated region (Fig. 7.7). The particles are pure metallic silver and are formed from Ag^+ ions in the glass. This phenomenon was called the "photoinduced surface deposition" (PSD) of silver. The PSD phenomenon is a purely optical phenomenon and any thermal effects are secondary [49].

7.2.1 Light Intensity, Photon Energy and Temperature Dependencies

Figure 7.8 shows the change in morphology of the Ag deposits as a function of the intensity of the illumination light. In panel (a) disc-like deposits are seen and their size decreases with increasing light intensity. In panel (b) which is representative of relatively large deposits, the peripheral region is higher than the central region, so that the particles have the appearance of a flower. Numerous small deposits are additionally observed between the flower-like deposits. The nucleation of small deposits seems to occur after the growth of the flower-like deposits.

A photon energy higher than the optical bandgap is needed to induce PSD and the photosensitivity of the PSD effect increases with an increase in the photon energy [50]. These results suggest that photoexcitation of electron–hole pairs is vital for the PSD to take place.

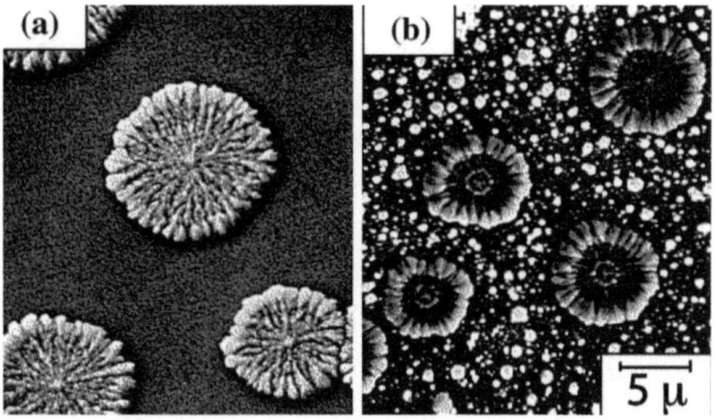

Fig. 7.8 Effect of light intensity for $Ag_{45}As_{15}S_{40}$ bulk glass [49]. Illumination with the light (a Hg lamp) of **a** 80 mW/cm^2, **b** 200 mW/cm^2 and **c** 530 mW/cm^2 was performed for 15 min at 21°C. Reprinted from Kawaguchi et al. [48] with permission from Elsevier

The effect of temperature on PSD phenomenon was examined in a temperature range of 0–50 °C [49]. The temperature dependence of the size of the Ag deposits shows a maximum at around 30 °C, but the density decreases with increasing temperature. Thus, the photosensitivity, proportional to (density)×(area) shows a maximum at around 20 °C. The PSD effect could not be observed at low temperatures (e.g. 77 K) and at temperatures above 80°C [49].

7.2.2 *Reversibility*

Figure 7.9 shows the alternate photodeposition and thermal erasure. Figure 7.9a is a SEM photograph of an as-prepared film. The particles in Fig. 7.9b are the Ag deposits resulting from the first illumination. The reversibility of the process is illustrated in Fig. 7.9c, d. In Fig. 7.9c, the disappearance of the deposits can be recognised, but very small white particles and dark spots of relatively large size are observed. The white particles are possibly the residue of Ag deposits and the dark spots may be due to an inhomogeneous Ag distribution caused by thermal erasing.

It is of interest to note that the photodeposted silver can be fixed on the surface and preserved from subsequent dissolution by covering with a thin gold layer: while the deposited silver could be reversibly dissolved in the uncovered region, in the covered region it stayed in the surface [43] (Fig. 7.10). This result should be compared to the effect of a gold layer in the dissolution process when excess silver was located in contact with the gold layer (cf. Fig. 7.4).

7.2.3 *Mechanism of Photosurface Deposition*

Similar to the photodiffusion, the phenomenon is explained by considering a flux of Ag^+ ions in the deep region towards the illuminated surface to counterflow the hole and electron fluxes. One can also argue that since the photodissolution of Ag is a photo-enhanced chemical reaction, the amount of the dissolved silver is determined

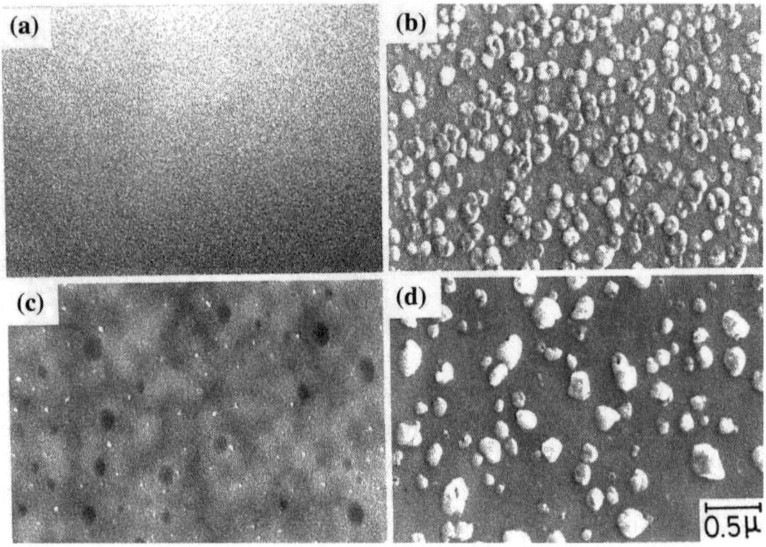

Fig. 7.9 SEM photographs of as-prepared and variously treated $Ag_{61.5}(Ge_{0.3}S_{0.7})_{38.5}$ films [44]: **a** as-prepared film (1400 Å thick); **b** after the first illumination (a Hg lamp, 160mW/cm^2, 10 min, 20°C); **c** after annealing of (**b**) (150°C, 10 min, in Ar gas); **d** after illumination of (**c**) (a Hg lamp, 160 mW/cm^2, 10 min, 20°C). Reprinted with permission by Japan Society of Applied Physics from Kawaguchi and Maruno [44]

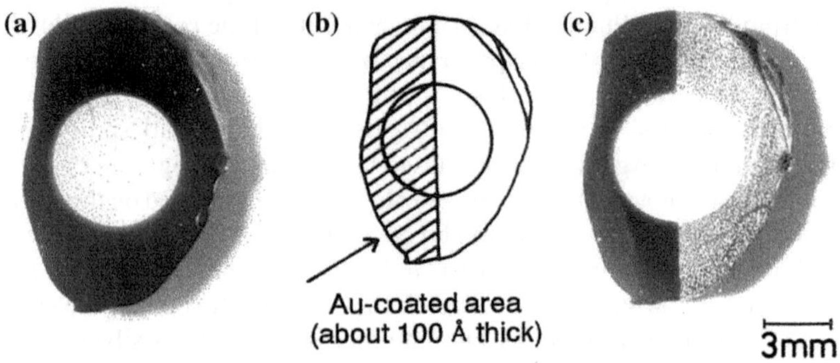

Fig. 7.10 Fixation of Ag pattern by covering of the substrate with a semitransparent Au film [43]. **a** a white circle was recorded on $Ag_{45}As_{15}S_{40}$ bulk glass by the PSD effect. The regions have been subsequently covered with a Au film **b**. After exposure to Hg lamp light (2 h, 30 mW/cm^2) silver on the uncoated region was dissolved while no change was observed for the coated region. Reprinted with permission by Japan Society of Applied Physics from Kawaguchi and Maruno [43]

by the electrochemical potential of the system. By changing the position of the chemical potential through exposure to light or by changing temperature one can change the amount of silver stably dissolved within the chalcogenide which can lead to either dissolution of additional silver or by the effusion of excessive silver.

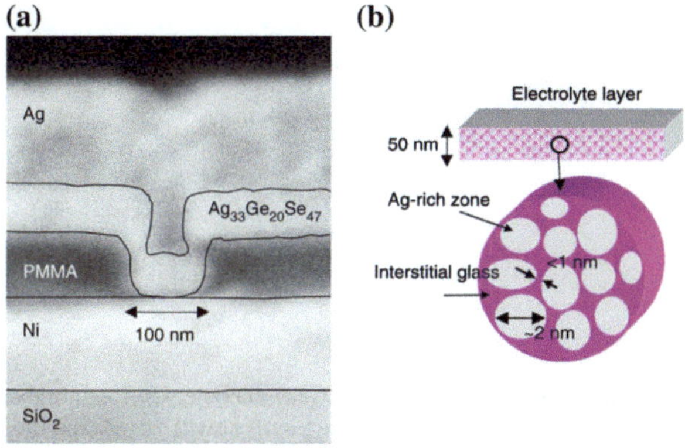

Fig. 7.11 **a** Scanning electron microscope image of a PMC memory cell cross-section fabricated using PMMA as the dielectric. The AgGeSe electrolyte and silver anode are formed in a 100 nm via in the dielectric. **b** Schematic of the nanostructure of the AgGeSe solid electrolyte. Electrolyte is nanophase separated into a dispersed conducting silver-rich phase and a germanium-rich/silver-poor resistive glassy interstitial phase. The average size of the silver-rich phase is around 2 nm whereas the average interstitial gap is less than 1 nm. Reprinted from Kozicki et al. [51] with permission from Elsevier

7.2.4 Programmable Metallisation Cell

The fact that silver particles can be dissolved and effused from Ag-rich chalcogenide glasses also has a potential application in non-volatile memory in so-called programmable metallisation cells (PMC) (Fig. 7.11a) that are based on the electrochemical control of nanoscale quantities of metal in thin films of solid electrolyte [51]. Ge_xSe_{1-x} host glass with x close to 0.25 is one of promising candidates. It has been argued that the electrolyte formed by silver dissolution into a thin film of chalcogen-rich glass is a nanostructured material [52], consisting of a finely dispersed low-resistivity silver-rich phase and an interstitial germanium-rich glassy phase that exhibits high resistivity (Fig. 7.11b). In the case of the AgGeSe system, the resistivity is around 2 mΩcm for the silver-rich phase and is estimated to be as high as 10^8 Ωcm for the germanium-rich phase. For a silver-saturated ternary, the glassy interstices between the silver-rich regions have an average width of less than 1 nm but the materials high resistivity makes the electrolyte resistance relatively high.

At the nanoscale, each silver-rich region acts as a local supply of ions. For each excess ion that enters one of these regions from the anode side, another ion will simultaneously leave on the cathode side and move into the interstitial zone there. Once in this glassy material, the high local electric field will cause the ions to move toward the adjacent downstream silver-rich region. The electron current from the cathode flows into the electrolyte and the supply of both ions and electrons in the interstitial zones results in electrodeposition of metallic silver similar to the process described earlier in this chapter. The nanoscale conductive electrodeposits bridge the

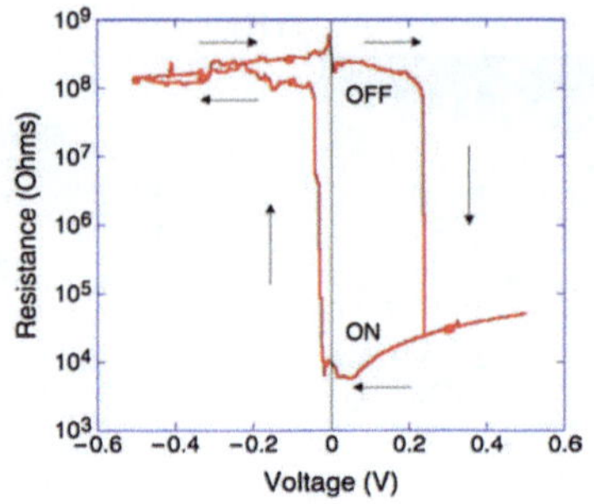

Fig. 7.12 Resistance versus voltage curve of a 100 nm diameter device fabricated using a 50 nm thick Ag-doped GeSe solid electrolyte. Programming current was 10 μA and the write (switch on) and erase (switch off) thresholds are +240 and −80 mV respectively. Off resistance is lower than expected due to poor step coverage in the small-geometry. Reprinted from Kozicki et al. [51] with permission from Elsevier

interstitial regions reducing the resistance between the electrodes. Upon reversal of the applied voltage the metallic silver dissolves and the high-resistance state of the glass is restored. A resistance–voltage characteristic of a PMC device is shown in Fig. 7.12.

References

1. A.V. Kolobov, S.R. Elliott, Adv. Phys. **40**, 625 (1991)
2. T. Wagner, M. Frumar, in *Phito-induced Metastability in Amorphous Semiconductors*, ed. by A.V. Kolobov (Wiley-VCH, Weinheim, 2003), p. 160
3. M.T. Kostyshin, E.V. Mikhailovskaya, P.F. Romanenko, Fiz. Tverd. Tela **8**, 571 (1966)
4. D. Goldschmidt, P.S. Rudman, J. Non-Cryst. Solids **22**, 229 (1976)
5. J. Rennie, S.R. Elliott, C. Jeynes, Appl. Phys. Lett. **48**, 1430 (1986)
6. A. Buroff, Ph.D. Thesis, Sofia, 1980
7. K. Chatani, I. Shimizu, H. Kokado, E. Inoue, Jpn. J. Appl. Phys. **16**, 389 (1977)
8. J. Oldale, Ph.D. Thesis, Cambridge University, 1989
9. R. Ishikawa, Solid State Commun. **30**, 99 (1979)
10. S.R. Elliott, J. Non-Cryst. Solids **130**, 85 (1991)
11. T. Yaji, S. Kurita, J. Appl. Phys. **54**, 647 (1983)
12. M.T. Kostyshin, Y.V. Ushenin, Sov. Phys. Semicond. **16**, 11 (1982)
13. T. Shirakawa, I. Shimizu, H. Kokado, E. Inoue, J. Photogr. Sci. Eng. **19**, 139 (1975)
14. P.J.S. Ewen, A. Zakery, A.P. Firth, A.E. Owen, Philos. Mag. B **57**, 1 (1988)
15. T. Wagner, G. Dale, P.J.S. Ewen, A.E. Owen, V. Perina, J. Appl. Phys. **87**, 7758 (2000)
16. T. Wagner, E. Marquez, J. Fernandez-Pena, J. Gonzalez-Leal, P.J.S. Ewen, S.O. Kasap, Philos. Mag. B **79**, 223 (1999)
17. G. Holmquist, J. Pask, J. Am. Chem. Soc. **62**, 183 (1979)
18. T. Wagner, A. Mackova, V. Peina, E. Rauhala, A. Seppala, S. Kasap, M. Frumar, M. Vlcek, J. Non-Cryst. Solids **299**, 1028 (2002)
19. A.V. Kolobov, B.T. Kolomiets, V.M. Lyubin, M.A. Tagirdzhanov, Solid State Commun. **54**, 379 (1985)

20. V. Lyubin, M. Klebanov, A. Arsh, N. Froumin, A.V. Kolobov, J. Non-Cryst. Solids **326**, 189 (2003)
21. V.M. Lyubin, A.V. Kolobov, J. Non-Cryst. Solids **90**, 489 (1987)
22. M. Yamaguchi, I. Shimizu, E. Inoue, J. Non-Cryst. Solids **47**, 341 (1982)
23. G.E. Bedelbaeva, A.V. Kolobov, V.M. Lyubin, Sov. Phys. Semicond. **25**, 119 (1991)
24. A. Matsuda, M. Kikuchi, J. Jpn. Soc. Appl. Phys. Suppl. **42**, 239 (1973)
25. A.E. Owen, A.P. Firth, P.J.S. Ewen, Philos. Mag. B **52**, 347 (1985)
26. T. Wagner, M. Frumar, in *Proceedings of International Conference on Non-Crystalline Semiconductors, Part II*, (Uzhgorod, 1989), p. 240
27. T. Wagner, M. Frumar, V. Suskova, J. Non-Cryst. Solids **128**, 197 (1991)
28. A.V. Kolobov, G.E. Bedelbaeva, Philos. Mag. B **64**, 21 (1991)
29. K. Chatani, I. Shinizu, E. Inoue, J. Soc. Photogr. Sci. Jpn. **41**, 21 (1978)
30. N. Funakoshi, S. Zembitsu, T. Kasai, Jpn. J. Appl. Phys. **20**, L649 (1981)
31. A. Buroff, E. Nebauer, P. Süptitz, Phys. Stat. Sol. (a) **48**, K109 (1978)
32. E.A. Lebedev, P. Süptitz, I. Willert, Phys. Stat. Sol. (a) **28**, 461 (1975)
33. Y. Kawamoto, M. Agata, S. Tsuchihashi, J. Ceram. Assoc. Jpn. (in Japanese) **82**, 502 (1974)
34. A.P. Firth, P.J.S. Ewen, A.E. Owen, in *The Structure of Non-Crystalline Materials*, ed. by P.H. Gaskell, J.M. Parker, E.A. Davis (Taylor and Francis, London, 1983), p. 286
35. T. Wagner, M. Frumar, J. Non-Cryst. Solids **116**, 269 (1990)
36. J. Malinowski, A. Buroff, Contemp. Phys. **19**, 99 (1978)
37. G. Kluge, Phys. Stat. Sol. (a) **101**, 105 (1987)
38. A.V. Kolobov (ed.), *Photo-Induced Metastability in Amorphous Semiconductors* (Wiley-VCH, Winheim, 2003)
39. E. Marquez, T. Wagner, J. Gonzalez-Leal, A. Bernal-Oliva, R. Prieto-Alcon, R. Jimenez-Garay, P. Ewen, J. Non-Cryst. Solids **274**, 62 (2000)
40. A.V. Kolobov, S.R. Elliott, A.T. Steel, Phys. Rev. B **41**, 9913 (1990)
41. Y. Kawamoto, M. Nishida, J. Non-Cryst. Solids **20**, 393 (1976)
42. T. Kawaguchi, in *Photo-Induced Metstability in Amorphous Semiconductors*, ed. by A.V. Kolobov (Wiley VCH, Weinheim, 2003), p. 182
43. T. Kawaguchi, S. Maruno, Jpn. J. Appl. Phys. **33**, 4521 (1994)
44. T. Kawaguchi, S. Maruno, Jpn. J. Appl. Phys. **33**, 6470 (1994)
45. T. Kawaguchi, S. Maruno, K. Tanaka, J. Non-Cryst. Solids **164**, 1231 (1993)
46. S. Maruno, J. Non-Cryst. Solids **59**, 933 (1983)
47. S. Maruno, T. Kawaguchi, J. Appl. Phys. **46**, 5312 (1975)
48. T. Kawaguchi, K. Tanaka, S.R. Elliott, in *Handbook of Advanced Electronic and Photonic Materials*, vol. 5, ed. by H.S. Nalwa (Academic Press, San Diego, 2001), p. 91
49. T. Kawaguchi, S. Maruno, J. Appl. Phys. **77**, 628 (1995)
50. T. Kawaguchi, S. Maruno, S.R. Elliott, J. Non-Cryst. Solids **202**, 107 (1996)
51. M.N. Kozicki, M. Mitkova, M. Park, M. Balakrishnan, C. Gopalan, Superlattices Microstruct. **34**, 459 (2003)
52. M.N. Kozicki, M. Mitkova, J. Zhu, M. Park, Microelectron. Eng. **63**, 155 (2002)

Part III
Phase-Change Materials

Chapter 8
Structure of the Crystalline Phase

As already mentioned (Sect. 1.2.3), materials that exhibit reversible crystallisation–amorphisation process with parameters (such as switching rate, stability, durability, etc.) that can be used for memory applications can be largely divided into two major groups. One major class of phase-change materials is represented by quasi-binary GeTe-Sb_2Te_3 (GST) alloys with the binary GeTe material also being a potential candidate for memory applications. The other major class is represented by Sb-rich SbTe alloys, typically close to the Sb_2Te composition. A subclass of these materials is formed by Ag- and In-doped Sb_2Te (AIST alloys) with the total concentration of the dopants close to 10%.

This chapter is dedicated to the structure of the crystalline phase of phase-change materials, mainly to the Ge-Sb-te alloys. It should be noted that GST alloys can crystallise into two different structures, namely the stable trigonal phase (alternatively referred to as the hexagonal phase) and a metastable cubic phase. The structures of these two phases are described separately.

8.1 Structure of the Stable Phase

8.1.1 Binary GeTe

It is natural to start the description with the binary GeTe which is the simplest material in the GST system. GeTe is a narrow bandgap semiconductor and also a ferroelectric with the simplest conceivable structure containing just two atoms in the primitive cell. In the low-temperature ferroelectric phase GeTe possesses a rhombohedral structure with space group R3m. This structure can be viewed as a rock salt structure slightly distorted along $\langle 111 \rangle$ direction with a subsequent shear relaxation along the $\langle 111 \rangle$ direction. The driving force for the formation of the rhombohedral phase has been a subject of various studies in the past [1]. In this phase, Ge and Te atoms are sixfold coordinated to each other with subsets of three shorter (2.83 Å) and three longer

A. V. Kolobov and J. Tominaga, *Chalcogenides*, Springer Series in Materials Science 164, DOI: 10.1007/978-3-642-28705-3_8,

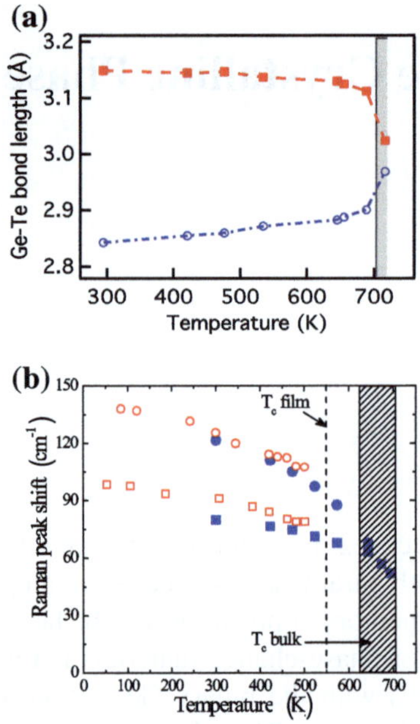

Fig. 8.1 **a** Experimental evidence for the displacive nature of the phase transition in GeTe. Ge–Te bond lengths calculated from unit cell data obtained from neutron diffraction data on bulk GeTe samples (after Chattopadhyay et al. [3]). **b** Raman phonon mode frequencies, filled symbols represent data on capped GeTe sputtered thin films; open symbols stand for data from Ref. [4] on bulk GeTe single crystals. The hatched area spans over the temperature range in which the critical temperature values T_c for the ferroelectric phase transition of GeTe have been reported in the literature. The *dashed line* represents the T_c of thin films, Ref. [8]. Reprinted with permission from Fons et al. [5]. Copyright 2010 by the American Physical Society

(3.15 Å) bonds often described as a Peierls distortion [2] due the reduced coupling between the *p*-type orbitals that constitute the basis for bonding in GeTe.

Based on diffraction studies, mainly neutron diffraction [3], it was concluded that GeTe undergoes a displacive ferroelectric-to-paralectric transition at the Curie temperature, T_C, around 705 K when the structure changes to the rock salt structure (space group Fm3m) with a concomitant disappearance of the Peierls distortion. Experimental bond lengths calculated from the unit cell data obtained from neutron diffraction [3] show an apparent symmetrical convergence of the two bond lengths present in the rhombohedral phase to the average value near T_C (Fig. 8.1, upper panel). Based upon the observation of phonon mode softening with temperature, the displacive nature of the transitions has been also suggested by Raman scattering [4], although the highest temperature studied was ca. 150 K lower than the estimated T_C.

As discussed earlier (Sect. 3.1.1), Bragg diffraction is not sensitive to stochastic local distortions and as a result can miss important details of the local structure. It is more appropriate to employ X-ray absorption fine structure (EXAFS) observations to directly observe the changes in the Ge-Te bond length with temperature. The characteristic time of EXAFS measurements is 10^{-15} s, i.e. EXAFS is capable of taking a true snapshot of the structure.

The results of the fitting of the Ge-Te distances for the shorter and longer bonds from the measured EXAFS data are shown in Fig. 8.2 (upper panel) [5]. The above results demonstrate that *locally* the structure remains distorted above T_c in essentially the same manner as it is distorted at lower temperatures. The fact that GeTe becomes paraelectric macroscopically means that the local distortions become stochastic at T_c, i.e. the ferroelectric-to-paraelectric transition is of the order–disorder type [6]. Schematic representation of displacive and order–disorder transitions is shown in Fig. 8.3.

Further grounds for the unchanged local structure come from consideration of the bond stiffness as a function of temperature [5]. Figure 8.2 (lower panel) shows the temperature dependence of the mean-square relative displacement (MSRD). The extent to which the MSRD increases with temperature as well as its absolute value are determined by the bond strength usually represented by the Einstein temperature Θ_E that is related to the MSRD, σ, through the following equation:

$$\sigma^2 = \frac{\hbar^2}{2\mu k_B \Theta_E} \coth\left(\frac{\Theta_E}{2T}\right) + \sigma_0^2 \tag{8.1}$$

Here, μ is the reduced mass, k_B is Boltzmann's constant, and σ_0 is static disorder. The Einstein temperature is typically around 300 K for tetrahedrally bonded semiconductors, around 120–150 K for Se and Te that form helical chains with only two first-nearest neighbours, and in a 30–60 K range for clathrates [7]. Should Ge atom become located within a soft-mode potential in the center of the rock-salt cell, there should be a noticeable increase in MSRD above T_c.

One can see, however, that the MSRD for GeTe in the entire temperature range can be fitted with a *single* value of $\Theta_E \approx 180$ K demonstrating again that there is no change in the Ge-Te bond strength (i.e. the first derivative of the interatomic potential) and hence the local potential relief remains unchanged as the material is heated above T_c.

The authors conclude that despite the currently accepted consideration that GeTe exhibits a ferroelectric transition of displacive nature, the recent experimental results provide evidence for an order–disorder nature of the transition since local distortions do not disappear at the Curie temperature [5].

This conclusion has been confirmed by a PDF analysis of total scattering when the apparently (based upon Bragg diffraction data) cubic phase could be well fitted by a model with Ge atom positions distributed among four different sites displaced along the four $\langle 111 \rangle$ directions as well as their inverses such that the rhombohedral displacement averaged to zero [9, 10]. Use of time-domain terahertz spectroscipy also "provided strong evidence against the purely displacive nature of the phase transition in GeTe [10]".

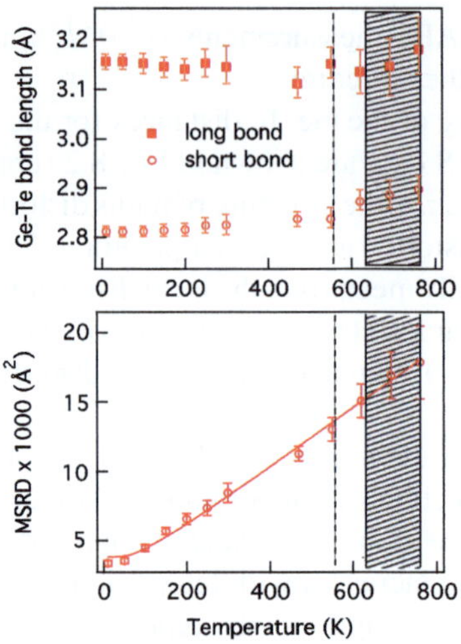

Fig. 8.2 Temperature dependence of the shorter and longer Ge-Te bonds across the phase transition (*upper panel*) and a similar dependence for MSRD for the shorter Ge-Te bonds (*lower panel*) reproduced after [5]. As in Fig. 8.1, the hatched area spans over the temperature range in which the critical temperature values T_c for the ferroelectric phase transition of GeTe have been reported in the literature. The dashed line represents the T_c of thin films, Ref. [8]. One can see that properties change *monotonically* across T_c. See text for more details. Reprinted with permission from Fons et al. [5]. Copyright 2010 by the American Physical Society

Fig. 8.3 Displacive (*left*) versus order–disorder (*right*) transitions in locally distorted materials accounting for the ferroelectric-to-paraelectric transition

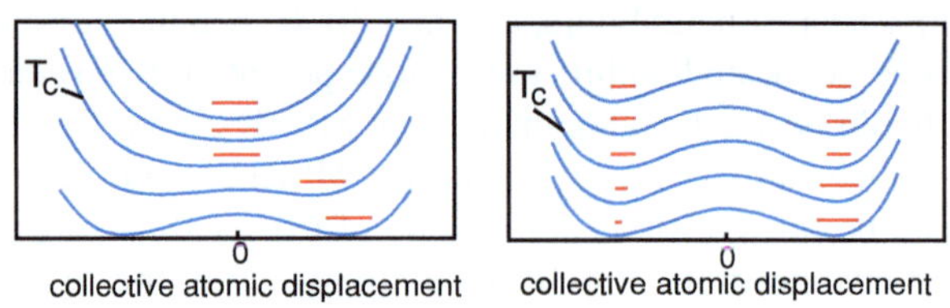

8.1.2 $Ge_2Sb_2Te_5$ and Related Materials

The first report on the structure of GST dates back to the late 1960s when Petrov et al. [11] investigated the structures of $Ge_2Sb_2Te_5$ and $GeSb_4Te_7$. Analysis of electron-diffraction results allowed the authors to conclude that the studied crystals possessed a hexagonal cell with the space group $P\overline{3}m1$ and the lattice constants $a = 4.20 \pm 0.02$ and $c = 16.96 \pm 0.06$ Å and $a = 4.21 \pm 0.02$ and $c = 23.65 \pm 0.08$ Å for $Ge_2Sb_2Te_5$ and $GeSb_4Te_7$, respectively. The structure of $Ge_2Sb_2Te_5$ was described as a layered structure with the stacking sequence of Te-Sb-Te-Ge-Te-Te-Ge-Te-Sb-Te. This sequence is sometimes refereed to as the Petrov sequence.

It should be noted that the studied samples were prepared by vacuum deposition of earlier prepared alloys on rock salt crystals and subsequent annealing for 3–4 h at 150–170°C (for $Ge_2Sb_2Te_5$), i.e. the conditions that are usually used to generate the

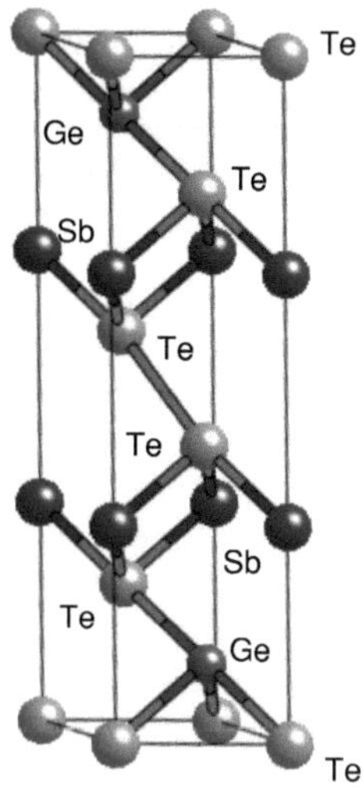

Fig. 8.4 The stable structure of $Ge_2Sb_2Te_5$ with the Kooi stacking sequence

metastable cubic phase. It may be possible that the annealing time was sufficiently long to transform the metastable cubic phase to the stable trigonal phase even at this rather low temperature.

The stable phase of $Ge_2Sb_2Te_5$ has been subsequently re-examined by Kooi et al. [12] also by electron diffraction and additionally by high-resolution transmission electron microscopy. Very similar lattice parameters were obtained ($a = 0.42$ and $c = 1.72$ nm) but the stacking sequence was found to be different, namely, Te-Ge-Te-Sb-Te-Te-Sb-Te-Ge-Te, i.e. with the Ge and Sb positions exchanged. This later sequence is sometimes referred to as the Kooi sequence. Ab-initio modelling [13] supported the Kooi sequence but it should be noted that the obtained energy difference between the Kooi and Petrov stacking sequences is rather small (-3.787 vs. -3.767 eV/atom). The structure of hexagonal $Ge_2Sb_2Te_5$ with the Kooi stacking sequence is shown in Fig. 8.4 and the obtained structural parameters for the stable $Ge_2Sb_2Te_5$ phase are summarised in Table 8.1 [14].

A later study using synchrotron radiation found that the Ge/Sb layers are randomly occupied by these two elements [14]. In this work the sample was a 300 nm thick film prepared by sputtering. The film was crystallised by laser illumination, scraped from the substrate and subsequently heated to 823 K for 5 min which transformed it to the stable trigonal phase. It may be possible that the differences in the distribution of Ge and Sb are caused by significantly shorter crystallisation times so that the structure has not reached equilibrium.

The properties of various coumpounds in the GeTe-Sb_2Te_3 system have been studied in detail by Shelimova et al. and a summary of the obtained results can be found in [15].

While most often materials from the GeTe-Sb_2Te_3 system are used in phase-change memories, another important class for practical applications is the Sb-rich

Table 8.1 Structural parameters for the Petrov [A] and Kooi [B] structures

[A] atom	Site	x	y	z	[B] atom
Te1	1(a)	0	0	0	Te1
Sb	2(d)	2/3	1/3	0.106	Ge
Te2	2(d)	1/3	2/3	0.212	Te2
Ge	2(c)	0	0	0.317	Sb
Te3	2(d)	2/3	1/3	0.421	Te3

Reprinted from Matsunaga et al. [14] by permission from IUCr.
The space group is $P\bar{3}m1$: $a = 4.20$, $c = 16.96$ Å

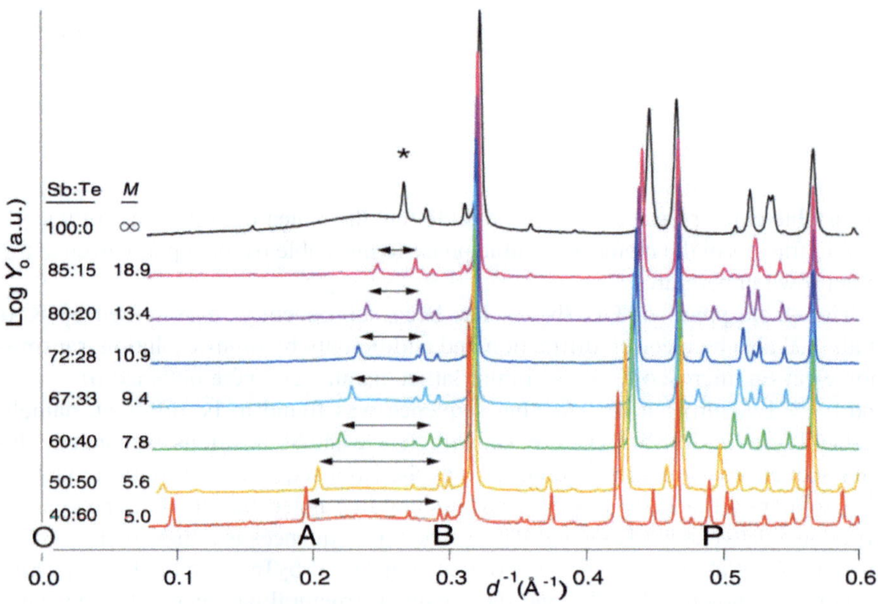

Fig. 8.5 X-ray powder diffraction profiles of a series of Sb-Te samples at 300 K. The *vertical axis* is plotted in logarithmic scale to emphasize the weak peaks and each profile is offset in the direction of this axis. The composition and stacking period M of each specimen are represented. Reprinted from Kifune et al. [16] by permission from IUCr

Sb-Te binary system. Based on systematic X-ray diffraction studies (Fig. 8.5) it was concluded that the structure of Sb-rich crystals was built up from Sb_2Te_3 building blocks separated by a certain number of Sb layers (Fig. 8.6) [16], in agreement with the conclusions drawn in [17].

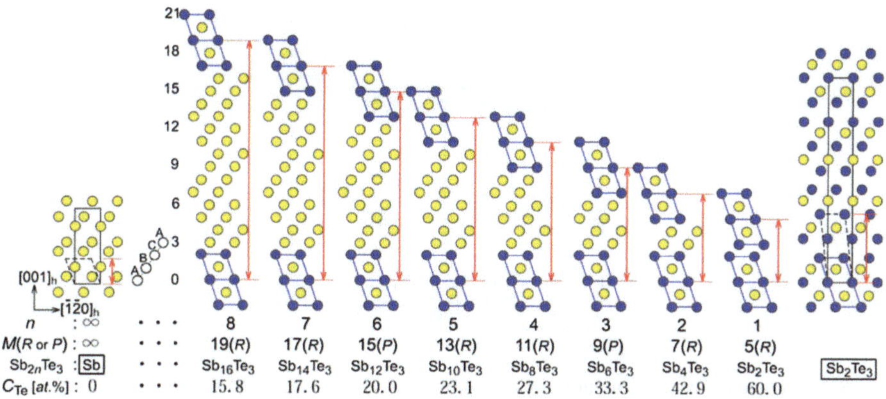

Fig. 8.6 $[100]_h$ projection of crystal structure models of the SbTe binary system. *Yellow solid circles* indicate Sb atoms and *blue circles* represent Te atoms. In the structures of Sb and Sb_2Te_3, each atom shows a projection of real atomic position and *rectangles* represent each unit cell. Except for these two structures, each atom shows a projection of ideal atomic position without displacement along the c axes. The stacking period M, chemical formula and Te concentration are shown under each model. Reprinted from Kifune et al. [16] by permission from IUCr

8.2 Structure of the Metastable Phase

8.2.1 Long-Range Order (Global Structure)

$Ge_2Sb_2Te_5$ and $GeSb_2Te_4$ Alloys

Yamada [18] was the first to report that the structure of a thin layer crystallised from an amorphous phase possesses a structure that is different from that of the stable trigonal phase. This issue was further studied using X-ray diffraction by Nonaka et al. [19] and by Yamada and Matsunaga [20]. The results of these studies agree well with each other and can be summarised as follows.

Based on results, the crystal structure of $Ge_2Sb_2Te_5$ was identified as the rock salt (NaCl) structure (Fm$\bar{3}$m). From Rietveld refinement it was concluded that the anion face-centred cubic (fcc) sublattice is fully occupied by Te atoms with Ge and Sb atoms randomly located on the cation sublattice. The stoichiometry of the structure requires that there are vacancies on the Ge/Sb sites and their presence has been confirmed by structure refinement. The proposed structure is shown in Fig. 8.7 and a summary of the obtained structural parameters is given in Table 8.2. The data analysis was performed assuming an isotropic thermal factor B_0 (the same for the Ge and Sb atoms) and the obtained rather large values suggested large vibration amplitudes, especially for Ge atoms.[1]

[1] The isotropic thermal factor B_0 is related to the mean atomic displacement u by the following equation: $\langle u^2 \rangle = B_0/8\pi^2$.

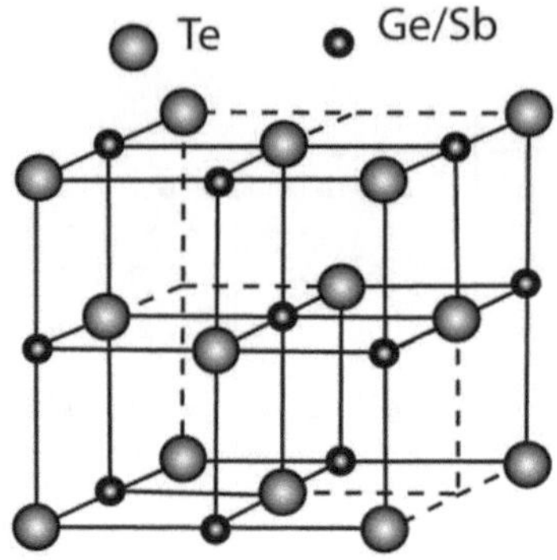

Fig. 8.7 Metastable cubic structure of $Ge_2Sb_2Te_5$

Table 8.2 A summary of the analytical results of the crystal structure of cubic $Ge_2Sb_2Te_5$

Lattice constant, Å			(°)			
a	*b*	*c*	*α*	*β*	*γ*	
6.0117(5)	6.0117	6.0117	90.000	90.000	90.000	
			Co-ordinates			Isotropic atomic displacement parameter (Å^2)
Site	Element	Occupancy	*x*	*y*	*z*	
4a	Ge	0.400	0.0	0.0	0.0	3.2(5)[a]
4a	Sb	0.400	0.0	0.0	0.0	3.25(5)[a]
4b	Te	1.000	0.5	0.5	0.5	1.1(3)[a]
R_{wp} (*R*-weighted pattern) (%)		R_R (*R*-structure factor) (%)		*S* value (goodness of fit)		
12.51		4.69		1.94		

Reprinted from Nonaka et al. [19] with permission from Elsevier
[a] A number in parentheses is an estimated standard deviation

It is worth mentioning that using the hexagonal representation the rhombohedral (distorted cubic) phase can be viewed as a six-layer structure with the hexagonal *c* axis going along the cubic [111] axis (Fig. 8.8). The layer sequences are very similar for the metastable cubic and stable trigonal phases and the differences between the structures are that the metastable phase contains vacancies while the stable phase does not and also that the stable phase contains pairs of Te-Te adjacent layers in the stacking sequence while in the metastable phase the Te and Ge/Sb layers alternate [21]. It may be interesting to note here that the stable phase has also been described as containing a layer of vacancies between the adjacent Te-Te layers [22].

It was subsequently shown that the rock salt structure is characteristic of a large range of Ge-Sb-Te alloys [8, 21, 23, 24]. The authors performed structure analysis (Rietveld refinement) and concluded that the Ge/Sb sites in various Ge-Sb-Te alloys

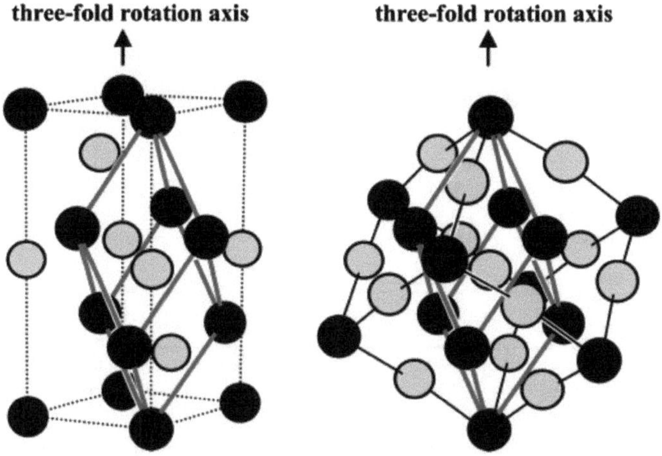

Fig. 8.8 Structure of the distorted NaCl-type low-temperature phase shown with both the rhombohedral (*gray thick lines*) and hexagonal (*dotted lines*) unit cells (*left*). *Black circles* indicate Te atoms, whereas *grey* ones represent Ge/Sb atoms and vacancies. The NaCl-type cubic structure of the high-temperature phase, framed by thin *solid lines*, is slightly deformed in the low-temperature phase (*right*). Reprinted from Matsunaga et al. [21] with permission from American Chemical Society

are randomly occupied by Ge/Sb and vacancies. The lattice parameter increases with an increase in the Sb_2Te_3 content [21] (Fig. 8.9) which can naturally be explained by the fact that Sb atoms have a larger size than Ge atoms.

Since some sites on the Ge/Sb sublattice are vacant, a question naturally arises whether they can be filled by adding Ge and/or Sb into the material composition. Such studies have been performed and it was concluded that addition of extra Ge/Sb atoms did not change the vacancy concentration but rather segregated at the grain boundaries [20, 25]. Theoretical studies of the effect of stoichiometric defects were performed by Caravati et al. [26]. It was found, in particular, that in the crystalline $Ge_2Sb_2Te_5$ deficiencies (excess) of Ge/Sb species shift the Fermi level inside the valence (conduction) band. In contrast, doping had no effect on the position of the Fermi level in the amorphous phase when it stayed pinned in the middle of the bandgap [26].

While it was argued [27] that addition of Ge serves to enhance the material's performance, the provided evidence was not sufficient. First, the observed improved crystallisation behaviour has only been demonstrated for as-deposited films. At the same time, it is known that as-deposited Ge-chalcogenides typically possess a large fraction of homopolar bonds [28] and may crystalise differently than the device-relevant amorphous phase as demonstrated, for example, for the binary GeTe [29]. No evidence for enhanced materials's performance in subsequent cycles has been reported. Second, the rock salt structure of the crystallised phase might well coexist with the excess Ge staying in the amorphous phase and hence invisible to Bragg diffraction similar to the case of the binary GeTe film when about 10% of Ge atoms

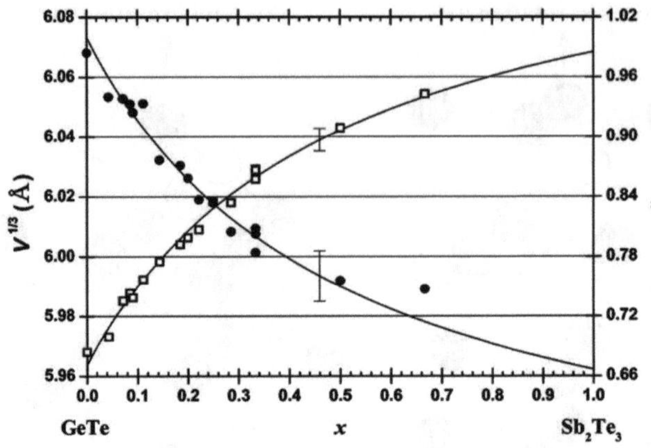

Fig. 8.9 Composition dependence of g parameter, which indicates the site occupancy of Ge+Sb atoms at cation sites (*closed circles*) and that of the cubic root unit cell volume ($V^{1/3}$) (*open squares*). The *horizontal axis* x indicates the fraction of Sb_2Te_3 in the GeTe-Sb_2Te_3 pseudobinary system. Reprinted from Matsunaga et al. [21] with permission from American Chemical Society

form an amorphous Ge phase generating vacancies on Ge sites in the crystallised phase [30].

Ge-rich Alloys

In GeTe-rich GST alloys the situation is more complex [8]. Bragg diffraction studies demonstrated that in $Ge_6Sb_2Te_9$ the structure maintains the NaCl-like cubic phase from 92 K (the lowest temperature used) essentially until the melting point. At the same time, the $Ge_8Sb_2Te_{11}$ (and compositions with higher Ge content) at low temperature possess rhombohedral structure which changes to the rock salt structure at higher temperatures (Fig. 8.10) [31]. The temperature of the observed rhombohedral-to-cubic transition decreases with the increased Sb content. The compositional boundary between the rhombohedral and cubic phases was determined to lie between GeTe:Sb_2Te_3 =8:1 and 6:1 [8].

Carbon- and Nitrogen-Doped GeTe

Doping GeTe by nitrogen and/or carbon has been shown to increase the crystallisation temperature [32] which is very important for certain so-called embedded applications, i.e. in the automotive industry when a memory chip is located close to a hot engine. Bragg diffraction studies have also shown that with increasing N and C content the structure gradually changes from rhombohedrally distorted to cubic [33]. In combination with the previous results this observations leads one to the conclusion that any doping of GeTe (alloying with Sb_2Te_3 may also be considered

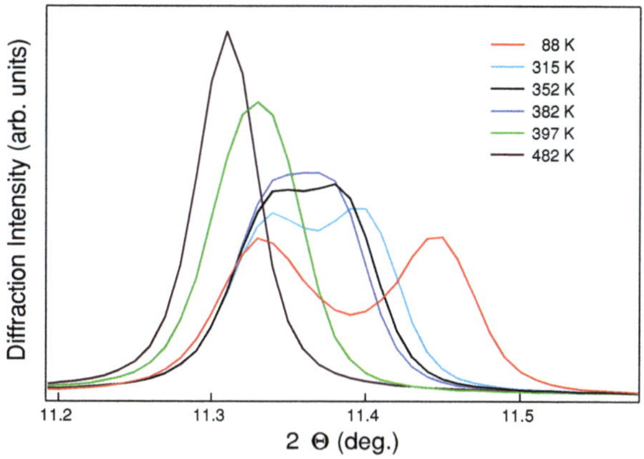

Fig. 8.10 Temperature evolution of X-ray diffraction of $Ge_8Sb_2Te_{11}$ showing a thombohedral-to-cubic transformation in the global structure. Reprinted with permission from Krbal et al. [31]. Copyright 2011 by the American Physical Society

as Sb-doping for the purpose of this discussion) leads to a change of its structure from rhomboherdal to cubic. At the same time, one should not forget that Bragg diffraction probes the average, or global, structure. The local structure may—and does (cf. Sect. 8.2.2)—respond differently to doping.

Ag-In-Sb-Te Alloys

Another class of phase-change materials are Ag- and In-doped Sb_2Te (AIST), the materials used in DVD-RW optical discs. The Ag and In contents may slightly vary but usually their total concentration is about 10 at.%. AIST was found to crystallise into an A7 structure (Fig. 8.11) with sites randomly occupied by the constituent atoms. The shorter and longer interatomic distances in the AIST were found to be 2.95 and 3.34 Å, respectively [34]. Although this structure is stable from room temperature to about 600 K [34], it is informative to compare it with the metastable cubic phase of GST since both these materials that are used in optical memories possess structures with subsets of shorter and longer bonds. The presence of the bond length asymmetry may thus be a general attribute of phase-change materials.

Variation of the Lattice Parameters upon External Stimuli

Temperature Dependence

The temperature dependence of the lattice parameter and the isotropic thermal factor B_0 for $Ge_2Sb_2Te_5$ are shown in Fig. 8.12 [35]. The obtained values for the thermal expansion coefficient was found to be $dV/VdT = 6.24 \times 10^{-5}\,K^{-1}$ for $Ge_2Sb_2Te_5$.

Figure 8.13 compares the temperature dependence of the mean volume per atom in the stable and metastable forms of $GeSb_2Te_4$ [23]. The volume per atom is smaller in the stable phase as one would expect considering the fact that there are no vacancies in this structure. It is interesting to note that the volume expansion is larger for the

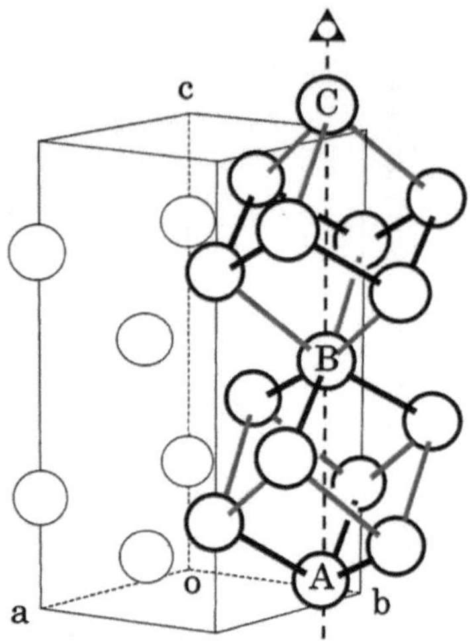

Fig. 8.11 A7-type crystal structure shown schematically in perspective. The unit cell is shown by means of the hexagonal lattice. *Open circles* show atomic positions. Interatomic bonds, illustrated with thick *black* and *gray lines*, show the two kinds of basic cells distorted in the direction parallel to the threefold rotation-inversion axis. Atoms labelled A, B and C exist on the same axis. The *black* and *grey lines* represent the shorter and longer bonds, respectively. Reprinted with permission from Matsunaga et al. [34]. Copyright 2001 by the American Physical Society

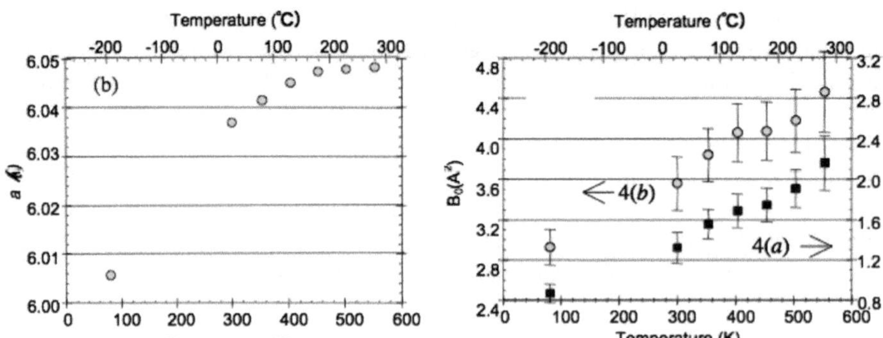

Fig. 8.12 Temperature dependence of the lattice parameter (*left*) and isotropic thermal factor B_0 (*right*) for cubic $Ge_2Sb_2Te_5$. Reprinted with permission after Yamada and Matsunaga [35]

stable phase suggesting that vacancies may serve to "absorb" the lattice expansion. This conclusion is in agreement with the fact that the bulk modulus for the trigonal phase is somewhat higher than that for the cubic phase (44 and 39 GPa, respectively) [36, 37] (Sect. 10.1.2).

It was also suggested that vacancies present in the crystalline phase serve to reduce the volume expansion upon melting and subsequent amorphisation during

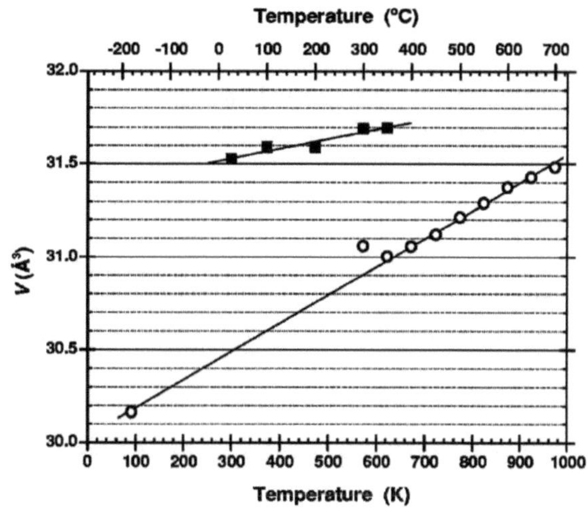

Fig. 8.13 Temperature dependence of the mean volume per single atom. *Black filled squares* and *open circles* show the metastable and stable phases, respectively. The estimated errors are smaller than the marks. Reprinted with permission from Matsunaga and Yamada [23]. Copyright 2004 by the American Physical Society

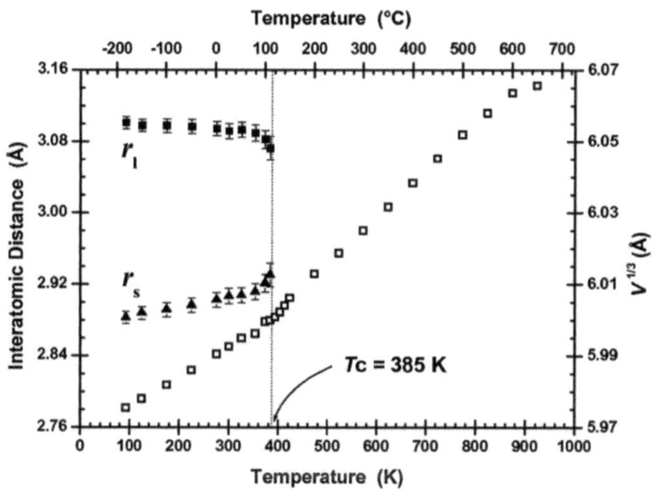

Fig. 8.14 Temperature variation of the interatomic distances and volume for the rhombohedral and rock salt phases obtained from X-ray diffraction measurements. Reprinted with permission from Matsunaga et al. [8]. Copyright (2008) by the American Institute of Physics [8]

the recording process thus reducing the generated stresses and ensuring long life time of devices [20].

The variation of the shorter and longer bond lengths and that of the lattice parameter (cubic root of the volume) as a function of temperature for the GeTe-rich composition $Ge_8Sb_2Te_{11}$ (a composition used in Blu-ray discs) obtained from Bragg diffraction is shown in Fig. 8.14 for both rhombohedral and rock salt phases [8].

Pressure Dependence

Variation of the unit cell volume upon application of hydrostatic pressure for the crystalline cubic and trigonal phases of $Ge_2Sb_2Te_5$ is shown in Fig. 10.9 [36, 37]. Fitting to the Birch-Murnaghan equation of state yields values for the bulk moduli of 39 ± 2 and 44 ± 2 GPa, respectively. The bulk modulus of $GeSb_2Te_4$ has a similar value [36].

Upon higher compression of the metastable cubic phase, both $Ge_2Sb_2Te_5$ and $GeSb_2Te_4$ become amorphous. The pressure-induced transformations of phase-change materials is the subject of a dedicated section (Sect. 10.1.1).

8.2.2 Short-Range Order (Local Structure)

Experimental Studies

EXAFS studies performed by various groups [38, 39] have demonstrated that the Ge-Te bond length in the metastable $Ge_2Sb_2Te_5$ is 2.83 ± 0.01 Å, i.e. significantly shorter than might be expected based on the rock salt structure as determined by XRD and the obtained lattice parameter of slightly over 6.00 Å. Based on these results, it was argued [39] that the structure of metastable $Ge_2Sb_2Te_5$ does not possess the rock salt symmetry but is locally distorted. The local structure around the Ge species is very similar to that of the binary GeTe that possesses a rhombohedral structure (that can be alternatively described as distorted rock salt structure) with subsets of three shorter and three longer bonds (Sect. 8.1.1). The distortion of the Ge atom location has been subsequently confirmed experimentally using scattering measurements [40, 41]. The Sb-Te bond length was also found to be shorter than half the lattice parameter (2.94 Å [39]). The distortion present in GST alloys is usually referred to as Peierls distortion [2].

While the obtained bond lengths are significantly shorter than those expected from the obtained lattice parameter and the rock salt symmetry, they are at the same time significantly longer than the sum of the corresponding covalent radii, which suggests that the bonds are not purely covalent.

One might expect that the distortion in the crystalline phase would disappear at higher temperatures, as typical for Peierls distortions [2]; however, this is not the case. The bond lengths remain either essentially unchanged with temperature (e.g. in GeTe, Fig. 8.2) or the short Ge–Te and Sb-Te bond lengths both *decrease* with increasing temperature in $Ge_2Sb_2Te_5$ (Fig. 8.15), i.e. Ge/Sb atoms *continuously move away* from the ideal octahedral symmetry site within the Te fcc lattice increasing the existing rhombohedral distortion [42].

The Einstein temperatures for the Ge–Te bonds are higher in the amorphous phase (179 and 274 K for the crystalline and amorphous phases, respectively) which provides numerical values for the comparison of the bond strengths.

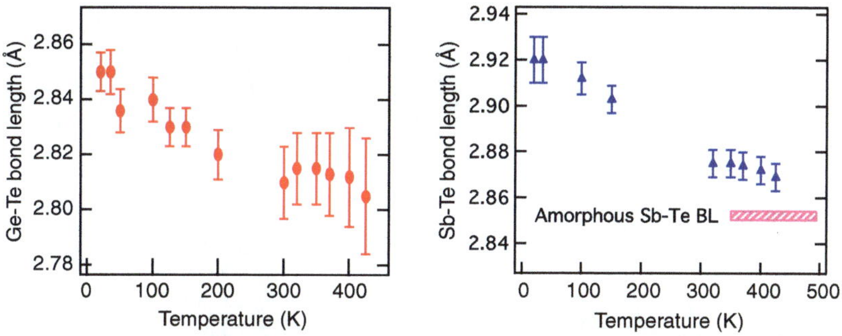

Fig. 8.15 Temperature variation of shorter Ge–Te and Sb-Te bond lengths in the metastable cubic phase of $Ge_2Sb_2Te_5$ determined from an EXAFS analysis. The behaviour is atypical; namely, the Ge–Te and Sb-Te bonds become even shorter with increasing temperature, i.e. the local distortion increases with temperature. Reprinted by permission from Macmillan Publishers Ltd.: Nature Chem. [42], copyright (2011)

When applied to $Ge_8Sb_2Te_{11}$, the composition that according to Bragg diffraction studies exhibits a rhombohedral-to-cubic transition at 385 K [8], EXAFS results have demonstrated that despite a change in the average structure detected by diffraction, locally the structure remains unchanged across the transition temperature (Fig. 8.16) similar to the case of the binary GeTe described above (Sect. 8.1.1). A similar inconsistency between the Bragg diffraction and EXAFS results has been observed for N- and C-doped GeTe: while XRD results clearly exhibit a rhombohedral-to-cubic transition, the local structure remains unchanged, i.e. rhombohedrally distorted [33]. A similar result, i.e. the preservation of the Ge atoms distortion across the Curie temperature was also observed in $Pb_{1-x}Ge_xTe$ where the ferroelectrioc transition has been attributed to long-range ordering of the dipoles [43].

It should be noted that the above EXAFS results are *not in contradiction* with the XRD data since Bragg diffraction probes an ensemble averaged structure and for this reason cannot distinguish between cases when the Ge atoms are loosely bonded and strongly oscillate around the centre of the rock salt structure sites (a displacive transition) and when Ge atoms are randomly displaced from those sites (an order–disorder transition) (Sect. 3.1.1). In both cases XRD "sees" a broad distribution of the Ge atoms around the centre of the rock salt cell with a large isotropic thermal factor.

XAFS, on the other hand, is a local probe. As such it is strongly sensitive to local distortions that are unaccessible to Bragg diffraction. It is only natural that the early XRD data were interpreted as a rock salt structure with a large B_0. A local structure probe-like EXAFS was needed to refine the local structure. At the same time it should be stressed that the average crystal structure can only be probed by a technique such as diffraction. It is the *combined complementary use of both techniques* that is crucial to adequately describe the structure of GST.

A possible explanation for the transition between the rhombohedral and cubic structures as a function of temperature and/or composition seen by XRD may be

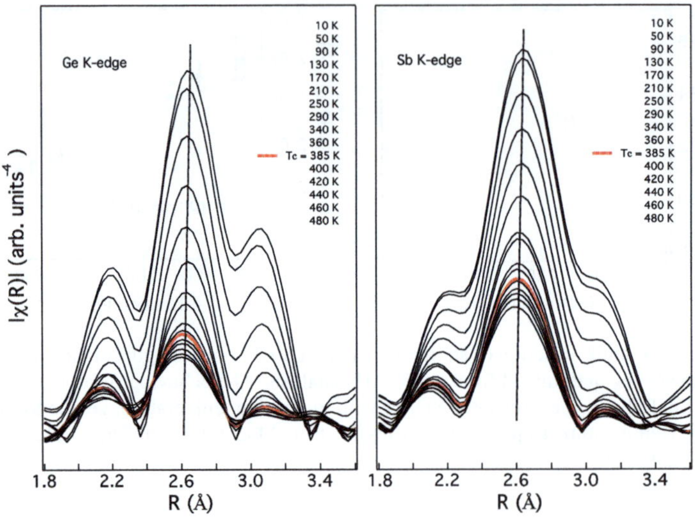

Fig. 8.16 Fourier transformed EXAFS spectra. The temperature dependence of the magnitude of χ(R) for the Ge and Sb K-edges of the $Ge_8Sb_2Te_{11}$ crystalline phase. The apparent peak positions are offset from the true bond lengths by the phase-shift value (properly accounted for during the fitting process [44]). The spectra highlighted in red represent the ferroelectric to paraelectric transition observed via Bragg difraction [8]. The *vertical black lines* serve to show the invariance of Ge–Te and Sb-Te bond lengths across $T_c = 385\,\text{K}$ clearly demonstrating that the 3 + 3 bond splitting is maintained across T_c. Reprinted with permission from Krbal et al. [31]. Copyright 2011 by the American Physical Society

the following. In the binary GeTe at low temperature, the Ge atoms are displaced in a correlated (coherent) manner due to the long-range forces giving rise to an XRD-detectable rhombohedral structure. As the concentration of dopants increases (in parallel with the concentration of vacancies), the long-range order in the Ge sublattice is lost. As a result, the Ge and dopant atoms (Sb, C, N and, possibly, other) become randomly displaced and the distortions are no longer XRD-detectable since on the average the atoms are located around the centre of the cell. Higher concentrations of dopants have a stronger effect which accounts for why alloys like $Ge_2Sb_2Te_5$ possess the "cubic" average structure even at low temperatures.

8.2.3 Resonance Bonding

As follows from the above, the most appropriate description of the GST structure is a *distorted* rock salt structure. For historical reasons, however, the word "distorted" is often dropped. For the same reason and also for the sake of simplicity in this volume we shall also refer to the structure of metastable GST as rock salt, or cubic, unless the role of distortions is important in the context used.

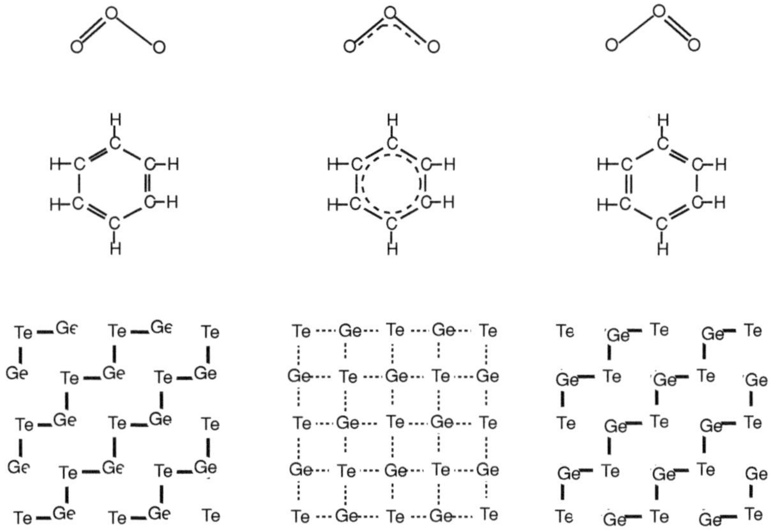

Fig. 8.17 Resonance between two different Lewis structure with pure covalent bonds shown in the *right* and *left-side* frames for ozone (*upper panel*), benzene (*middle panel*) and GeTe (*lower panel*) results in the the formation of resonantly bonded phases shown in central frames

The rock salt-like arrangement of atoms in the crystalline phase requires the formation of six bonds by each participating atom while at the same time, the number of valence electrons located on Ge and Sb atoms is lower. A way to ensure six bonds is by virtue of sharing the valence electrons among several bonds with less than two electrons per bond on the average through resonance bonding, the concept that is described below.

There are numerous examples in inorganic and organic chemistry when more than one Lewis structure can be drawn for a molecule and none of them represents the molecule entirely. The real structure in such cases is descibed by a resonance hybrid which is a linear combination of the possible Lewis structures. One such example is ozone whose two possible structures are shown in the upper panel of Fig. 8.17, right and left frames. Both these structures have one single (longer) bond and one double (shorter) bond. Experimentally, both bond length are identical and the way to ensure this is called resonance [45] between the structure shown in panels on the right and left with the resulting structure shown in the central frame. The dashed line in the resonantly bonded structure indicates that electrons are delocalised among more than two atoms.

Another well-known example of resonance bonding is in benzene. Benzene also represents a special problem in that, to account for all the bonds, there must be alternating double carbon bonds (Fig. 8.17 middle panel right and left frames show the Kekulé representation of benzene). At the same time, all six carbon–carbon bonds in benzene are of the same length of 1.40 Å, i.e. longer than a double bond (1.35 Å) but shorter than a single bond (1.47 Å). This intermediate distance is consistent with electron delocalization: the electrons for C–C bonds are distributed equally between each of the six carbon atoms (Fig. 8.17 central frame shows the Thiele representation of benzene). Because the only orbitals occupied by π electrons are those with net bonding character [46], delocalisation of π electrons stabilises the benzene ring. A characteristic feature of resonance bonding is a large bond polarisability.

A similar idea can also be applied to solids and, in particular, to IV–VI crystals. Lucovsky and White [47] were the first to discuss this possibility and its consequences. Similar to the cases described above, one can imagine two different structure for IV–VI crystals such as GeTe (right and left frames in panels Fig. 8.17, lower panel). It was argued that the formation of resonance bonding with the resulting structure shown in the central frame lowers the total energy of the system. The conclusion drawn from diffraction experiments that at elevated temperatures GeTe has the rock salt structure with a large isotropic thermal factor suggesting that Ge atom moves in a soft potential allowing for large atomic displacement supports this idea.

In particular, in the cited work [47] resonance bonding in IV–VI and chalcogen (Se and Te) crystals was considered and it was suggested that "we might expect materials that exhibit resonance bonding in the crystalline phase to have different properties in the amorphous phase". It was also stated that long-range order is crucial for resonance bonding to exist. If long-range order is lost, the system will be unable to achieve resonant bonding [47]. This approach has been subsequently elaborated in [48–50] to describe the unusually large property contrast in GST alloys and the fast switching rate.

At the same time, it has been demonstrated recently that the apparent rock salt structure of GeTe is a result of site averaging effect of Bragg diffraction. Local probes such as EXAFS and total scattering with subsequent pair-distribution function analysis demonstrated that locally the distortions persist in both pure and doped GeTe. The presence of subsets of shoter and longer bonds was also demonstarted for Ge-Sb-Te alloys.

There are thus two major differences between resonance bonding in molecules and phase-change alloys. While in the former, the bond lengths (and strengths) are identical, in the latter, there are shorter and longer bonds making resonance difficult. The second important difference is that in molecules resonance bonding exists *in addition* to the conventional covalent σ bonds. As a result, the bond lengths get shorter, and stronger, then single covalent bonds. In this situation, even when the resonance interaction is broken, the structure remains stable because the primary covalent bonds persist. In phase-change materials, on the other hand, the longer bonds are formed *at the expence* of the shorter bond that are consequently longer than normal covalent bonds. In this case, when the resonance interaction is broken there will be no bonds between certain pairs of atoms making the structure locally unstable.

How does the bond length asymmetry affect the electron density distribution? The plot of the charge difference between the simulated relaxed GeTe model at 0 K and isolated pseudoatoms shown in Fig. 8.18 demonstrates that the electron density is only significant along the short bonds. In contrast, the electron density pile-up along the long bonds is *significantly* lower demonstrating that there is a pronounced bonding

Fig. 8.18 Difference in electron charge density for GeTe at 0 K and isolated pseudoatoms. A significant bonding charge can be seen midway along the shorter bonds which corresponds to covalent bonding. At the same time, the bonding localised along the longer bonds is *significantly* lower clearly demonstrating a pronounced bonding energy difference between the short and long bonds. Reprinted by permission from Macmillan Publishers Ltd.: Nature Chem. [42], copyright (2011)

energy hierarchy between the short and long bonds [42]. The existing bonding energy hierarchy should manifest itself in the different response of the short and long bonds to thermal and/or electronic excitations. The fact that the long-range order is associated with the weak resonance bonding suggests that the crystalline phase is intrinsically fragile.

Figure 8.18 also demonstrates that referring to such a structure as octahedral, i.e. each atom having six first nearest neighbours is, strictly speaking, incorrect. A more appropriate way to describe the coordination number would be 3 + 3 as used, for example, in [51]. In order to account for the fact that bonds are strongly asymmetric, it has been recently suggested to use the effective coordination concept when determining the coordination number [52] where higher weighting is attributed to shorter bonds over longer bonds in order to take into account structural distortions. Using this approach the effective coordination numbers (ECN) have been determined as ECN = 5.1(Te), 5.1(Ge) and ECN = 5.55 (Ge), 5.73 (Sb), 4.37 (Te) for GeTe and for the metastable crystalline phase of $Ge_2Sb_2Te_5$, respectively [52].

It is interesting to note that the description of benzene as a mixture of the two structures was given a firm foundation in quantum theory as a resonance hybrid in terms of delocalised π electrons. The view that delocalised electrons provide essentially the correct description for this type of system appears to be generally accepted [46]. In 1986, however, the delocalised picture of benzene has been contested by Cooper, Gerratt and Raimondi who argued that the π electrons in benzene are almost certainly localised, and the aromatic properties of benzene originate from the symmetric coupling of the electron spins around the carbon ring framework rather than from electron delocalisation [53]. The correctness of the 'resonance bonding' term has also been disputed in [54]. While this criticism has to be kept in mind, below in this volume we shall use the term 'resonance bonding' to refer to bonding that is delocalised among three of more atoms.

8.2.4 Point Defects, Vacancies and Distortions

Vacancies as an Intrinsic Attribute of the Cubic Phase

Vacancies present in the structure deserve a dedicated paragraph. One might imagine that presence of vacancies is a simple consequence of the GST stoichiometry and the vacancies are structural imperfections, or defects, in the rock salt structure. Upon a closer look, however, the situation is more complex. As already mentioned, attempts to fill the vacancies by adding excess Ge or Sb atoms have shown that vacancies segregate at grain boundaries [20, 25]. This result was the first indication that vacancies may not be defects.

Kolobov et al. [39] suggested that vacancies serve to stabilise the structure by ensuring the appropriate charge balance. The role of vacancies was a subject of several consequent studies and where it was also concluded that vacancies, whose concentration varies as $x/(1+2x)$ for the $GeTe_{1-x} - (Sb_2Te_3)_x$ system [21], are not electronic defects but are an intrinsic feature of the GST rock salt structure [21, 27, 48]. Vacancies were also argued to account for the *p*-type conductivity of the crystalline phase of GeTe [55].

Wuttig et al. [27] have demonstrated that introduction of vacancies into Ge/Sb sublattice decreases the total energy of the system (Fig. 8.19). The authors attribute this finding to the anti-bonding nature of the Ge-Te and Sb-Te states in the vicinity of the Fermi level. A related conclusion, namely that an increase in the vacancy number leads to an increase in the bonding energy (for $Ge_2Sb_2Te_5$) has been reached in [56].

Are Vacancies Ordered?

Experiments

Polycrystals. In all experimental XRD studies, the distribution of Ge/Sb/*v* (where *v* stands for vacancy) within the cation sublattice was found random [19, 20]. It cannot be excluded, however, that vacancy ordering requires significantly longer times than those used to generate the crystalline cubic phase experimentally. (The fact that chemical ordering of the material continues after the crystal phase was reported in [57]).

Epitaxial Layers. Of special interest is vacancy distribution in epitaxial layers (Sect. 8.3.2) since the latter are obtained under conditions close to thermodynamical equilibrium. One would therefore expect some ordering on the non-Te sublattice of the cubic structure that contains Ge, Sb and vacancies. If the vacancies interact, they could either attract, in which case the equilibrium structure would consist of a dense crystal without vacancies and vacuum, or repel, in which case the equilibrium structure for common interaction potentials would be an ordered array of vacancies. No superstructures corresponding to vacancy ordering were found in the extensive synchrotron-based X-ray diffraction studies for epitaxial GST grown on GaSb(100):

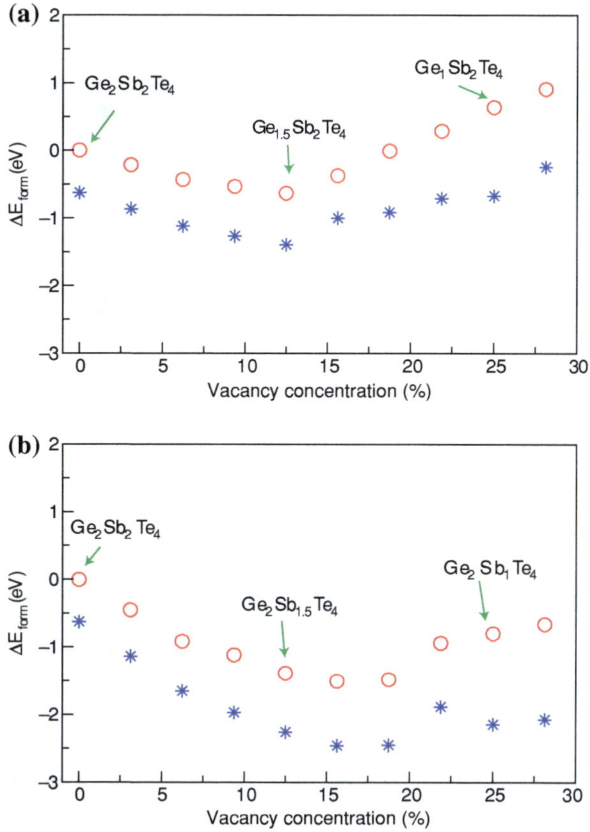

Fig. 8.19 Formation energies (*open circles*) for Ge (**a**) and Sb (**b**) vacancies for different concentrations of vacancies. The corresponding stoichiometries are marked above. Negative values for the formation energies mean that the respective compound is energetically preferred. On removing Ge atoms the energy of the crystal gets lower in the order of half an eV per supercell. The distortions (*stars*) lead to a further reduction, also of the order of half an eV. Reprinted by permission from Macmillan Publishers Ltd.: Nature Mater. [27], copyright (2007)

all reflections originated from the basic periodicity of the GST crystal [58] suggesting that vacancies do not interact in GST, which seems strange as they locally distort the lattice, involving a significant elastic energy.

Ab Initio Studies

At the same time, ab initio computer simulations [13] comparing various stacking sequences for $Ge_2Sb_2Te_5$ suggested that vacancies should be ordered and form a vacancy-rich layer. It was argued that the most stable configuration of the cubic of the cubic $Ge_2Sb_2Te_5$ phase possesses a stacking sequence of Te-Ge-Te-Sb-Te-v-

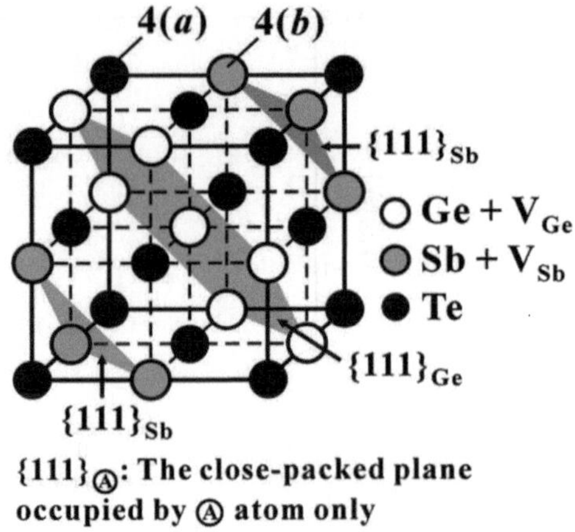

Fig. 8.20 Ordered arrangement of vacancies in cubic $Ge_2Sb_2Te_5$. Reprinted with permission from Park et al. [60]. Copyright 2005 by the American Institute of Physics

Te-Sb-Te-Ge in the [111] direction, where v represents an ordered vacancy layer. This stacking sequence is similar to the Kooi stacking sequence [12] for the stable $Ge_2Sb_2Te_5$ phase. At the same time, it should be noted that the observed energy differences between the Petrov and Kooi sequences are rather small (−3.797 eV/atom vs. −3.779 eV/atom [13]) and as a result probabilities of obtaining various stacking sequences experimentally become essentially indistinguishable at room temperature (and above) which may also account for the random vacancy distribution observed experimentally. Subsequently, the same authors investigated other materials in the GST system and concluded that the vacancy layer is always located within the SbTe block (-Sb-Te-v-Te-Sb-) [59].

The ordering of Ge and Sb atoms have also been concluded by Park et al. [60] based on HR-TEM studies. These authors proposed the structure shown in Fig. 8.20 as the one providing the best agreement with the experiment. The difference between this structure and the one proposed by Sun et al. [13] is twofold. First, in the model of Sun et al. the vacancies form vacancy layers while in the model of Park et al. there are no vacancy layers. Second, in the model of Sun et al. the structure can be viewed as a sequence of GeTe and Sb_2Te_3 grouped layers with all vacancies confined within the Sb_2Te_3 layers while in the model of Park et al., the Ge and Sb layers alternate and vacancies are randomly distributed within both the Ge and Sb layers.

Distribution of vacancies has also been studied by Eom et al. [56]. While for $Ge_2Sb_2Te_5$ these authors deduced random distribution of vacancies, for $GeSb_2Te_4$, they found that "the most stable structure is the vacancy layer normal to the [111] direction" which is very simillar the conclusion of Sun et al. for $Ge_2Sb_2Te_5$.

Point Defects

Regarding point defects, the available information is very limited. Thus Ge atoms located on tetrahedral symmetry sites have been reported based on HR-TEM results

[61, 62] and their presence was also suggested by X-ray holography studies [63] of the crystalline cubic GST phase. At the same time, it should be noted that careful Rietveld analysis or powder diffraction data did not detect such atoms [19, 20]. Nor could they be found in epitaxially grown $Ge_2Sb_2Te_5$ films when the authors specially examined the possibility of the existance of such sites [58].

Distortions

The distortions in the crystal structure of cubic GST similar to those in the binary GeTe were first determined experimentally by EXAFS [38, 39]. Their presence in the crystal structure is usually attributed to the Peierls effect similar to the case of pure group V elements such as Sb whose structure can be viewed as obtained from a simple cubic lattice by displacing every second atom in the direction of a space-diagonal and also slightly altering the angles. Such distortions result in reducing the translational symmetry and opening a gap in the density of states near the Fermi level. The IV–VI crystals such as GeTe that also possess on an average five valence electrons per atom are also often described as Peierls distorted. In phase-change materials the situation is similar, the three orthogonal *p*-orbitals allow for the formation of the rock salt structure that is unstable against local distortions. A detailed study of Peierls distortions in covalently bonded semiconductors has been performed by Gaspard et al. [2].

Wuttig et al. [27] have shown that distortions decrease the total energy of GST, their effect being more strongly pronounced in structures that contain vacancies (Fig. 8.19). These authors also concluded that it is mostly the Te subsystem that reduces its free energy upon structural relaxation.

While Peierls distortion is expected to decrease with temperature, this is not the case in GST alloys where the shorter Ge–Te bonds remain essentially unchanged [64] or get even shorter [42] as the temperature is increased.

While the existing experimental diffraction results and those of ab inito DFT simulations all agree that the crystalline phase is well described as a distorted rock salt structure, it may be worth noting that XANES simulations performed on such structures obtained by either relaxation of the rock salt phase at 0 K or by annealing of the melt-quenched amorphous phase *in silico* [57] do not agree (Fig. 8.21a) with the experimental results for the crystalline phase $Ge_2Sb_2Te_5$ shown in panel (b), in particular, the slopes in the energy region II have opposite signs. At the same time, simulated XANES spectra for clusters with octahedral coordination around the central atom but containing a certain number of purely covalent fragments without resonance bonding (Fig. 8.21c) give a near-perfect match to the experimental results suggesting that the degree of disorder in the experimentally obtained metastable crystalline phase may be larger than generally believed and, in particular, that numerous non-resonantly bonded fragments are present in the crystalline phase.

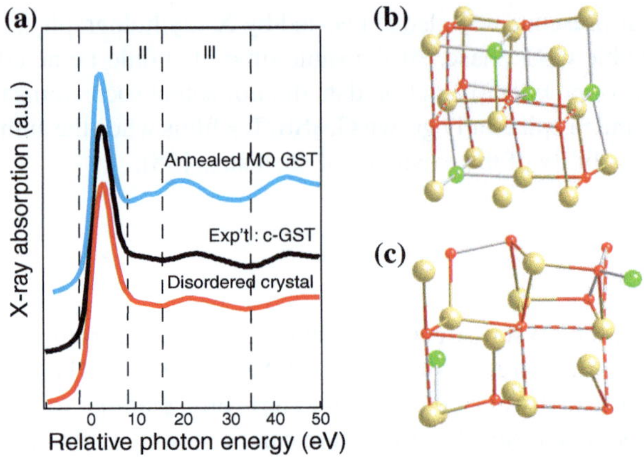

Fig. 8.21 Experimental and simulated XANES spectra for the metastable cubic phase of $Ge_2Sb_2Te_5$ (**a**). Theoretical spectra were simulated for the *in-silico* annealed rock salt phase of Ref. [57] (**b**) and for a cluster that contains numerous covalently bonded fragments without resonance bonding between them (**c**). In the models, the notations are: Ge-small (*red*) atoms, Sb-middle size (*green*) atoms and Te-large (*yellow*) atoms

8.3 Engineered Structures

8.3.1 GeTe-Sb_2Te_3 Layered Structures

While the analysis of experimental data suggests that Ge, Sb and vacancies are randomly distributed, computer studies of the spatial distribution of Ge and Sb species on the Ge/Sb/v sublattice demonstrated that the layered structure with the Te-Ge-Te-Sb-Te-v-Te-Sb-Te-Ge sequence in the [111] direction possesses the lowest energies [13]. Subsequently, the same authors investigated other materials in the GST system and concluded that the vacancy layer is always located within the SbTe block (-Sb-Te-v-Te-Sb-) [59]. In addition to possessing the lowest energy, such structures are characterised by two other features. First, they possess a well-defined axis along the [111] direction suggesting that the Ge atom switching during the phase-change process should proceed along this direction and hence the transition should have much lower entropic losses. Second, all Ge atoms possess an identical local structure which suggests that the phase-change process in such a material should be characterised by a much narrower energy distribution. Both these features seem advantageous for memory design.

Experimentally, the DSI group was the first to spatially separate GeTe and Sb_2Te_3 layers with individual layer thicknesses of a few nm [65–67]. It was shown that the layered structure was preserved after repeated cycles and devices made from the layered structure demonstrated better switching characteristics, such as low programming current [66], compared to composite materials. Making use of the different

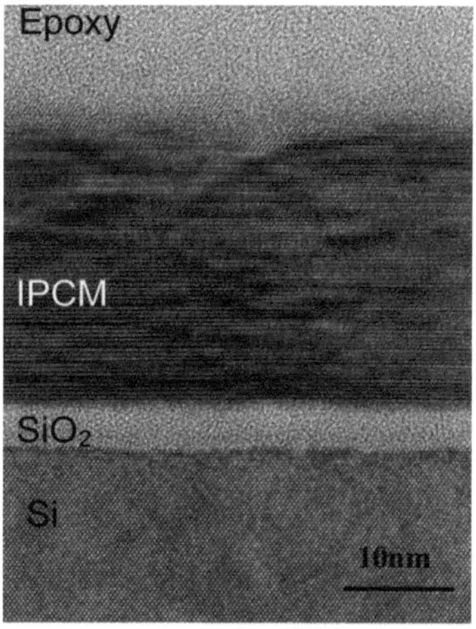

Fig. 8.22 A TEM image of an atomically flat layered GeTe–Sb_2Te_3 structure obtained by helicon-wave sputtering

crystallisation temperatures for the constituent materials, the authors demonstrated that multilevel recording can be achieved using such structures [67].

Layered structures consisting of alternating GeTe and Sb_2Te_3 blocks with sub-nanometer thicknesses have been prepared using helicon-wave sputtering (Fig. 8.22) using three different targets [68, 69]. One can see that despite the fact that the samples were grown using a mass-production sputtering method, the structure has a very strong orientation along the $\langle 111 \rangle$ axis of the "cubic" phase.

The layered structures possess switching characteristics significantly superior to those obtained using conventional magnetron sputtering with a composite target and pave the way to design novel phase-change memory devices (Sect. 12.2.2).

8.3.2 Epitaxial Growth of Phase-Change Materials

Successful growth of epitaxial GST films in the metastable cubic phase by molecular beam epitaxy has been reported in [70–72]. The initial experiments were performed on GaSb(001). The growth mode strongly depends on the substrate temperature. At low temperature an amorphous film grows, this is followed by polycrystal growth and incubated epitaxial growth. In a rather narrow temperature range following this, epitaxial films grow (Fig. 8.23). Despite the rough morphology, the layers show a good crystallinity and strongly dominant epitaxial orientation. Subsequently, the authors performed successful growth of GST on GaSb, InAs and Si substrates [73]. These results are briefly described below in this section.

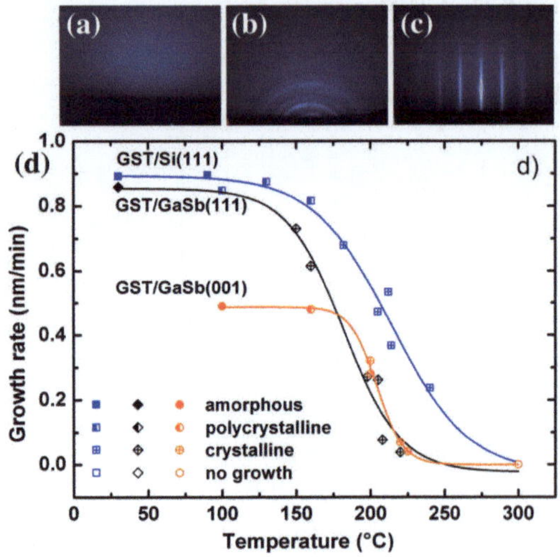

Fig. 8.23 In situ RHEED patterns acquired during growth, showing: **a** diffuse scattering from an amorphous layer. **b** Laue rings due to a polycrystalline GST layer and **c** crystal truncation rods originating from the diffraction of a single crystalline GST epilayer. **d** GST growth rate for the above sturcture versus substrate temperature. After Rodenbach et al. [73]

Structure of Epitaxial Phase-Change Layers

In agreement with the results obtained for the layered growth using helicon-wave sputtering, the epitaxial GST layers show a strong preference to grow with (111) facets which was attributed to the tendency of GST to form (111)-oriented Ge, Sb and Te planes in its trigonal structure as well as to the symmetry of the distortion: growth on a (111) plane retains the full rotational symmetry in the (111) plane and the distortion can relax normal to it. The films grown on GaSb and InAs show the same trend.

Figure 8.24a, d, g shows the RHEED pattern of the starting surface and its reconstruction for each substrate. Figure 8.24b, e, h displays the patterns at an intermediate stage of growth. The patterns of final surfaces are shown in Fig. 8.24c, f, i. One can see that growth on (001) surfaces results in a spotty transmission pattern characteristic of a rough surface with crystallites that are small enough that the electron beam can penetrate. A smooth surface, on the other hand, produces a streaky RHEED pattern as is clearly observed for films grown on (111) oriented substrates. A comparison of the diffraction patterns obtained for the two different substrate orientations demonstrates that the growth of GST on (111) oriented substrates is superior, without a strong dependence on the substrate used or the lattice mismatch [73].

The crystal structure becomes apparent when combining RHEED patterns in different directions as shown in Fig. 8.25. The circular in-plane pattern is constructed from a scan measured during continuous rotation of the substrate. Quasi-3D rendering reveals the body-centred cubic reciprocal lattice of the grown layer. This is in agreement with the expected real-space face-centred NaCl structure expected for the metastable cubic phase. Streaks extending from the reflections in plane along the $\langle 110 \rangle$ and out of plane along the $\langle 111 \rangle$ axes imply relaxation and facet formation

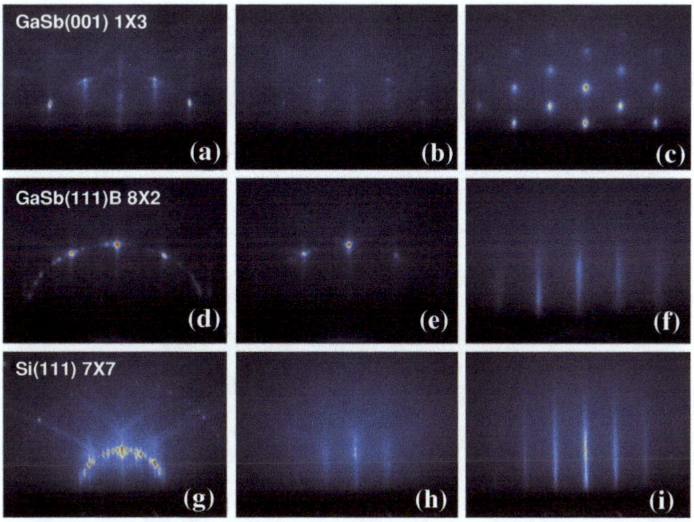

Fig. 8.24 Comparison of RHEED patterns before, during, and at the end of the growth of GST on (**a–c**) GaSb(001), GaSb(111)B (**d–f**) and on (**g–i**) Si(111). After Rodenbach et al. [73]

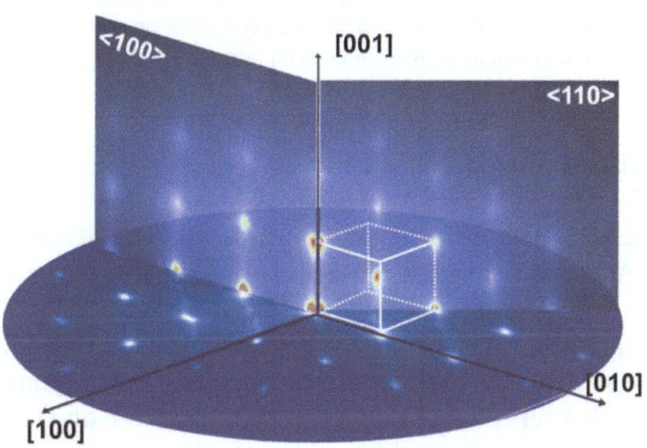

Fig. 8.25 Reciprocal lattice of the $Ge_2Sb_2Te_5$ film grown on GaSb(100) constructed from different RHEED scans. The reciprocal unit cell is indicated by the shaded cube. After Rodenbach et al. [73]

along the ⟨111⟩ directions, indicating that the Peierls distortion observed by EXAFS is preferentially directed along ⟨111⟩ [71].

Of special interest is the successful growth of epitaxial GST layers on Si(111) despite a rather large lattice mismatch of about 10% [73]. This fact can be explained by the presence of vacancies in the structure and their ability to "absorb" the stress. In addition, the obtained result demonstrates perfect compatibility of phase-change materials with silicon technology. It is interesting to note that single-crytal growth of GST could also be obtained by sputtering (which is a mass-production technique) GST on a hot Si (111) substrate [74].

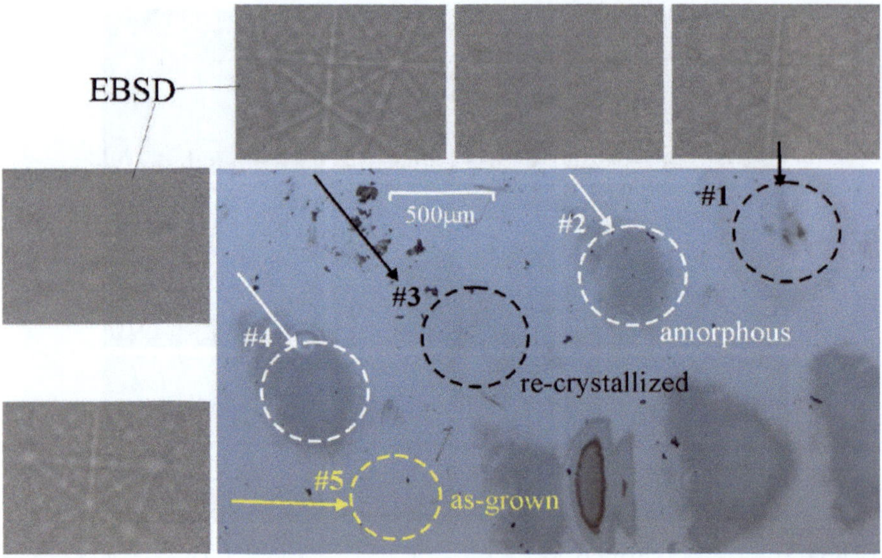

Fig. 8.26 Optical microscope image of an epitaxial $Ge_2Sb_2Te_5$ layer on GaSb(001) amorphised and recrystallised by laser pulses, together with EBSD patterns of selected regions. Electron backscattering Kikuchi line patterns indicate the crystalline phase of the material. Recrystallisation reproduces the initial orientation. Reprinted with permission from the author from [58]

Laser-Induced Phase Change

Results of laser switching experiments on epitaxial $Ge_2Sb_2Te_5$ using a Nd-doped YAG laser with 60 ps pulse duration [58] are shown in Fig. 8.26. The strong contrast in optical properties between the two phases shows up as a darker colour of the amorphised regions. The regions enclosed by black and white dashed circles have been switched back and forth between the two phases multiple times demonstrating that repeated switching of epitaxial GST on GaSb(100) is possible. Reversible switching of epitaxial films between the crystalline and amorphous phases have also been confirmed by Raman scattering [58].

The authors further took electron backscatter diffraction (EBSD) that measures the Kikuchi line pattern produced by the electron beam in scanning electron microscopy. Due to its small sampling depth, it can detect the crystallinity and grain orientation in thin films. EBSD of the different regions shown in Fig. 8.26 clearly confirms the amorphous and crystalline character of the different regions. Interestingly, the recrystallised regions are not polycrystalline, as would be expected for heterogeneous nucleation. Instead, the epitaxial pattern of the as-grown epitaxial state is recovered, indicating that the amorphised material retains memory of the epitaxial arrangement, i.e. a *latent long-range order* exists in the amorphous phase obtained from the ordered single crystalline phase.

Recrystallisation of amorphised epitaxial phase-change layers back to the single-crystal phase has also been demonstrated using nanobeam X-ray diffraction (P. Rodenbach et al., unpublished).

References

1. K.M. Rabe, J.D. Joannopoulos, Phys. Rev. B **36**, 6631 (1987)
2. J.P. Gaspard, A. Pellegatti, F. Marinelli, C. Bichara, Philos. Mag. B **77**, 727 (1998)
3. T. Chattopadhyay, J. Boucherle, H. Von Schnering, J. Phys. C **20**, 1431 (1987)
4. E. Steigmeier, G. Harbeke, Solid State Commun. **8**, 1275 (1970)
5. P. Fons, A.V. Kolobov, M. Krbal, J. Tominaga, K.S. Andrikopoulos, S.N. Yannopoulos, G.A. Voyiatzis, T. Uruga, Phys. Rev. B **82**, 155209 (2010)
6. E.A. Stern, Y. Yacoby, J. Phys. Chem. Solids **57**, 1449 (1996)
7. B. Sales, B. Chakoumakos, D. Mandrus, J. Sharp, J. Solid State Chem. **146**, 528 (1999)
8. T. Matsunaga, H. Morita, R. Kojima, N. Yamada, K. Kifune, Y. Kubota, Y. Tabata, J.J. Kim, M. Kobata, E. Ikenaga, K. Kobayashi, J. Appl. Phys. **103**, 093511 (2008)
9. T. Matsunaga, P. Fons, A.V. Kolobov, J. Tominaga, N. Yamada, Appl. Phys. Lett. **99**, 231907 (2011)
10. F. Kadlec, C. Kadlec, P. Kužel, J. Petzelt, Phys. Rev. B **84**, 205209 (2011)
11. I.I. Petrov, R.M. Imamov, Z.G. Pinsker, Sov. Phys. Cryst. **13**, 339 (1968)
12. B.J. Kooi, J.Th.M. De Hosson, J. Appl. Phys. **92**, 3584 (2002)
13. Z. Sun, J. Zhou, R. Ahuja, Phys. Rev. Lett. **96**, 055507 (2006)
14. T. Matsunaga, N. Yamada, Y. Kubota, Acta Cryst. B **60**, 685–691 (2004)
15. L.E. Shelimova, O.G. Karpinskii, P.P. Konstantinov, M.A. Kretova, E.S. Avilov, V.S. Zemskov, Inorg. Mater. **37**, 342 (2001)
16. K. Kifune, Y. Kubota, T. Matsunaga, N. Yamada, Acta Cryst. B **61**, 492 (2005)
17. V. Agafonov, N. Rodier, R. Céolin, R. Bellissent, C. Bergman, J.P. Gaspard, Cryst. Struct. Commun. **47**, 1141 (1991)
18. N. Yamada, MRS Bull. **21**, 48 (1996)
19. T. Nonaka, G. Ohbayashi, Y. Toriumi, Y. Mori, H. Hashimoto, Thin Solid Films **370**, 258 (2000)
20. N. Yamada, T. Matsunaga, J. Appl. Phys. **88**, 7020 (2000)
21. T. Matsunaga, R. Kojima, N. Yamada, K. Kifune, Y. Kubota, Y. Tabata, M. Takata, Inorg. Chem. **45**, 2235 (2006)
22. J.L.F. Da Silva, A. Walsh, H. Lee, Phys. Rev. B **78**, 224111 (2008)
23. T. Matsunaga, N. Yamada, Phys. Rev. B **69**, 104111 (2004)
24. T. Matsunaga, R. Kojima, N. Yamada, K. Kifune, Y. Kubota, M. Takata, Appl. Phys. Lett. **90**, 161919 (2007)
25. S. Privitera, E. Rimini, C. Bongiorno, R. Zonca, A. Pirovano, R. Bez, J. Appl. Phys. **94**, 4409 (2003)
26. S. Caravati, M. Bernasconi, T. Kühne, M. Krack, M. Parrinello, J. Phys. Condens. Matter **21**, 255501 (2009)
27. M. Wuttig, D. Lüsebrink, D. Wamwangi, W. Welnic, M. Gillessen, R. Dronskowski, Nature Mater. **6**, 122 (2007)
28. K. Tanaka, in *Handbook of Advanced Electronic and Photonic Materials*, vol. 5, ed. by H.S. Nalwa (Academic Press, San Diego, 2001), p. 120
29. S. Raoux, H.Y. Cheng, M.A. Caldwell, H.S.P. Wong, Appl. Phys. Lett. **95**, 071910 (2009)
30. A.V. Kolobov, J. Tominaga, P. Fons, T. Uruga, Appl. Phys. Lett. **82**, 382 (2003)
31. M. Krbal, R.E. Simpson, J. Tominaga, A.V. Kolobov, P. Fons, T. Matsunaga, N. Yamada, T. Uruga, Phys. Rev. B **84**, 104106 (2011)

32. V. Sousa, L. Perniola, A. Fantini, G.B. Beneventi, E. Gourvest, S. Loubriat, A. Bastard, A. Roule, A. Persico, H. Feldis, A. Toffoli, D. Blachier, S. Maitrejean, B. Hyot, J.F. Nodin, C. Jahan, G. Reimbold, T. Billon, B. Andre, B.D. Salvo, F. Boulanger, S. Lhostis, P. Mazoyer, D. Bensahel, P. Zuliani, R. Annunziata, in *European Phase Change and Ovonics Science Symposium*, Milano, 2010
33. X. Biquard, M. Krbal, A.V. Kolobov, P. Fons, R.E. Simpson, B. Hyot, B. André, J. Tominaga, T. Uruga, Appl. Phys. Lett. **98**, 231907 (2011)
34. T. Matsunaga, Y. Umetani, N. Yamada, Phys. Rev. B **64**, 184116 (2001)
35. N. Yamada, T. Matsunaga, in *Studies of High-Speed Phase-Change Materials Using Synchrotron Radiation*. European Phase Change and Ovonics Science Symposium, Lugano, 2003
36. A.V. Kolobov, J. Haines, A. Pradel, M. Ribes, P. Fons, J. Tominaga, Y. Katayama, T. Hammouda, T. Uruga, Phys. Rev. Lett. **97**, 035701 (2006)
37. M. Krbal, A.V. Kolobov, J. Haines, P. Fons, C. Levelut, R. Le Parc, M. Hanfland, J. Tominaga, A. Pradel, M. Ribes, Phys. Rev. Lett. **103**, 115502 (2009)
38. B. Hyot, X. Biquard, L. Poupinet, in *European Phase Change and Ovonics Science Symposium*, Säntis, 2001
39. A.V. Kolobov, P. Fons, A.I. Frenkel, A.L. Ankudinov, J. Tominaga, T. Uruga, Nature Mater. **3**, 703 (2004)
40. S. Shamoto, N. Yamada, T. Matsunaga, T. Proffen, J.W. Richardson, J.H. Chung, T. Egami, Appl. Phys. Lett. **86**, 081904 (2005)
41. S. Shamoto, K. Kodama, S. Iikubo, T. Taguchi, N. Yamada, T. Proffen, Jpn. J. Appl. Phys. **45**, 8789 (2006)
42. A.V. Kolobov, M. Krbal, P. Fons, J. Tominaga, T. Uruga, Nature Chem. **3**, 311 (2011)
43. Q.T. Islam, B.A. Bunker, Phys. Rev. Lett. **59**, 2701 (1987)
44. B. Ravel, M. Newville, J. Synchrotron Radiat. **12**, 537 (2005)
45. L. Pauling, *The Nature of the Chemical Bond and the Structure of Molecules and Crystals: An Introduction to Modern Structural Chemistry* (Cornell University Press, Ithaca, 1960)
46. P.W. Atkins, *Physical Chemistry*, 6th edn. (Oxford University Press, Oxford, 1998)
47. G. Lucovsky, R. White, Phys. Rev. B **8**, 660 (1973)
48. J. Robertson, K. Xiong, P.W. Peacock, Thin Solid Films **515**, 7538 (2007)
49. K. Shportko, S. Kremers, M. Woda, D. Lencer, J. Robertson, M. Wuttig, Nature Mater. **7**, 653 (2008)
50. B. Huang, J. Robertson, Phys. Rev. B **81**, 081204R (2010)
51. R. Mazzarello, S. Caravati, S. Angioletti-Uberti, M. Bernasconi, M. Parrinello, Phys. Rev. Lett. **104**, 085503 (2010)
52. J.L.F. Da Silva, J. Appl. Phys. **109**, 023502 (2011)
53. D. Copper, J. Gerratt, M. Raimondi, Nature **323**, 699 (1986)
54. R. Kerber, J. Chem. Educt. **83**, 223 (2006)
55. A.H. Edwards, A.C. Pineda, P.A. Schultz, M.G. Martin, A.P. Thompson, H.P. Hjalmarson, C.J. Umrigar, Phys. Rev. B **73**, 045210 (2006)
56. J. Eom, Y. Yoon, C. Park, H. Lee, J. Im, D. Suh, J. Noh, Y. Khang, J. Ihm, Phys. Rev. B **73**, 214202 (2006)
57. J. Hegedüs, S.R. Elliott, Nature Mater. **7**, 399 (2008)
58. R. Shayduk, Molecular beam epitaxy of GeTe–Sb_2Te_3 phase change materials studied by X-ray diffraction. Ph.D. Thesis, Humboldt-Universität, Berlin, Germany, 2009
59. Z. Sun, S. Kyrsta, D. Music, R. Ahuja, J. Schneider, Solid State Commun. **143**, 240 (2007)
60. Y.J. Park, J. Lee, M. Youm, Y. Kim, H. Lee, J. Appl. Phys. **97**, 093506 (2005)
61. E.T. Kim, J.Y. Lee, Y.T. Kim, Appl. Phys. Lett. **91**, 101909 (2007)
62. X.Q. Liu, X.B. Li, L. Zhang, Y.Q. Cheng, Z.G. Yan, M. Xu, X.D. Han, S.B. Zhang, Z. Zhang, E. Ma, Phys. Rev. Lett. **106**, 025501 (2011)
63. S. Hosokawa, T. Ozaki, K. Hayashi, N. Happo, M. Fujiwara, K. Horii, P. Fons, A.V. Kolobov, J. Tominaga, Appl. Phys. Lett. **90**, 131913 (2007)
64. P. Fons, A.V. Kolobov, M. Krbal, J. Tominaga, K.S. Andrikopoulos, S.N. Yannopoulos, G.A. Voyiatzis, T. Uruga, Phys. Rev. B **82**, 155209 (2010)

65. T.C. Chong, L.P. Shi, W. Qiang, P.K. Tan, X.S. Miao, X. Hu, J. Appl. Phys. **91**, 3981 (2002)
66. T.C. Chong, L. Shi, R. Zhao, P. Tan, J. Li, H. Lee, X. Miao, A. Du, C. Tung, Appl. Phys. Lett. **88**, 122114 (2006)
67. T.C. Chong, L.P. Shi, X.Q. Wei, R. Zhao, H.K. Lee, P. Yang, A.Y. Du, Phys. Rev. Lett. **100**, 136101 (2008)
68. J. Tominaga, R. Simpson, P. Fons, A. Kolobov, in *European Phase Change and Ovonics Science Symposium*, Milano, 2010
69. R.E. Simpson, P. Fons, A.V. Kolobov, T. Fukaya, M. Krbal, T. Yagi, J. Tominaga, Nature Nanotech. **6**, 501 (2011)
70. R. Shayduk, W. Braun, J. Cryst. Growth **311**, 2215 (2009)
71. W. Braun, R. Shayduk, T. Flissikowski, M. Ramsteiner, H.T. Grahn, H. Riechert, P. Fons, A.V. Kolobov, Appl. Phys. Lett. **94**, 041902 (2009)
72. W. Braun, R. Shayduk, T. Flissikowski, H.T. Grahn, H. Riechert, P. Fons, A. Kolobov, in *Materials and Physics for Non-Volatile Memories*, ed. by Y. Fujisaki, R. Waser, T. Li, C. Bonafos. Materials Research Society Symposium Proceedings, vol. 1160, 2009, p. 177
73. P. Rodenbach, R. Calarco, K. Perumal, F. Katmis, M. Hanke, A. Proessdodf, W. Braun, A. Guissani, A. Trampert, H. Riechert, P. Fons, A.V. Kolobov (in press)
74. G. Cohen, S. Raoux, Y. Zhu, M. Hopstaken, H. Wildman, S. Maurer, A. Giussani, R. Calarco, *MRS Spring Meeting (symposium F)*, (San Francisco, 2012)

Chapter 9
Structure of the Amorphous Phase

As have already been described above, despite numerous recent efforts, even the structure of the crystalline phase of GST is still not completely resolved. The situation with the amorphous phase is naturally even more complex. In early works the structure of the amorphous phase was tacitly assumed to be completely disordered. Extensive recent studies, performed both experimentally and using ab initio computer simulations, revealed several characteristic features of the amorphous phase. The results obtained in these studies form the main body of the present chapter.

Experimental studies of the amorphous structures are complicated by several circumstances. First, it is generally more difficult to study the amorphous phase because the number of suitable techniques is more limited. Second, in order to obtain the amorphous phase of GST ultra-high quenching rates are required that can only be realised in device structures. Attempts to generate melt-quenched GST by splat cooling have failed: no amorphous phase could be found is the splat-cooled samples with the cooling rates as high as 10^5–10^6 K/s (Kolobov and Pradel, unpublished).

This leaves experimentalists with two options. One is to use as-deposited films to model the melt-quenched state and the other one is to use device structures to fabricate the required amorphous phase. Both these approaches have shortcomings. The as-deposited phase, on the one hand, may not be the same as the one obtained from quenching the melt and as-deposited films of Ge-based chalcogenides in particular are known to contain a large concentration of homopolar bonds [1]. Device structures, on the other hand, do not provide sufficiently large amounts of the material required for accurate measurements, the latter being further complicated by the presence of other phases in the device structures.

An appealing alternative is to study the structure of the amorphous phase using computer simulations. Recent advances in computer hardware and software development have turned ab initio studies into a highly efficient practical tool. Furthermore, the very short crystallisation times of phase-change materials are comparable with the time range accessible by DFT-based molecular dynamics which allows for direct simulations of the crytallisation process. Use of computer modelling to analyse the structure of the amorphous phase of phase-change memory alloys has been described

A. V. Kolobov and J. Tominaga, *Chalcogenides*, Springer Series in Materials Science 164, DOI: 10.1007/978-3-642-28705-3_9,

in [2]. At the same time, computer simulations are based on certain approximations and, additionally, in order to optimise the cost of simulations (to reduce the relaxation time) the melt is usually generated at 2,000–3,000 K, i.e. significantly higher than the experimental melting point and this very high temperature may introduce additional disorder into the structure that is not present in experimental samples or in devices during the phase change process that takes place at much lower temperatures.

It should also be mentioned that the above issues generate a consequential problem: the experimental results obtained for as-deposited samples are often compared with the simulation results obtained on melt-quenched samples although the structures of as-deposited and melt-quenched phases are not identical. This issue should be kept in mind whenever such a comparison is made.

In what follows we discuss the results of both experimental studies and computer simulations starting with the experimental studies.

9.1 Experimental Studies

9.1.1 Binary GeTe

Of special interest may be the structure of amorphous binary GeTe that is the simplest phase-change material and an end point of the quasibinary GeTe–Sb_2Te_3 tie-line. Despite years of studies by various groups, there is still a controversy regarding the structure of amorphous GeTe. The structural X-ray diffraction studies made by Betts, Bienenstock and Ovshinsky [3] back in the 1970s revealed that the interatomic distance and the coordination number of a-GeTe are in poor agreement with those of c-GeTe. From a detailed analysis of the peak, these authors concluded that a random covalent model with a 4(Ge):2(Te) local coordination was the most appropriate as a local coordination model for a-GeTe. Subsequent electron diffraction studies [4, 5] obtained similar radial distribution functions. The 4(Ge):2(Te)-coordinated structure was also suggested on the basis of an EXAFS measurements around both Ge and Te K-edges [6–8]. The Ge–Te bond length (2.59 Å) was also found to be consistent with the 4(Ge):2(Te) bonding geometry [8]. Raman scattering and far-infrared absorption spectra for a-GeTe [9] were equally interpreted in terms of the presence of $GeTe_4$ tetrahedra. These results strongly supported a random covalent network model of the 4(Ge):2(Te)-coordinated atoms, and excluded the possibility of a c-GeTe microcrystalline structure. Combined photoemission and inverse photoemission studies [10] also favoured the 4(Ge):2(Te) coordination.

On the other hand, neutron scattering [11] and Mossbauer spectrometry of ^{125}Te nuclei [12] studies suggested the 3(Ge):3(Te)-coordinated local structure, demonstrating that the local structure of GeTe is still far from being resolved. One explanation of the different conclusions could be that the samples were prepared differently in different studies. Since the melt-quenched bulk amorphous phase is unachievable, the experimental studies are necessarily performed on as-deposited films whose structure may significantly vary depending on the preparation conditions.

Recent DFT simulations [13] have shown that the amorphous phase of GeTe obtained directly from the crystalline phase without going through the melt possess predominantly 3(Ge):3(Te) building blocks. Tetrahedrally coordinated Ge atoms are limited to Te_3Ge–$GeTe_3$ tetrahedra.

9.1.2 *Quasibinary GeTe-Sb_2Te_3 Alloys*

X-ray Absorption Fine Structure

The first report on the local structure of the amorphous phase belongs to the LETI group in France [14]. These authors performed EXAFS studies on as-deposited and thermally crystallised layers of GeTe and GST and obtained the bond lengths characteristic of the two states. In particular, they found the Ge–Te bond to be 2.62 Å and the Sb–Te bond to be 2.84 Å (as compared to 2.83 and 2.94 Å, respectively in the crystalline phase), i.e. the bonds are *shorter* in the amorphous phase. This work, however, only reported the obtained numerical values and stayed short of drawing further conclusions.

Kolobov et al. [15] were the first to go beyond the simple measurements and their results are described below in more detail. It should be noted that in this work laser-crystallised and laser-amorphised states of GST were studied, i.e. exactly the structures that are used in optical memory devices (the measurements were performed on real-device structures). XAFS measurements were performed at room temperature using conversion electron yield at the K-edges of all constituent elements.

For the amorphous phase, it was found that the Ge–Te and Sb–Te bonds become shorter and the structure possesses more local order than the crystalline phase as evidenced by more intense and narrower peaks in the spectra corresponding to the amorphous phase (Fig. 9.1). It is informative to note that the obtained Ge–Te and Sb–Te bond lengths are very close to the sum of the corresponding covalent radii for the elements ($r_{Ge} = 1.22$ Å, $r_{Sb} = 1.38$ Å, $r_{Te} = 1.35$ Å [16]).[1] No evidence for the presence of Ge–Ge or Ge–Sb bonds was detected in the laser-amorphised phase.

The observed bond shortening and increased local order are highly unusual for covalent solids when, due to anharmonicity of the interatomic potential, disorder usually results in longer and weaker bonds and suggests that the local structures in the two cases are significantly different. It should also be noted that despite the bond shortening the density of the amorphous phase is about 5% lower than that of the crystalline phase.

The conclusion that a significant change occurs in the local structure finds more support in experimentally measured XANES spectra [15] shown in Fig. 9.2. As was discussed earlier, XANES has the advantage (which is also a complication for the data analysis) that it is sensitive to the three-dimensional spatial arrangement of scat-

[1] The covalent radius for Ge is given assuming sp^3-hybridisation.

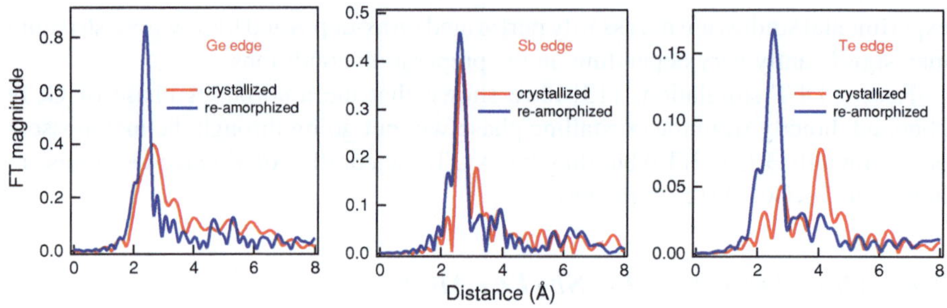

Fig. 9.1 Fourier transformed EXAFS spectra for Ge, Sb and Te K-edges for laser-crystallised and laser-amorphised $Ge_2Sb_2Te_5$. Reprinted by permission from Macmillan Publishers Ltd.: Nature Mater. [15], copyright (2004)

tering atoms around the central (absorbing) atom [17, 18]. The observed differences thus provide evidence that the local structures (i.e. the coordination numbers, nature of bonding with the nearest neighbours and the interatomic distances) are, indeed, different in the two cases.

The decreased bond length in the amorphous state concomitant with the density decrease is reminiscent of molecular solids, e.g. selenium, when different kinds of bonding exist in a solid and their interplay determines the local structure. Thus, in the case of crystalline selenium, the covalently bonded atoms within helical chains are held together by weaker van-der-Waals interactions (occasionally also referred to as resonance bonds [19, 20]) which further deepens the analogy between GST and molecular solids such as selenium. Upon the loss of the long-range order, the int*er*-chain interaction decreases and the int*ra*-chain forces dominate the local structure. As a result, the first nearest neighbour distances get shorter (2.32 Å in the amorphous state as opposed to 2.36 Å in the crystalline state). At the same time, the first nearest neighbour EXAFS peak has a larger intensity in the amorphous phase (Fig. 9.3) [21]. While it may be possible that the coordination number in amorphous selenium is larger than 2.0 [21] it has also been argued that this may be due to an *increased local order in the amorphous state* due to weakening of the int*er*-chain bonds [22].

Similar to the case of $Ge_2Sb_2Te_5$, despite the first nearest neighbour bond shortening, the density of amorphous selenium is lower than that of the crystal and in the case of Se this is due to the fact that the second nearest neighbour distances, i.e. int*er*chain distances, increase in the amorphous phase. This analogy between GST and Se EXAFS suggests that the transition between the crystalline and amorphous phases involves rupture of the weaker (longer) bonds that exist in the crystalline phase [23]. A case example of selenium has been recently used to model the large change in optical properties between the crystalline and amorphous phases in phase-change materials [24].

The overall observed changes in XANES upon the phase transition could be reproduced by FEFF simulations [25] when the Ge atoms were placed into tetrahedral symmetry sites [15] within the Te fcc lattice (Fig. 9.4). Based on this finding the

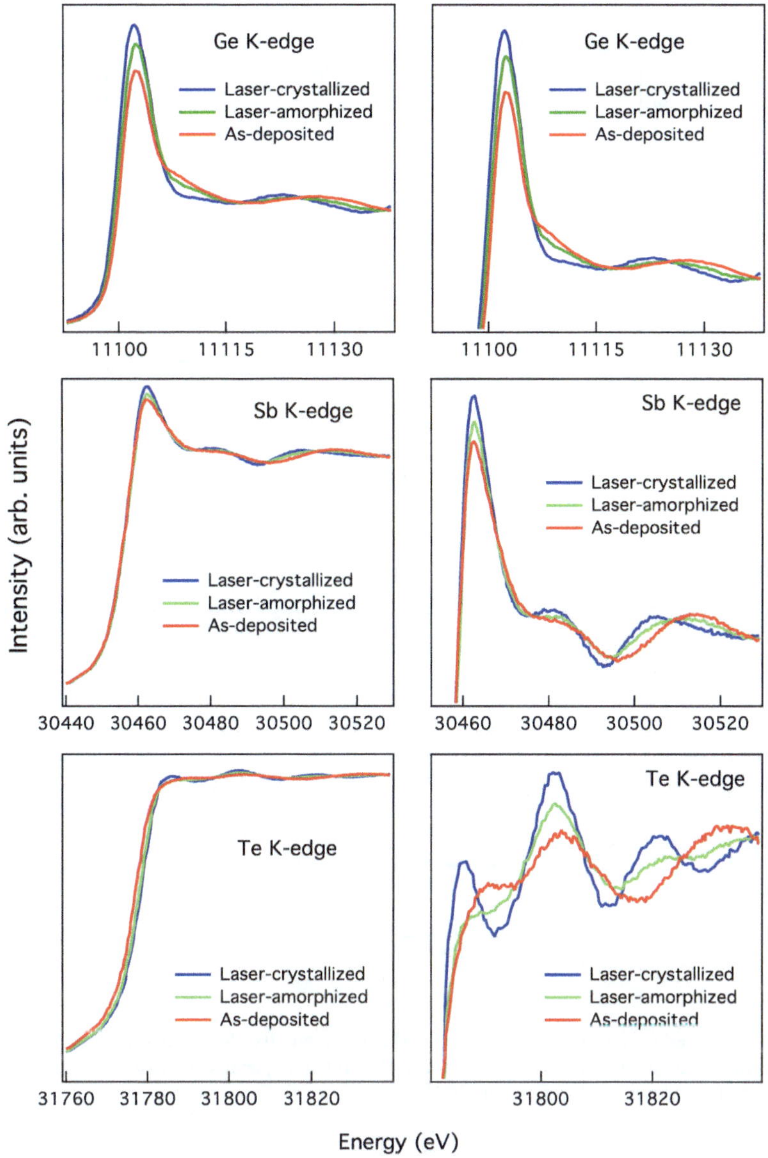

Fig. 9.2 XANES spectra for the as-deposited, laser-crystallised and laser-amorphised $Ge_2Sb_2Te_5$ samples. The left-side panels show full spectra; in the right-side panels the spectra have been scaled vertically for better visualisation of the difference. Reprinted with permission from Krbal et al. [39]. Copyright 2011 by the American Physical Society

authors argued that the phase transition is triggered by an umbrella-flip of Ge atoms [15] between the octahedral and tetrahedral symmetry sites within the Te fcc lattice

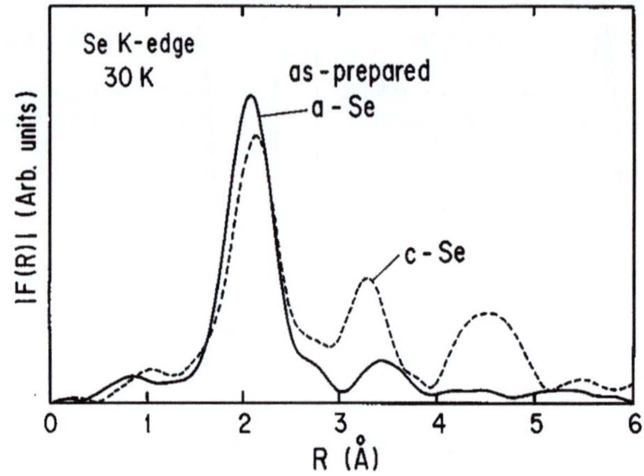

Fig. 9.3 Fourier transformed EXAFS spectra for as-deposited and crystallised selenium films. Reprinted with permission from Kolobov et al. [21]. Copyright 1977 by the American Physical Society

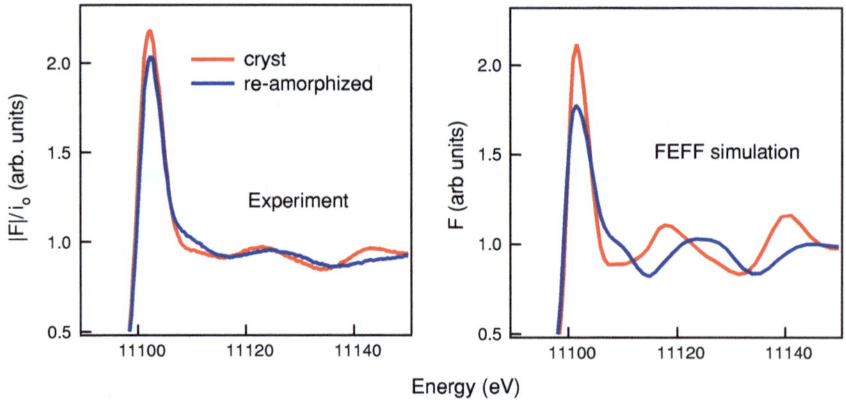

Fig. 9.4 Ge K-edge XANES spectra measured experimentally (*left*) and simulated using FEFF8 code assuming tetrahedral location of Ge atoms. Reprinted by permission from Macmillan Publishers Ltd.: Nature Mater. [15], copyright (2004)

(Fig. 9.5). This provided the first atomistic explanation why the transition is so fast (a single atom switches) and why the medium is so stable (the Te sublattice is essentially preserved).

A simple switch of a single Ge atom within an ordered Te fcc lattice as shown in Fig. 9.5 is, of course, a schematic oversimplification; in reality, the situation is much more complex. First, the Te atoms shown in the figure are not only bonded to the central Ge atom but also to other atoms in the solid. Additionally, a change in the Ge atom bonding geometry necessarily changes the force balance acting on

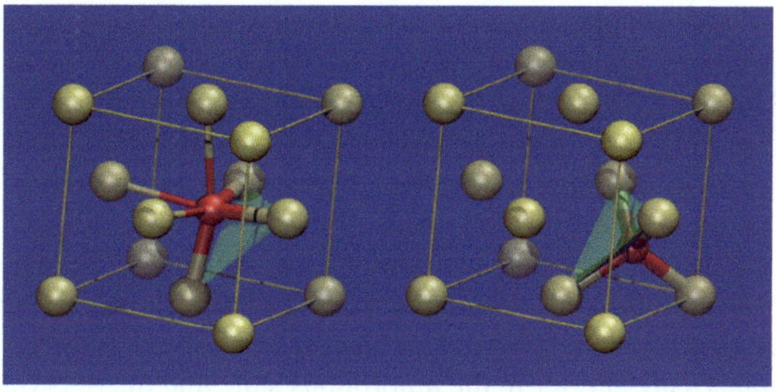

Fig. 9.5 Ge atom located on an octahedral (*left*) and tetrahedral (*right*) symmetry site within the Te fcc lattice. Reprinted by permission from Macmillan Publishers Ltd.: Nature Mater. [15], copyright (2004)

Table 9.1 Total coordination numbers for as-deposited amorphous $Ge_2Sb_2Te_5$ obtained from EXAFS analysis

Atom type	Total coordiantion number
Ge	3.9 ± 0.7 [30], 3.85 [27]
Sb	2.8 ± 0.5 [30], 3.12 [27]
Te	2.4 ± 0.6 [30], 1.99 [27]

other atoms. As a result, the Te atoms are displaced from their original fcc positions leading to loss of the long-range order as first suggested in [26].

Amorphous $Ge_2Sb_2Te_5$ has subsequently been studied using EXAFS by other groups [27–29]. An important difference, however, is that in those studies *as-deposited* amorphous films were studied. The obtained Ge–Te and Sb–Te bond lengths are in a good agreement with the results described above.

In addition to the Ge–Te and Sb–Te bonds found in previous works, for as-deposited samples, Ge–Ge bonds were identified by various groups with the bond length of 2.47–2.48 Å [27, 30]. It should be noted in passing that presence of "wrong" bonds, and especially Ge–Ge bonds, is also characteristic of as-deposted films of chalcogenide glasses [1]. As to the existence of Ge–Sb bonds in amorphous $Ge_2Sb_2Te_5$, such bonds were not observed in [31] but their presence was reported in [27, 32] and the bond length of 2.69 Å was reported [27]. Sb–Sb and Te–Te bonds were not reported from EXAFS studies.

Using the coordination numbers obtained from the EXAFS data analysis, the authors applied the bond constraint theory [33] to interpret the phase change in $Ge_2Sb_2Te_5$ and calculated the average number of constraints to be 3.07. According to the authors, the obtained number of constraints corresponds to a good glass former [29]. This conclusion is in stark contrast to the existing consensus that attributes high crystallisation rates of phase-change materials to the fact that they are marginal glass-formers [34].

The 8–N rule has been experimentally confirmed for all constituent elements for as-deposited GST. This conclusion follows from the coordination numbers summarised in Table 9.1 and also has been specifically noted in [27].

The reported EXAFS data allow one to draw the following conclusions about the structure of the amorphous phase of $Ge_2Sb_2Te_5$. The Ge–Te and Sb–Te bond lengths are 2.61–2.63 Å and 2.83–2.85 Å, respectively, i.e. significantly shorter than in the crystalline phase and very close to the sum of covalent radii of the constituent atoms. Additionally, Ge–Ge and Ge–Sb "wrong" bonds may be present in as-deposited samples. The constituent atoms obey the 8–N rule, i.e. the Ge, Sb and Te atoms are predominantly fourfold, threefold and twofold coordinated, respectively.

EXAFS studies performed on other materials within the GeTe–Sb_2Te_3 system revealed an essentially identical behaviour namely, in all studied cases the first nearest neighbour distances are shorter in the amorphous state and the 8–N rule is satisfied.

Scattering Studies

Bond shortening in the as-deposited amorphous phase of $Ge_2Sb_2Te_5$ has also been observed by high-energy X-ray scattering experiments [35]. Figure 9.6 shows the total correlation functions for the binary GeTe and GST. One can clearly see that the peak positions shift to shorter distances in the disordered (amorphous and liquid) phases in agreement with the EXAFS results.

The results were analysed using reverse Monte Carlo (RMC) technique (cf. Sect. 3.3.1). The RMC analysis suggested that the Ge(Sb)–Te–Ge(Sb) and Te–Ge(Sb)–Te angles were close to 90 degrees in disagreement with the tetrahedral coordination of the Ge atoms suggested from XAFS analysis. At the same time, the coordination numbers found for Ge (Sb) species of 3.7 (3.0) is in agreement with the previous conclusions. The existence of the "wrong" Ge(Sb)–Ge(Sb) bonds was not confirmed by the RMC analysis.

The authors further performed a ring statistics analysis (cf. Sect. 3.3.2); the results are reproduced in Fig. 9.7. For GeTe, they found a broad distribution of different ring sizes which is due to the formation of Ge–Ge bonds. At the same time, "the ring statistics of amorphous $Ge_2Sb_2Te_5$ is dominated by four- and six-fold rings analogous to the crystal phase". The authors conclude that such unusual ring statistics of amorphous $Ge_2Sb_2Te_5$ is the key for the fast crystallisation speed of the material. It should be noted that the obtained ring statistics in a–$Ge_2Sb_2Te_5$ is the result of the RMC simulation under the assumption that homopolar Ge–Ge and Sb–Sb bonds are not formed in a–$Ge_2Sb_2Te_5$. At the same time, different experimental EXAFS studies of as-deposited $Ge_2Sb_2Te_5$ [27, 30] reported the existence of the Ge–Ge bonds.

The obtained results were subsequently refined using ab initio simulations as a starting point for the RMC with the following conclusions drawn: (a) ring structures play an essential role in the phase change, (b) numerous cavities provide space for the phase change to occur without breaking many bonds and (c) over-coordinated Sb and Te and a small number of wrong bonds indicate that $Ge_2Sb_2Te_5$ is a marginal glass former [36].

Scattering studies have also been performed for Ag-In-Sb–Te alloys and the results are described below (Sect. 9.2.4).

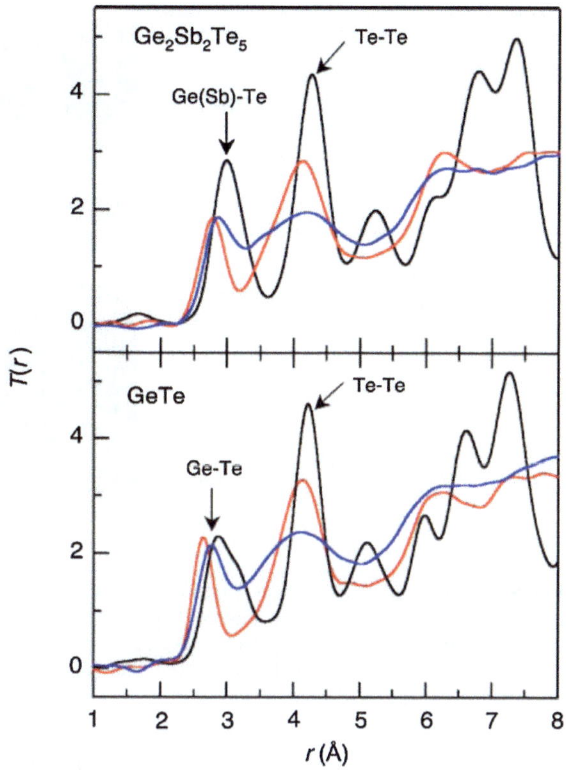

Fig. 9.6 Total correlation functions $T(r)$ of crystalline (*black*), liquid (*blue*) and amorphous (*red*) $Ge_2Sb_2Te_5$ and GeTe. Reprinted with permission from Kohara et al. [35]. Copyright 2006 by the American Institute of Physics

9.1.3 Differently Prepared Amorphous Phases: Polymorphism

Finally, we address the issue of differences in the amorphous structure obtained through different pathways. First of all, it should be noted that optical reflectivity of the amorphous phase obtained after the material has once been crystallised is intermediate between those of the as-deposited and crystalline phases suggesting that the two phases possess different structures [37]. Secondly, it should be noted that the crystallisation time is significantly shorter for the "recycled" amorphous phase [37, 38] which further suggests that the two phases are different.

Direct structural evidence for the differences can be seen from comparison of XANES spectra of as-deposited and laser-amorphised samples [39] as shown in Fig. 9.2. One can see that curves for the laser-amorphised sample are intermediate between the as-deposited sample and the crystalline samples. Raman scattering spectra (Fig. 9.8) exhibit the same trend [39]. This result may be compared with photodarkening in chalcogenide glasses when the absorption edge of annealed (i.e. equilibrated) films is located in between those for the as-deposited and illuminated states (cf. Fig. 4.1).

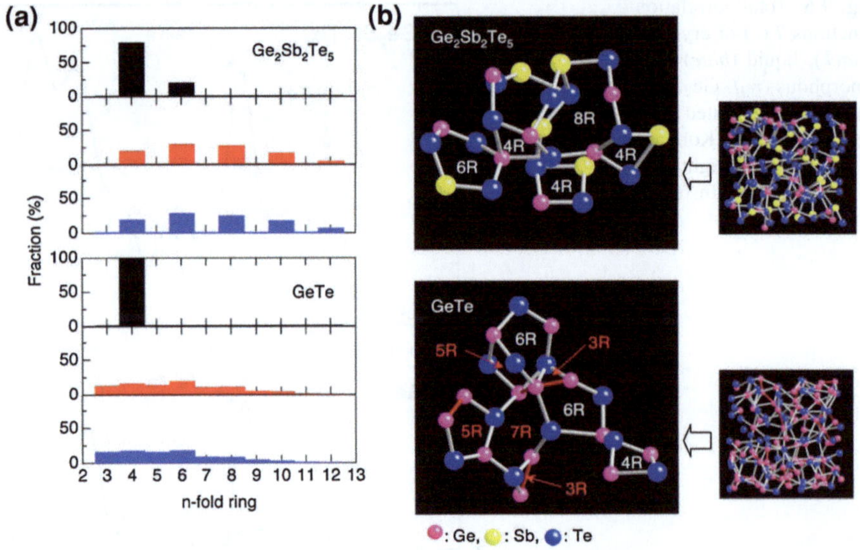

Fig. 9.7 Ring size distribution in $Ge_2Sb_2Te_5$ and GeTe (**a**). *Black* bars represent the distribution in the crystals and *red* and *blue* bars correspond to the amorphous structures obtained by RMC simulation started with NaCl-type structure (*red*) and with a random structure (*blue*). The obtained structures are shown in panel (**b**). Reprinted with permission from Kohara et al. [35]. Copyright 2006 by the American Institute of Physics

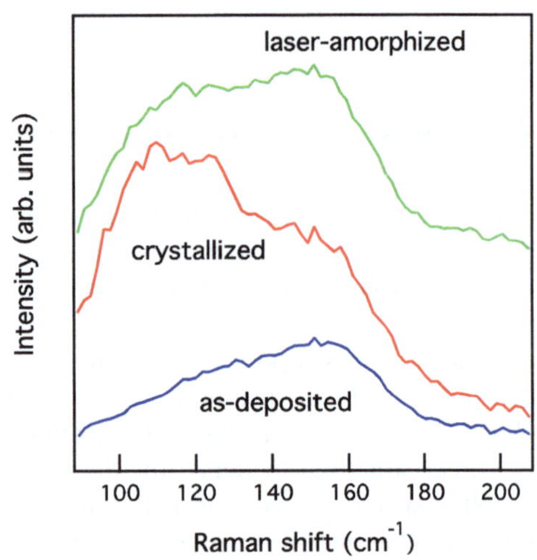

Fig. 9.8 Raman spectra of as-deposited and laser-amorphised $Ge_2Sb_2Te_5$ compared to that of the laser-crystallised layer. Reprinted with permission from Krbal et al. [39]. Copyright 2011 by the American Physical Society

One possible explanation for the laser-amorphised $Ge_2Sb_2Te_5$ to possess an intermediate structure would be that the sample was not uniformly laser amorphised and the intermediate spectra are due to the presence of a certain fraction of the crystalline phase. An alternative and more likely explanation may be that the nature

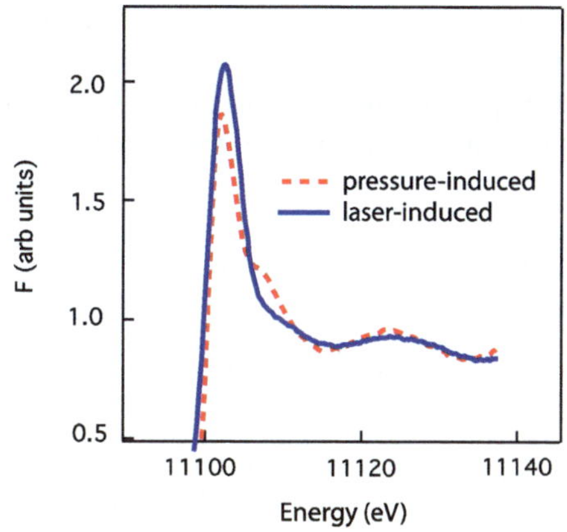

Fig. 9.9 An experimental XANES spectrum of $Ge_2Sb_2Te_5$ amorphised at 22 GPa compared with the spectrum for a laser-amorphised $Ge_2Sb_2Te_5$

of the crystallisation–amorphisation process is such that the strongly distorted Te lattice persists in the amorphous state [15, 40]. The observed intermediate range order in the amorphous GST supports this idea which is further supported by the fact that four-ring building blocks that form the metastable cubic phase of GST are also found in "melt-quenched" $Ge_2Sb_2Te_5$ [41]. Several other studies also found that the amorphous phase contains nuclei of the cubic phase [42–44].

One can imagine that in the as-deposited film obtained from the vapour phase the atoms are more likely to possess the "ideal" amorphous coordination, such that Ge atoms form $GeTe_4$ tetrahedra while Sb atoms are threefold coordinated. The coordination numbers found from EXAFS analysis on as-deposited samples that satisfy the 8–N rule support this proposal. On the other hand, the amorphous phase obtained from the solid metastable cubic GST phase is more likely to preserve a certain number of Ge- and Sb-containing fragments with the valence angles close to those in octahedrally coordinated sites in the crystalline phase. Such a conclusion has also been drawn from DFT and detailed XANES simulations when structures of as-deposited and melt-quenched $Ge_2Sb_2Te_5$ were compared from first principles [39, 45].

As described in Sect. 10.1.1, metastable cubic GST can also be rendered amorphous by application of hydrostatic pressure. XANES studies performed on a pressure-amorphised sample (Fig. 9.9), the results, and in particular the lower white line and an appearance of a pronounced shoulder following the white line, suggest that a significant amount of Ge atoms become tetrahedrally coordinated after compression (Krbal et al. unpublished).

Analysis of experimental data as well as computer modelling have shown that pressure amorphised GST contains a large amount of "wrong" Sb–Sb and Te–Te bonds that were not detected in laser-amorphised or as-deposited GST [46].

Finally, the amorphous GST obtained by ion implantation has been investigated [47, 48] and it was found that ion-irradiated samples crystallised significantly faster than the as-deposited sample. Based on Raman scattering studies, it was argued that Sb^+ ion implantation (120 keV) leads to destruction of tetrahedral Ge–Te bonds in the initial as-deposited phase [48]. At the same time, the existing ambiguities [49–51] about the location of the Raman feature arising from tetrahedral $Ge–Te_4$ configurations suggest that this conclusion may have to be revisited.

9.2 Computer Simulations

9.2.1 Around the Umbrella-Flip Model

The experimental observation of a significant change in the local order between the crystalline and amorphous phases triggered numerous computer studies that were initially aiming at supporting—or disproving—the umbrella-flip model.

The Spinel Approximation

Welnic et al. [52] were the first to apply ab initio computer simulations based on density functional theory for phase-change materials. Based on the results of experimental XAFS studies which concluded that Ge atoms are located on tetrahedral symmetry sites and Sb atoms on octahedral symmetry sites within the Te fcc lattice, these authors came up with an elegant idea to model the local structure of GST by using the spinel structure. For a reader unfamiliar with crystallography we note here that the spinel is a ternary crystal with the composition of $MgAl_22O_4$. In this structure, Mg ions are tetrahedrally coordinated and Al ions are octahedrally coordinated. Use of the spinel structure with Ge atoms on Mg sites, Sb atoms on Al sites and Te atoms on O sites enabled the authors to substantially reduce the cell size and thus perform cost-efficient calculations.

The authors first compared the total energy of the rocksalt structure with that of spinel and found that the (distorted) rocksalt structure possesses a very similar energy to that of the spinel structure (Fig. 9.10). Furthermore, they found that the energy difference between the two states is in very good agreement with the value obtained experimentally using differential scanning calorimetry (DSC).

The authors also found that the Ge–Te bond length in the spinel structure was shorter than that in the rocksalt structure (2.81 Å as opposed to 2.86 Å) in agreement with the conclusions drawn from XAFS. Numerical discrepancies between the experimental and simulated values were attributed to the simulation method that tends to produce longer interatomic distances. At the same time, the Sb–Te distance was found to increase in the simulations (from 3.00 Å in the rocksalt to 3.10 Å in the spinel structure) while in the experiments the Sb–Te bond shortening was observed.

Calculations of the band structure have shown band opening at the Γ point upon transformation from the rocksalt to the spinel phase in agreement with the experiment. It was suggested that the smaller values obtained in simulations (as opposed to those measured experimentally) are characteristic of DFT simulations [52].

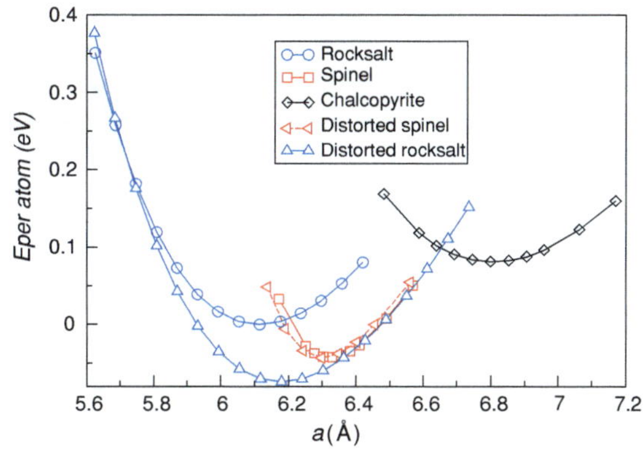

Fig. 9.10 Electronic ground state energy per atom plotted versus lattice parameter for the rocksalt, spinel and chalcopyrite structure. The lattice parameter corresponds to a cubic lattice containing 1 Ge, 2 Sb and 4 Te atoms. The total energy of the relaxed, undistorted rocksalt phase was set to zero in this plot, the crucial and physically relevant data are the energy differences between the structures. The distorted rocksalt structure has the lowest ground state energy and can be identified as the crystalline phase. Reprinted by permission from Macmillan Publishers Ltd.: Nature Mater. [52], copyright (2005)

While subsequent computer simulations casted doubt in the correctness of the use of the spinel analogue of the amorphous phase (as well as the umbrella-flip model deduced from XANES studies), it was nonetheless the first demonstration that the presumed change in the Ge bonding geometry towards a purely covalent bonding does reproduce the change in properties observed experimentally.

The results of computer simulations for $Ge_2Sb_2Te_5$ were also reported in [53, 54] and the energy barrier between the octahedral and tetrahedral sites was found to be 0.4 eV which is significantly smaller than the experimentally determined barrier of 2.3 eV (see e.g. [55]) for the crystallisation process.

Effect of Site Occupancy in Ge/Sb Sublattice

Lang et al. [54] were the first to explicitly point out that upon a Ge atom switching into a tetrahedral symmetry site within the Te fcc lattice, in addition to the four Te atoms, the Ge atom is at least initially, i.e. before any diffusion-mediated changes occur, coordinated to three more sites, each containing either a Sb or Ge atom or a vacancy. They have calculated the energies for transitions of an octahedrally coordinated Ge atom into a site that has *only* four Te neighbours and also into a site that additionally has two Sb atoms.

The results of simulations are shown in Fig. 9.11. For the case when the switched Ge atom only has Te neighbours (adjacent sites on the Ge/Sb sublattice are vacant), a barrier of 0.4 eV was found between the octahedral and tetrahedral sites. On the other hand, for the case with two Sb atoms additionally present (on the Ge/Sb sublattice),

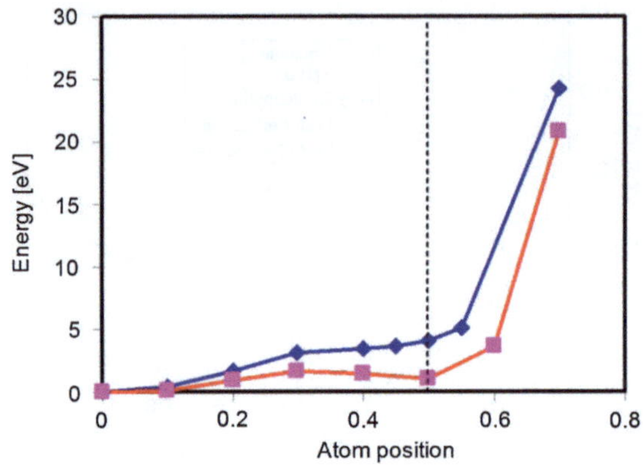

Fig. 9.11 Energy variation for a Ge atom motion between an octahedral site and a "tetrahedral" site within the cubic phase of $Ge_2Sb_2Te_5$ when Ge only has four Te neighbours (*squares*) and when it additionally has two Sb neighbours (*diamonds*) in the "amorphous" state. The atom position at 0 corresponds to the octahedral site and the position marked by a *dashed line* to that of a tetrahedral site. Reprinted with permission from Lang et al. [54]. Copyright 2007 by the American Physical Society

there is no barrier. The structures were subsequently relaxed by molecular dynamics for a simulation time of 2 ps for temperatures of 400 and 500 K. At both temperatures the purely tetrahedral Ge location was stable. From this result the authors conclude that the umbrella flip is "only favourable if there are no Sb atoms surrounding the site to which the Ge atom moves".

This result is in agreement with a later finding of Da Silva et al. [56] who calculated the potential energy path along the rocksalt [111] direction for GeTe as a function of Ge shift between the octahedral and tetrahedral sites. It was found that the tetrahedral Ge positions in crystalline GeTe (assuming zincblende crystal structure) are not a local minimum. The zincblende structure relaxes without an energy barrier to graphite-like or "long-Ge–Te" structures, which have lower energies than the high symmetry zincblende phase and concluded that the intrinsic instability of Ge at ideal tetrahedral sites drives them to move away and adopt a variety of lower symmetry coordination environments as illustrated by Fig. 9.12.

It should be also kept in mind that presence of vacancies disturbs the symmetry of the octahedral site and the octahedrally coordinated atoms are likely to be displaced not completely randomly but the displacement direction should correlate with the location of the adjacent vacancies [57]. As a consequence, the Ge atom is likely to flip in a preferred direction which increases its possibility to find itself in the most stable site. Such a scenario is in agreement with the results of [58, 59] where it was found that only a fraction of the Ge atoms acquire tetrahedral bonding geometry in the amorphous phase.

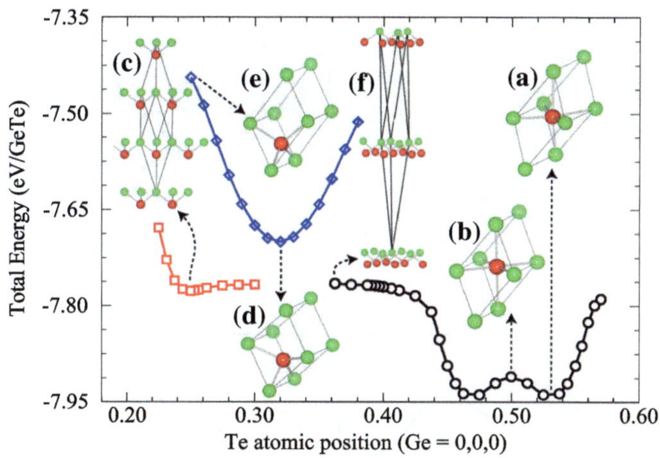

Fig. 9.12 Potential energy path for atomic displacements of Ge atoms along the rocksalt [111] direction of GeTe. **a** Distorted rocksalt structure. **b** Perfect rocksalt structure. **c** Long GeTe bond zincblende structure. **d** Graphite-like structure. **e** Perfect zincblende structure. **f** Rocksalt-layer structure. Reprinted with permission from Da Silva et al. [56]. Copyright 2009 by the American institute of Physics

Long-Range Order Effect

To investigate the effect of the long-range order loss resulting from the lattice relaxation in the amorphous state Kolobov et al. [60] performed a simplistic calculation using the bond-orbital method [61] for a motion of a Ge atom between the octahedral and tetrahedral sites within eight first nearest Te neighbours (Fig. 9.13) for two cases. In one case, all eight Te atoms were fixed at the ideal positions on the fcc sites (ordered), in the other case, the four atoms with which the bonds have been broken after the flip were moved away by 0.1 Å in agreement with the experimentally observed Sb–Te bond shortening upon amorphisation (disordered). One can see from the figure that loss of the long-range order has a drastic effect: while the octahedral location is more stable in the ordered structure, it is the tetrahedral location that is more stable in the disordered structure.

9.2.2 Melt-Quenched Amorphous Phase

A significant step forward was the *in silico* generation of a melt-quenched amorphous phase. Such simulations were performed by several groups and their main results are reviewed below.

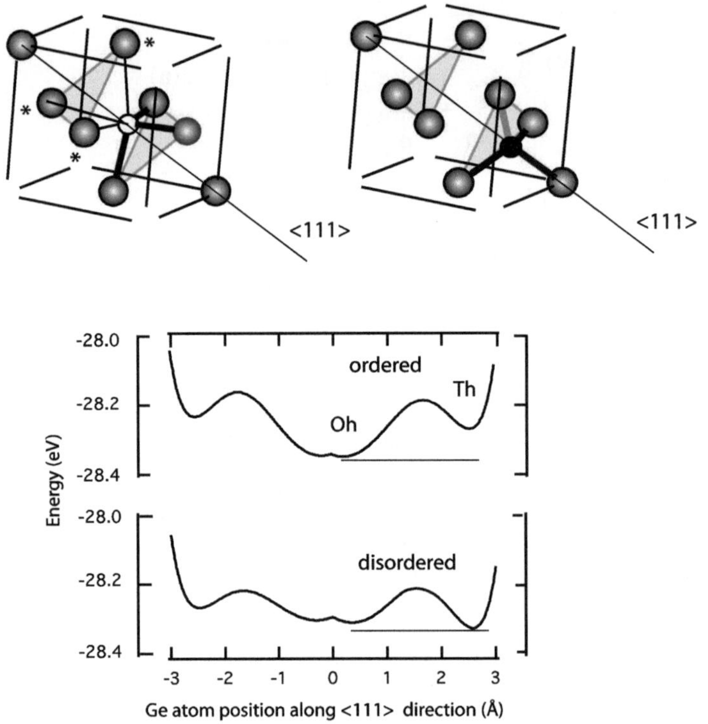

Fig. 9.13 Motion of a Ge atom between octahedral and a "tetrahedral" sites within the first nearest neighbour eight Te atoms located on the fcc lattice. The lower panel shows energy diagrams for perfectly ordered and disordered Te sublattice. Reprinted from Kolobov et al. [60] with permission from IOP Publishing Ltd.

Caravati, Bernasconi, Kühne, Krack and Parrinello Model

The first simulation of the melt-quenched amorphous $Ge_2Sb_2Te_5$ was reported by Caravati et al. [58]. The liquid structure (270 atoms) generated at 2,300 K was equilibrated for 6 ps and then quenched in 16 ps and further equilibrated for 18 ps at 990 K. In order to generate a model of a–$Ge_2Sb_2Te_5$, the liquid has been brought to 300 K in 18 ps. Subsequently these studies were extended [62].

The calculated X-ray scattering function is in agreement with the XRD data of [35] as shown in Fig. 9.14 [62], although some differences are visible at larger Q values ($Q > 4$). The authors find the average coordination numbers of 3.8 (Ge), 4.0 (Sb) and 2.9 (Te) with the bonding cut-off chosen as the outer edge of the appropriate partial pair correlation functions. Together with the angle distribution function (Fig. 9.15) these results can be summarised as follows. Ge and Sb atoms are mostly fourfold coordinated and form bonds preferentially with Te atoms. A rather large fraction of Ge–Ge, Sb–Sb, Ge–Sb as well as Te–Te bonds were found. In particular, 38% of Ge atoms were bonded with at least one other Ge or Sb atom.

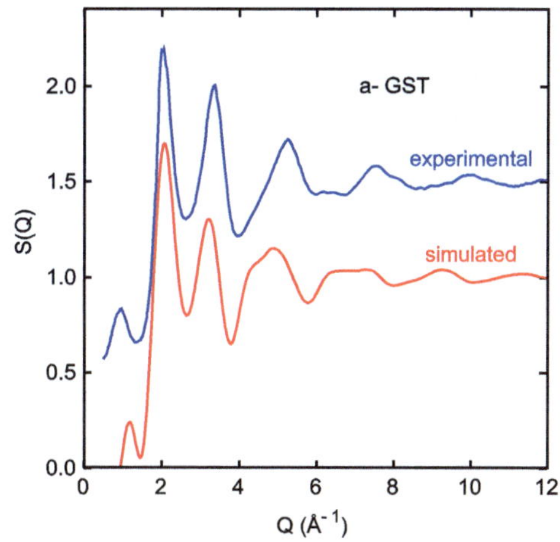

Fig. 9.14 Comparison of the simulated and experimental X-ray scattering factor S(Q) of amorphous $Ge_2Sb_2Te_5$. Reprinted from Caravati et al. [62] with permission from IOP Publishing Ltd.

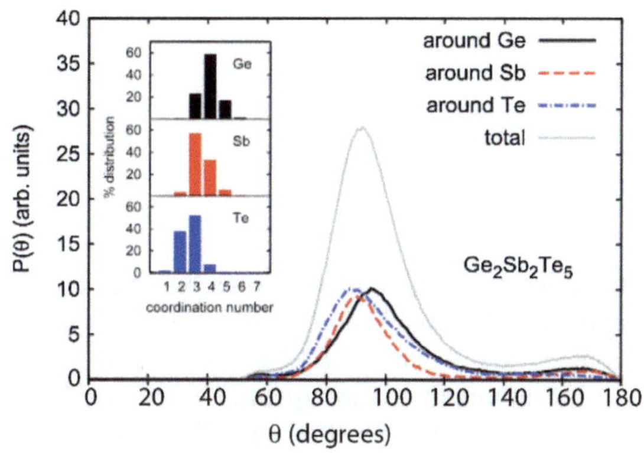

Fig. 9.15 Angle distribution function (total and resolved for different central atoms). Insert: distribution of coordination numbers of different species. Reprinted from Caravati et al. [62] with permission from IOP Publishing Ltd.

The latter result is in disagreement with the experimental EXAFS data [15] and may be caused by the significantly higher than the experimental melting point temperature used to melt GST *in silico* that resulted in extra disordering and/or insufficient time for equilibration at the lower temperature.

For threefold coordinated Te atoms, only angles at 90° are found as well as for Sb atoms while only a fraction of Ge atoms are in defective octahedral sites. A large fraction of Ge atoms are found on tetrahedral sites. A signature of the tetrahedral geometry is visible as a shoulder at 110° in the angle distribution functions for Ge atoms. Furthermore, the fourfold coordinated Ge distribution is found to be bimodal with peaks corresponding to defective octahedral and tetrahedral sites. For Te and Sb atoms no evidence of tetrahedral geometry was found.

The authors further analysed the distribution for fourfold coordinated Ge atoms and found that the presence of bonds with Ge or Sb favours a tetrahedral geometry. Only a few tetrahedral Ge atoms were found to be bonded to Te. On the other hand,

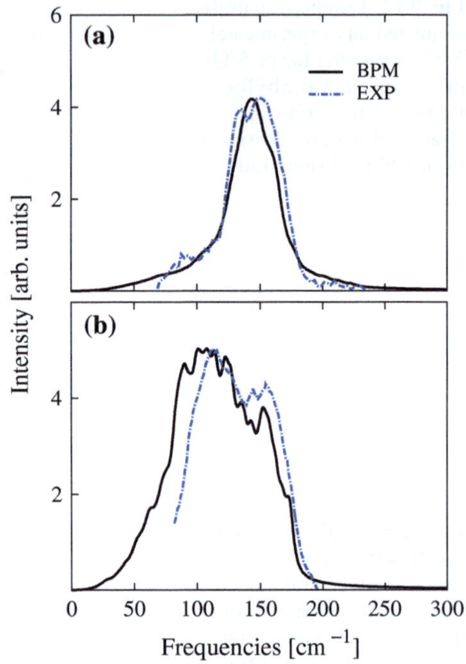

Fig. 9.16 Reduced Raman spectra of amorphous (**a**) and crystalline (**b**) $Ge_2Sb_2Te_5$ computed with bond polarisability model (*solid lines*) for unpolarised light compared with the experimental results (*dashed lines*) of [48]. Reprinted with permission from Sosso et al. [51]. Copyright 2011 by the American Physical Society

all Ge atoms with more than one "wrong" bond (to Ge or Sb) are in a tetrahedral geometry. This conclusion is in contrast with the one made in [54] that the tetrahedral geometry is only stable if a Ge atom is exclusively bonded to Te atoms.

The authors have subsequently calculated Raman spectra for the amorphous and cubic phases of $Ge_2Sb_2Te_5$ using the bond polarisability model [51]. The results of the calculations are compared with the experiment in Fig. 9.16. The authors conclude that the feature associated with $TeGe_4$ tetrahedra is located around 190 cm^{-1} [50].

Akola and Jones Model

Akola and Jones [40, 59] performed similar simulations using significantly larger cells (460 atoms as opposed to 56 atoms of Ref. [58]) and carried out molecular dynamics simulations of hundreds of picoseconds that is comparable to the amorphisation times observed experimentally (ca. 1 ns). A comparison of the simulated and experimental structure factors is given in Fig. 9.17. One can see that while the general shape of the spectrum is well reproduced by simulations, the exact shape and positions of some of the features slightly differ between the experiment and simulations in a way that is very similar to the results reported in [58] for a much smaller cell. It should be mentioned that the comparison of the *in silico* melt-quenched amorphous phase has been done to the experimental data corresponding to the as-deposited films.

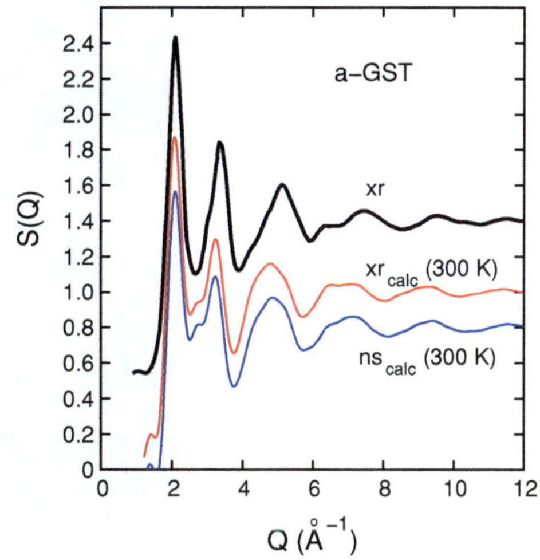

Fig. 9.17 The structure factor S(Q) of amorphous $Ge_2Sb_2Te_5$. The experimental data (*upper curve*) are from [35]. Reprinted with permission from Akkola and Jones [40]. Copyright 2007 by the American Physical Society

The observed difference may thus be related to the differences that exist between the two amorphous phases.

The Ge–Sb (2.78 Å) and Sb–Te (2.93 Å) bond lengths were found to be shorter than in crystals but longer than those obtained from experimental studies. The most prominent topologies are Ge–Te_4, Sb–Te_3, and Te–Ge(Sb)$_3$. "Wrong" bonds such as Ge–Ge (0.4), Sb–Sb (0.6) and Ge–Sb (0.2) were also found. The total coordination numbers using a cut-off distance of 3.2 Å are 4.2 (Ge), 3.7 (Sb) and 2.9 (Te). As in the previous work [58] the cut-off distance used may be too large since the bonds in the amorphous $Ge_2Sb_2Te_5$ are purely covalent [20] and their lengths are typically 2.61 Å (Ge–Te) and 2.83 Å (Sb–Te) [14, 15].

About 60% of Ge atoms are found to be fourfold coordinated but only 34% are tetrahedrally bonded. Sb atoms were found not to form a tetrahedral bonding geometry. The order parameter a for the first nearest neighbours ($a = 1$ for a crystal) has been calculated to be around 0.7 for all three kinds of atoms.

The dihedral angles around Ge–Te (and Sb–Te) bonds show pronounced maxima at 0, 90 and 180 that are consistent with octahedral bonding geometry and imply similarities between the amorphous and crystalline phases. In line with this, the study found the dominance of square fragments in the structure of the a-GST. Denoting Te atoms as "A" and Ge/Sb atoms as "B", the authors introduce ABAB squares as building blocks for the a-GST (Fig. 9.18).

Another important finding was a strikingly strong long-range correlation of Te atoms of up to 10 Å (Fig. 9.19) which underscores the importance of large simulation cells; simulations performed on smaller cell may miss some important features. While a cubic arrangement of the Te atoms was not found, the strong Te correlations provide support to the previous suggestion [15] that the changes occur "within the Te lattice". The observed correlations are in agreement with the results of fluctuation electron microscopy [42, 63] when a characteristic dimension of the ordered region was found

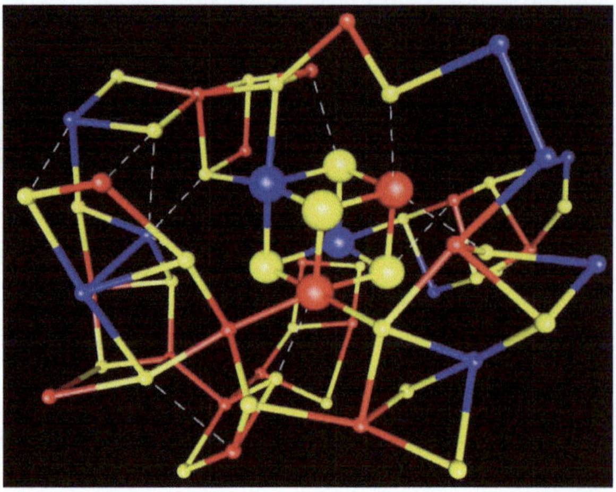

Fig. 9.18 A fragment of the amorphous structure showing an ABAB cube. Reprinted wih permission from Akkola and Jones [40]. Copyright 2007 by the American Physical Society

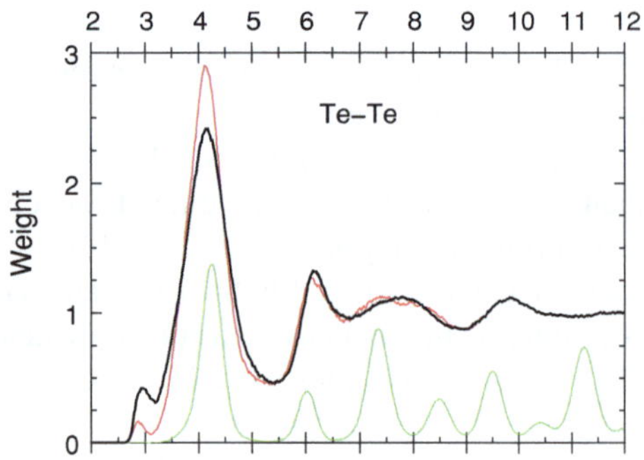

Fig. 9.19 Radial distribution functions of amorphous $Ge_2Sb_2Te_5$ and GeTe at 300 K. *Black*-a-$Ge_2Sb_2Te_5$, *red*-a-GeTe, *green*-$Ge_2Sb_2Te_5$ metastable crystal phase. Reprinted with permission from Akkola and Jones [40]. Copyright 2007 by the American Physical Society

to be ca. 1.3 nm for samples that appear entirely amorphous in electron diffraction measurements.

The authors have also studied the role of vacancies and voids, in the amorphous $Ge_2Sb_2Te_5$. The vacancies were found to be surrounded by Te atoms as in the crystalline form of GST and vacancies of different sizes have been detected (Fig. 9.20). In addition, long-range vacancy-vacancy correlations to at least 10 Å was observed, which is similar to the behaviour of Te atoms that are the preferred neighbours of vacancies. This is the first report of voids in the amorphous $Ge_2Sb_2Te_5$. Vacancies in GST were also studied in [64].

The rapid amorphous-to-crystalline transition is viewed by the authors as a vacancy-supported reorientation of ABAB squares.

In this work, the vibrational densities of states have also been calculated for both crystalline and amorphous phases and pronounced differences were observed between the spectra for the crystalline and amorphous states.

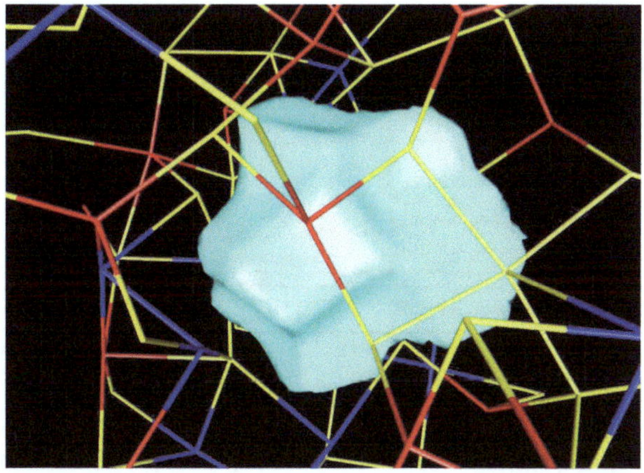

Fig. 9.20 A medium size cavity in amorphous $Ge_2Sb_2Te_5$. Reprinted wih permission from Akola and Jones [40]. Copyright 2007 by the American Physical Society

It is of interest to note that while most of the Ge atoms are not tetrahedral, the authors consider a change from sp^3 bonding to p^3 bonding to account for the differences between the crystalline and amorphous phases writing that "although most Ge atoms are not tetrahedral, the overall differences between the amorphous and ordered phases can be understood in terms of a change from sp^3 bonding (low coordination) to p^3 bonding (high coordination)" [40].

The same authors have subsequently studied the $Ge_8Sb_2Te_{11}$ composition used in Blu-ray discs. They found the total coordination numbers to be Ge:4.0, Sb:3.7 and Te:2.9, and the Ge–Ge partial coordination number is 0.7. Most atoms (particularly Sb) prefer octahedral coordination but 42% of Ge atoms are tetrahedral. Structural details, including ring statistics, local coordination and cavities were found to be very to those of $Ge_2Sb_2Te_5$ [65].

Hegedüs and Elliott Model

Hugedüs and Elliott [41] have performed a comprehensive study of the phase change in $Ge_2Sb_2Te_5$ thoughout the phase-change cycle. Starting with the liquid phase, they have found existence of four-membered rings (4-rings), analogous to ABAB blocks of Ref. [40], even at temperatures as high as 1073 K. The concentration of the 4 rings increases as the temperature goes down and this increase was found to be correlated with the size of the maximal cluster of *connected square rings*.

As mentioned earlier, the determination of coordination numbers in a structure where the nearest neighbours are located at different distances is not straightforward and may be influenced by the choice of the cut-off distance [66]. In order to avoid ambiguity associated with the choice the cut-off distance when determining the coordination number, Hegedüs and Elliott have chosen to investigate the evolution of interatomic distances around Ge atoms during slow and fast cooling of the melt (Fig. 9.21). Such an approach allows one to see if the number of the first nearest

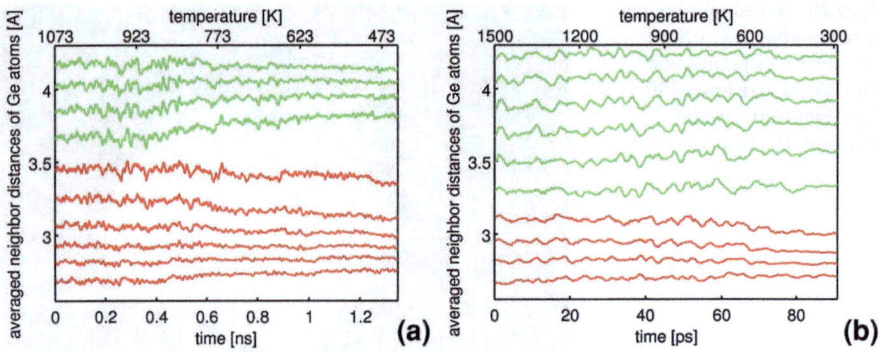

Fig. 9.21 Evolution of the interatomic distances around Ge atoms during slow (*left*) and rapid (*right*) quenches from the melt, eventually leading to crystalline and amorphous phases, respectively. Reprinted by permission from Macmillan Publishers Ltd.: Nature Mater. [41], copyright (2005)

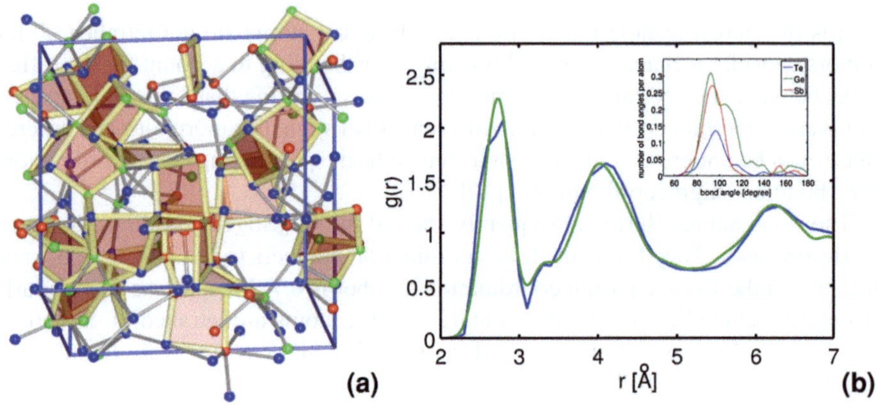

Fig. 9.22 Rapidly quenched amorphous model of $Ge_2Sb_2Te_5$ with a high concentration of crystal seeds in the form of connected near-regular four-fold rings (*left*) and the experimantal (*green*) and simulated (*blue*) PDF of the 90-atom relaxed amorphous model (*right*). The insert to the right panel shows the bond-angle distribution function. Reprinted with permission from Macmillan Publishers Ltd.: Nature Mater. [41], copyright (2005)

neighbours that tend to a similar value of the interatomic distance are six and four for the crystalline and amorphous phases, respectively. At the same time, it should be noted that even for the crystalline phase there are pronounced differences among various bond lengths within the six first nearest neighbours suggesting a rather large degree of structural disorder.

Their melt-quenched amorphous phase also contains a high density of 4-ring crystal seeds that seem to be homogeneously distributed (Fig. 9.22). The bond angle distribution functions obtained for this model demonstrate the existence of both octahedrally and tetrahedrally coordinated Ge atoms while for Sb atoms the bond angle distribution has a single peak at 90°.

The authors further found that the "melt-quenched" structure transforms into the rocksalt structure upon annealing *in silico* and the phase-change adol mechanism suggested by the authors consists of the ordering/disordering of the 4-rings.

9.2.3 Ab Initio XANES Studies

The results obtained using the DFT simulations described above clearly demonstrate that not all Ge atoms acquire tetrahedral symmetry positions in the amorphous phase. This could be either due to the fact that not all Ge atoms switch, or that the amorphisation mechanism is different from the umbrella flip. At the same time, none of those models have been tested against experimental XANES data which is one of the very few 3D-sensitive techniques that can be used for amorphous materials while such a comparison is crucial for validation of any suggested model.

To close this gap, Krbal et al. [39] performed detailed Ge K-edge XANES analyses on the MQ amorphous model by Hegedüs an Elliott [41] that contains 90 atoms in the simulation cell, including 20 Ge atoms. Thanks to the 3D sensitivity of XANES independent of the presence of long-range order, this technique allows one to discriminate between various existing models [17]. The results of the XANES calculations for this model are shown in Fig. 9.23. While XANES spectra calculated for each of the 20 Ge atoms in the cluster are all different, as expected for an amorphous material, the average XANES spectrum is in excellent agreement with experiment (middle panel of Fig. 9.23b).

In order to determine which atomic configurations are crucial to reproduce the experimentally measured XANES spectra, the authors have investigated the correlation between the local structure around each Ge atom and the corresponding XANES spectrum. An analysis of the amorphous model of GST [41] revealed that the local structure around Ge atoms can be grouped into three characteristic groups with respect to the first nearest neighbour geometry. First, there are purely tetrahedral (T_d) configurations (seven sites) with bond lengths typically around 2.58 Å, in agreement with the experimental EXAFS results.

Another characteristic group (seven sites) can be described as pyramids (P_y), with a Ge atom at the apex and with Te–Ge–Te angles very close to 90°. The Ge–Te bond lengths in this configuration are in the range 2.6–2.8 Å, with (one or two) next nearest neighbours located at distances around 3.0–3.5 Å. It should be stressed that the bond angles for the longer bonds deviate from 90°. This configuration can alternatively be referred to as a "four-coordinated defective octahedral site" [58] but we believe (as follows from the next paragraph) that the pyramidal description is more appropriate.

It should be noted here that the coordination number used in previous DFT studies to discuss the structure of the amorphous phase tacitly assumed that there are covalent bonds between the central atom and its neighbours up to a specific cut-off distance, usually around 3.2 Å. At the same time, the simple occurrence of a neighbour at a certain distance does not necessarily mean that there is a covalent bond between the atoms. In an amorphous solid, there is a continuous spectrum of interatomic

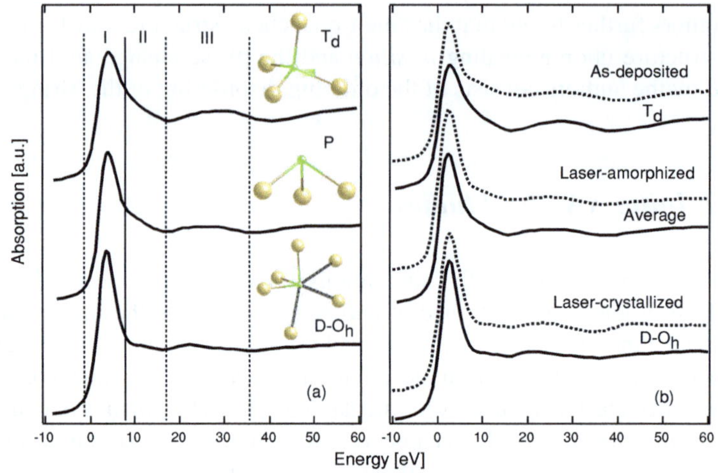

Fig. 9.23 **a** Simulated (averaged) Ge K-edge XANES spectra for the three main configurations found in the computer-generated amorphous GST model, namely, tetrahedral sites (T_d sites), pyramids (P) and distorted octahedra (D–O_h sites). The typical local configurations of the analysed groups are depicted in the top right-hand corner of the respective plot; Ge atoms are drawn in *green* and the larger Te atoms are drawn in gold. **b** Comparison of the experimental spectra for as-deposited and MQ amorphous and laser-recrystallised layers with the simulated XANES spectra of characteristic configurations. ("Average" refers to the averaged spectrum taken over all Ge sites.) The energy origin is set at the Ge K-edge. Reprinted with permission from Krbal et al. [39]. Copyright 2011 by the American Physical Society

distances and an appropriate selection of the cut-off distance is very important in order to determine properly the coordination numbers. Isosurface plots of the charge difference between the simulated MQ amorphous model and isolated pseudo-atoms clearly demonstrate that the electron density that can be associated with a covalent bond is only significant between atoms whose distance is less than 3 Å. This characteristic can be seen in Fig. 9.24 (left) which shows a Ge atom tetrahedrally coordinated by four Te atoms at a distance of approximately 2.57 Å; there is a significant pileup of bonding charge located midway along the bonds for each of the four neighbours. In contrast, the apparently fivefold coordinated fragment shown in Fig. 9.24 (right) *only* displays bonding isosurfaces for the *three* bond lengths less than 3 Å, implying an absence of significant bonding for atoms located at distances >3 Å.

These results suggest that when determining the coordination number, which tacitly implies the number of *covalent bonds*, only atoms located at distances less than ca. 3 Å should be considered. To include atoms that are located at distances just exceeding 3 Å that may interact with the central Ge atom, a notation Ge(3+n):Te(3) may be used which also includes the distorted octahedral configuration O_h, although the latter may possess different electronic properties due to "resonance" bonding. An alternative way to account for the fact that atoms located at shorter and longer distances contribute differently to bonding is to use the concept of effective coordination number [66], where an effective weight depending on the interatomic distance

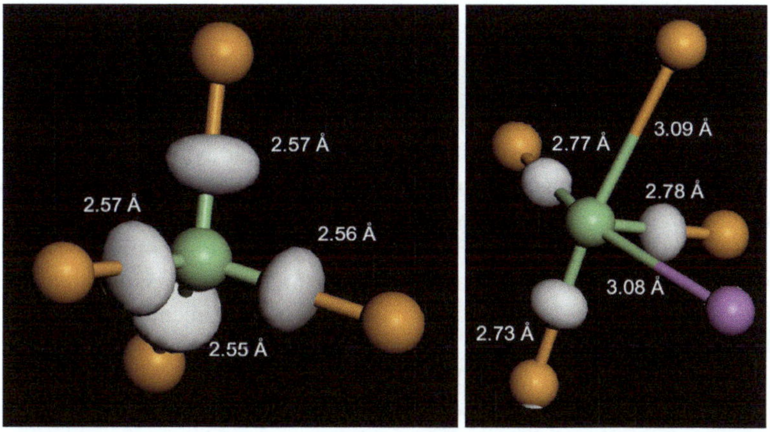

Fig. 9.24 Isosurface plots (*grey regions*) of the difference in electron charge density for the simulated amorphous model of Ref. [41] and isolated pseudo-atoms. Significant bonding charge can be seen midway along all four bonds of the tetrahedrally coordinated Ge atom (*green*) shown on the left, while for the fragment shown on the right, bonding charge isosurfaces are only present for bonds of length less than ~3 Å. Te (Sb) atoms are indicated as gold (*purple*). Reprinted with permission from Krbal et al. [39]. Copyright 2011 by the American Physical Society

is incorporated into the calculation of the coordination number. Considering the covalent backbone structure, the local Ge:Te coordination in the amorphous phase can be presented as a mixture of Ge(4):Te(2) and Ge(3):Te(3), and these T_d and P_y configurations form the main structural motifs of the covalently bonded amorphous phase.

Finally, there is a third group (six sites) that can be described as highly distorted octahedral (O_h) structures. In this group, the shorter Ge–Te distances are also in the range 2.6–2.8 Å, and typically two or three next nearest neighbours. While the further neighbours in this group are also located at distances around 3.0–3.5 Å similar to the previous group, *all* bond angles—not just those for the shorter bonds—are close to 90° which is indicative of strong resonance bonding.

The simulated (averaged) XANES spectra for these three types of configurations are shown in Fig. 9.23a. The authors consider three characteristic regions in the spectra that show pronounced differences between the amorphous and crystalline states. One is the so-called white line (I) that is higher in the crystalline state, another is a characteristic shoulder above the white line (II) that is more pronounced in the amorphous state, and the third is the location of the first feature above the white line (III) that is located at a higher energy in the crystalline phase. It can be seen that the spectrum for tetrahedrally coordinated Ge is quite similar to the experimentally observed spectra for as-deposited amorphous films (top panel, Fig. 9.23b).

At the same time, the authors found that the average XANES spectrum of the pyramidal configurations (middle panel, Fig. 9.23a) also reproduces the overall features of XANES for amorphous $Ge_2Sb_2Te_5$, namely, the white line becomes lower than for the crystalline phase, the shoulder II has a steeper slope and the peak in energy range III is shifted to larger energies compared to the crystalline sample. The observed similarity between the T_d and P_y spectra (the lower white line and the steeper shoulder in region II) is thus characteristic of a covalently—as opposed to resonantly—bonded structure. This result provides clear evidence that the observation of Ge sites with octahedral bonding angles through DFT simulations is also in agreement with the experimental XANES results [40, 41, 58, 64].

The XANES spectrum for octahedrally bonded Ge atoms (bottom panel) is characterised by a higher white line and a shallower slope at energies immediately above the white line (region II) and is rather similar to the experimentally measured spectrum for a laser recrystallised film (bottom panel, Fig. 9.23b).

A comparison of the simulated XANES spectra with the experimentally measured spectra for as-deposited amorphous, laser-amorphised and laser-crystallised $Ge_2Sb_2Te_5$ layers (Fig. 9.23b) led the authors to several important observations. First, the spectrum of as-deposited amorphous $Ge_2Sb_2Te_5$ is rather well approximated by the calculated spectrum for T_d Ge atoms in terms of the steeper slope in region II and both peak locations and intensities, in particular, in energy region III, suggesting that Ge atoms in as-deposited films predominantly have their usual sp^3 electron configuration with four nearest neighbours. This deduction is in agreement with some of the conclusions drawn previously from experimental XAFS and Raman measurements [27, 30, 49] as well as with recent ab initio simulations [45].

At the same time, the experimental XANES spectrum for the laser-amorphised layer exhibits a stronger white line (I), a stronger shoulder (II) and the higher energy feature (III) is located at a slightly different energy. This experimental spectrum is well reproduced (middle panel, Fig. 9.23b) by a mixture of the three above-mentioned configurations (T_d, P_y, and O_h) found in the model.

Ge(4):Te(2) and Ge(3):Te(3) Local Bonding Configurations and the 8–N Rule

The presence of T_d coordinated Ge atoms and twofold coordianted Te atoms in Ge(4):Te(2) configurations in the amorphous phase is not surprising considering the usual sp^3 electronic configuration for Ge and was predicted by the umbrella-flip model. The most intriguing question is what makes the group-IV element Ge form P_y configurations in the MQ amorphous phase since it would appear to be energetically unfavourable for sp^3-hybridised Ge atoms. The answer to this question derives from the definition of a covalent bond that is characterised by the sharing of a pair of electrons by the two bonded atoms (see e.g. [67]). Germanium has the $s^2p_x^1p_y^1p_z^0$ electronic configuration of its outer shell. In most materials, the orbitals hybridise and the electronic configuration changes to sp^3, allowing for the formation of four equivalent covalent bonds.

A rather special situation is created when Ge is bonded to a chalcogen. The electronic configuration of Te is $(s^2)p_x^1p_y^1p_z^2$, i.e. there are two unpaired *p* electrons and a non-bonding lone pair in the outer shell. The Te lone-pair *p* electrons have

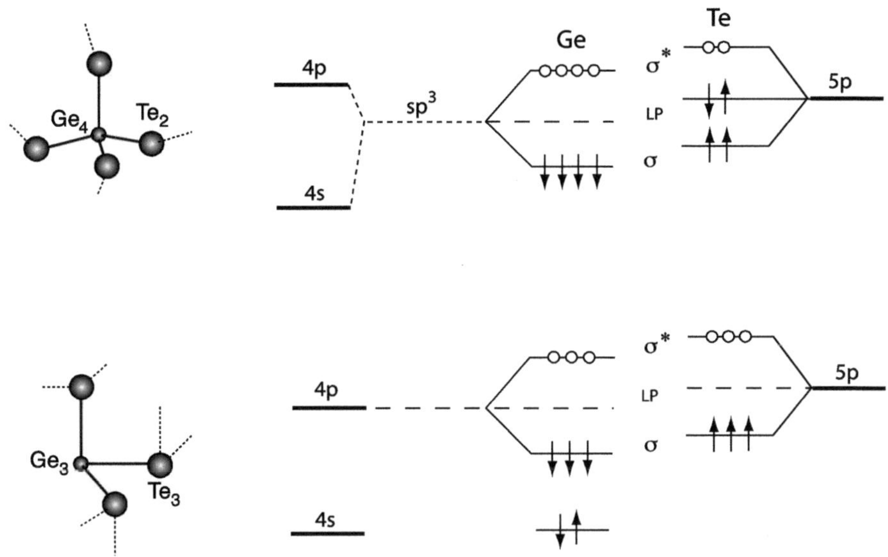

Fig. 9.25 Bonding schemes for Ge(4):Te(2) and Ge(3):Te(3) configurations and the corresponding energy diagrams (LP- lone-pair level taken as zero, σ corresponds to the bonding and σ^* to the anti-bonding states) for different atomic configurations. Note that the energy scales for the Ge and Te atoms are different

the option to interact with the empty *p*-orbital of an unhybridised Ge atom, forming a two-electron dative bond, which, once formed, is indistinguishable from usual covalent bonds [68]. A similar approach to explain the existence of Ge(3):Te(3) configurations was used in [69] where the authors also consider the formation of Ge–Te bonds that involve lone-pair electrons of Te.

Ge(4):Te(2) and Ge(3):Te(3) configurations are shown in Fig. 9.25 together with the corresponding energy diagrams. The comparison of the energy diagrams suggest that the Ge(4):Te(2) configuration is locally more stable which is also suggested by the shorter Ge–Te bonds in the latter despite a larger coordination number.

It is this very special combination of empty Ge p-orbitals and Te lone-pair electrons that allow for the creation of a *covalently bonded* network with coordination numbers that apparently deviate from the 8–N rule. We would like to note here that the original Mott rule, proposed to account for insensitivity of amorphous chalcogenides to doping [70], requires that atoms "use all their valence electrons in bonds with surrounding atoms or ions, modifying the local coordination to make this possible" and thus essentially *only* requires that all constituent elements have all saturated bonds, i.e. the outer shells are complete. From this perspective, both the Ge(3):Te(3) and Ge(4):Te(2) configurations described above satisfy the requirements imposed by the Mott rule for ideal covalent glass networks.

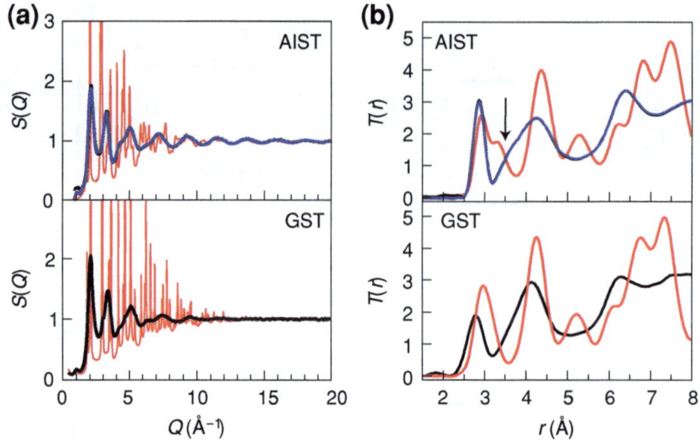

Fig. 9.26 Comparison of structure factors $S(Q)$ and total correlation functions $T(r)$ of AIST and GST. *Red lines* indicate experimental data for the crystalline phase, *black lines* correspond to experimental data for the amorphous phase, *blue lines* correspond to DF-RMC model. The DF-RMC and experimental results are practically indistinguishable. Reprinted by permission from Macmillan Publishers Ltd.: Nature Mater. [71], copyright (2011)

9.2.4 Ag-In-Sb–Te Alloys

Information about amorphous AIST alloys is rather limited. An early EXAFS study [23] suggested that in this class of materials the bond lengths are also shorter in the amorphous phase. Subsequently, Matsunaga et al. [71] performed a comprehensive study of amorphous AIST combining experimental (X-ray diffraction, high-resolution XPS, and EXAFS) and theoretical (density functional (DF) molecular dynamics simulations combined with reverse Monte Carlo (RMC) method) approaches. These results reveal that the bond shortening in the amorphous phase is significantly smaller in AIST than in GST (Fig. 9.26) suggesting that the structural rearrangement during the phase transition is also smaller.

The authors conclude that the amorphisation process of AIST consists of disordering of spatial orientations of the shorter and longer bonds present in the A7 structure of crystalline AIST (see Sect. 11.3 for more detail).

9.3 Liquid Phase

9.3.1 Experimental Studies

Scattering Experiments

The long standing consensus of the phase-change mechanism is that a short intense pulse melts the material that is subsequently quenched into the amorphous state. For this reason, knowledge of the structure and properties of the molten phase is crucial for better understanding the phase-change process. Certain similarities between structures of the two phases have been experimentally observed by Kohara et al. [35] using high-energy X-rays (cf. Fig. 9.6). A visual examination of the figure suggests that the structure of the liquid phase is essentially similar to that of the amorphous phase but possesses a larger degree of disorder. No further details of the structure of the liquid phase were reported.

The structure of liquid GST alloy of various compositions have been studied by neutron scattering [72]. The obtained total structure factors are displayed in Fig. 9.27. For a better comparison, the authors used normalised data according to $S_{\text{norm}}(q/q_1) = [S(q/q_1) - 1]/S(q/q_{\text{max}}) - 1] + 1$, where q_1 is the momentum transfer of the first maximum or shoulder and q_{max} the position of the overall maximum of $S(q)$. The analysis is based on establishing a relationship between the relative amplitudes and positions of the first and second peaks in the structure factor with the average number of sp valence electrons. Shown in the figure are the studied Te-alloys and also $GeSe_2$ that form a tetrahedral liquid as well as $GeTe_6$ that is argued to possess distorted octahedral order in the liquid state. We would like to note in passing that the attribution of the local structure of liquid $GeTe_6$ to octahedral in the cited work [73] is based on the coordination number (that varies between 2.65 and 3.36 as the cut-off distance changes from 3.17 to 3.31 Å) obtained from (1D) pair distribution function and on analogy with the Peierls distorted low-temperature crystalline GeTe, the fact that may be insufficient to justify the conclusion about the octahedral bonding geometry of liquid $GeTe_6$ since nothing is known about the number and spatial location of the next nearest neighbours. The description of this local geometry as pyramidal would seem to be more accurate.

The authors conclude that as the average number of the valence electrons decreases, the second peak becomes more intense and the distance between the first and second peaks becomes larger. The obtained results can be grouped into two groups: one with $S = 0.8$ and $r_2/r_1 = 1.4$ and the other one with $S = 1.3$ and $r_2/r_1 = 1.6$. The boundary between the two groups is at the average number of valence electrons of around 4.2. From these trends the authors conclude that all phase-change alloys possess octahedral geometry in the liquid state. At the same time, they add that "the actual bonding mechanism can only be understood through atomistic modelling". The coordination numbers are not reported in this work but they are reported in a different publication [74] by (almost) the same authors although the details of the analysis are not given. The average coordination number in liquid

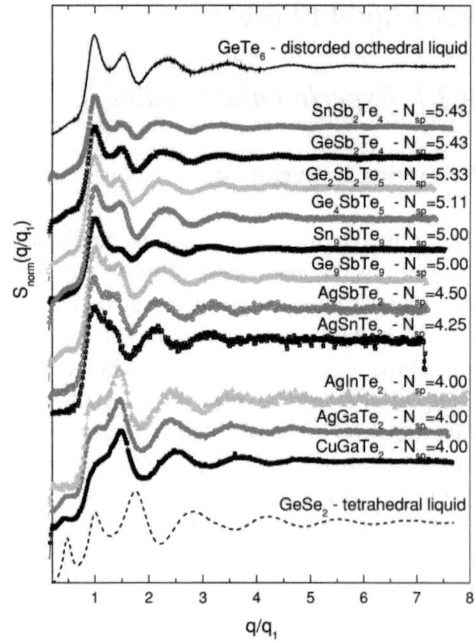

Fig. 9.27 Normalised structure factors $S_{\text{norm}}(q/q_1)$ as a function of the average number of valence electrons N_{sp} increasing from bottom to top. Reproduced from Steimer et al. [72] with permission by Wiley–VCH Verlag GmbH

GST was found to be about 3.2–3.5. As with the liquid $GeTe_6$ it is not obvious from this result that the local geometry can be described as octahedral since nothing is known about the number and spatial location of the remaining three neigbours. A description of this bonding geometry and pyramidal, or $3+n$, may be more accurate.

X-Ray Absorption studies

The local structure of the liquid phase was investigated using EXAFS [75]. Special care was taken to avoid irreversible changes due to oxidation and/or Te sublimation during melting. From the EXAFS analysis, it was concluded that Ge–Te (2.64 Å) and Sb–Te (2.85 Å) bonds are dominent in the liquid phase. Additionally, Ge–Sb correlations (3.04 Å) were also detected. An important point that should be noted is that the obtained Ge–Te and Sb–Te bond lengths are similar to those in the amorphous phase implying the bonds in the liquid phase preserve their covalent nature and hence suggesting that the liquid phase is semiconducting. This conclusion is in agreement with the conclusion drawn from electrical conductivity measurements of liquid GST (cf. Sect. 2.1.6) when the observed temperature dependence of conductivity also suggested semiconductor nature of the molten phase [76].

XANES spectra for the liquid and amorphous phases are directly compared in Fig. 9.28. One can see that while the features above the white line that have structural origin are essentially identical, the white-line intensity is different for the two cases,

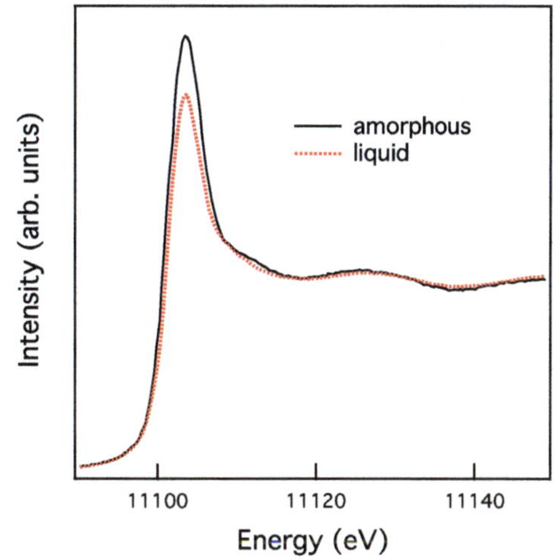

Fig. 9.28 Comparison of XANES spectra of $Ge_2Sb_2Te_5$ in the liquid and amorphous phases. Reprinted with permission from Kolobov et al. [75]. Copyright 2009 by the American institute of Physics

which likely reflects a lower density of unoccupied states in the liquid phase where thermal excitation of charge carriers across the rather narrow bandgap is efficient.

9.3.2 Computer Simulations

Binary GeTe: Reentrant Peierls Distortion

Since analysis of the experimental data or computer modelling for a ternary liquid is rather complicated, it may be informative to start with the binary GeTe alloy that possesses a very similar (distorted rock salt) structure in the crystalline state. Comprehensive studies of the liquid GeTe have been performed by Raty et al. [77, 78]. One interesting aspect of the results is that the obtained coordination numbers for the liquid phase vary with temperature from 3.90 to 5.03 suggesting that the structure of the liquid may be strongly temperature dependent. This result is also reminiscent of the melting of elemental Te. Just above the melting point Te preserves its semiconducting properties but as the temperature increases further it becomes a metal, indicating an increase in the coordination number usually associated with the semiconductor-metal transition upon melting of materials such as Si, Ge, and most III-V semiconductors [79].

A very interesting finding was an observation of local distortions in the liquid phase called "reentrant Peierls distortions" [77, 80]. At the same time, we would like to note that the disappearance of the local distortion in the high-temperature solid phase of GeTe was concluded based on Bragg diffraction that is not sensitive to stochastic local distortions (cf. Sect. 3.1.1). Subsequent XAFS and pair-distribution function analyses provided evidence that the distortions remain even in the high-temperature paralelectric phase [81, 82]. The fact that the shorter Ge–Te bonds

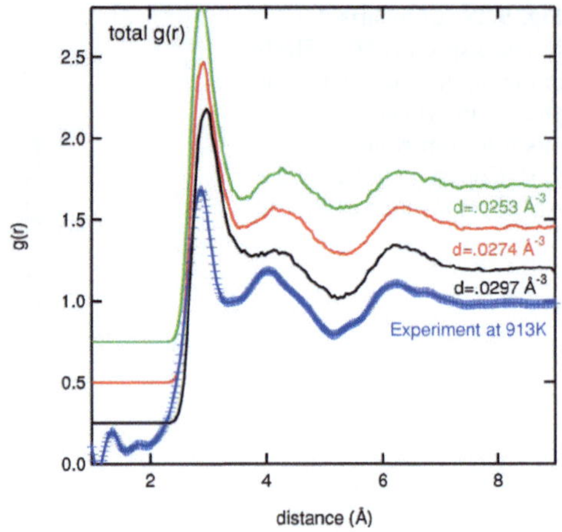

Fig. 9.29 Comparison of the total pair correlation function of $GeSb_2Te_4$ calculated at 0.0297 (*black*), 0.0274 (*red*) and 0.0253 (*green*) $Å^{-3}$ with the one measured experimentally at 973 K. Reprinted with permission from Bichara et al. [83]. Copyright 2007 by the American Physical Society

could be detected in the liquid phase is likely related to the fact that total scattering (and not Bragg diffraction) approach was used to characterise the latter. We propose that the term "revisualised Peierls distortion" may be more appropriate.

Liquid GST Alloys

The results of ab initio simulations [83] performed for $GeSb_2Te_4$ are shown in Fig. 9.29. The authors found that *all* atoms have an octahedral coordination, characterised by bond angle distributions peaked around 90–95°. We would like to reiterate here that while in the crystalline phase the coordination number of three may be identical to that of an octahedral bonding geometry, in the amorphous or liquid phases three first nearest neighbours with the bond angles close to 90° are more appropriately described as pyramidal configurations.

In a subsequent work [84] the predominantly defective octahedral structure of $GeSb_2Te_4$ has been confirmed that is "similar to the crystalline form on average but with important fluctuations round the mean". Most Sb atoms are found to be octahedrally coordinated while there are coexisting tetrahedral and fivefold coordination at octahedral sites for the Ge atoms, the Te atoms essentially being four-fold and threefold coordinated at octahedral sites with contributions from twofold, fivefold and sixfold coordination.

Of special interest is an observation of the unique melting behaviour of GST in that atoms form two-dimensional linear and entangled clusters while only vibrating around their original configurations in the perpendicular direction. The authors propose that "this unique melting behavior is responsible for the fast and reversible phase transition between amorphous and crystalline phases" [85]. A correlation in atomic motion as evidenced by "planes" found for atom trajectories at 973 K on a timescale of 20 ps was also reported in [41]. This "layering" of the liquid phase is reminiscent of a molecular crystal when upon melting weak van-der-Waals interactions break while the stronger covalent backbone persists. It is also of interest that

Hegedüs and Elliott [41] found the existence of four-membered rings even in the liquid phase.

Liquid AIST Alloys

Liquid AIST has been studied in [86]. The authors found that (a) there are signs of medium-range order, (b) Ag and In prefer to be located near Te rather than Sb atoms, and promote octahedral coordination in association with Te, (c) Ag is the most mobile element, has the shortest bonds (2.8–2.9 Å) and the highest coordination (5.5) and is anionic with the largest effective charge, and (d) there are few cavities (4 vol.%).

References

1. K. Tanaka, in *Handbook of Advanced Electronic and Photonic Materials*, vol. 5, ed. by H.S. Nalwa (Academic Press, New York, 2001), p. 120
2. W. Welnic, in *Phase Change Materials: Science and Applications*, ed. by S. Raoux M. Wuttig, (Springer Science + Business Media, Berlin, 2009), p. 17
3. F. Betts, A. Bienenstock, S.R. Ovshinsky, J. Non-Cryst, Solids **4**, 554 (1970)
4. D. Dove, M. Heritage, K. Chopra, S. Bahl, Appl. Phys. Lett. **16**, 138 (1970)
5. O. Uemura, Y. Sagara, M. Tsushima, T. Kamikawa, T. Satoh, J. Non-Cryst, Solids **33**, 71 (1979)
6. M. Wakagi, Y. Maeda, Phys. Rev. B **50**, 14090 (1994)
7. A.V. Kolobov, P. Fons, J. Tominaga, A.L. Ankudinov, S.N. Yannopoulos, K.S. Andrikopoulos, J. Phys. Condens. Matter **16**, S5103 (2004)
8. K. Hirota, K. Nagino, G. Ohbayashi, J. Appl. Phys. **82**, 65 (1997)
9. G. Fisher, J. Tauc, Y. Verhelle, in *Amorphous and Liquid Semiconductors*, ed. by J. Stuke, W. Brenig (Taylor and Francis, London, 1974), p. 1259
10. S. Hosokawa, Y. Hari, T. Kouchi, I. Ono, H. Sato, M. Taniguchi, A. Hiraya, Y. Takata, N. Kosugi, M. Watanabe, J. Phys. Condens. Matter **10**, 1931 (1998)
11. S.J. Pickart, Y.P. Sharma, J.P. de Neufville, J. Non-Cryst, Solids **34**, 183 (1979)
12. P. Boolchand, B.B. Triplett, S.S. Hanna, J.P. de Neufville, in *Mössbauer Effect Methodology*, vol. 9, ed. by J. Gruverman, C.W. Seidel, D.K. Dieterly (Plenum Press, New York, 1974), p. 53
13. A.V. Kolobov, M. Krbal, P. Fons, J. Tominaga, T. Uruga, Nature Chem. **3**, 311 (2011)
14. B. Hyot, X. Biquard, L. Poupinet, in *European Phase Change and Ovonics Science Symposium*, Säntis, 2001
15. A.V. Kolobov, P. Fons, A.I. Frenkel, A.L. Ankudinov, J. Tominaga, T. Uruga, Nature Mater. **3**, 703 (2004)
16. http://www.webelements.com
17. G. Smolentsev, A.V. Soldatov, M.C. Feiters, Phys. Rev. B **75**, 144106 (2007)
18. G. Smolentsev, A.V. Soldatov, M.C. Feiters, AIP Conf. Proc. **882**, 83 (2007)
19. G. Lucovsky, R. White, Phys. Rev. B **8**, 660 (1973)
20. K. Shportko, S. Kremers, M. Woda, D. Lencer, J. Robertson, M. Wuttig, Nature Mater. **7**, 653 (2008)
21. A.V. Kolobov, H. Oyanagi, Ka. Tanaka, K. Tanaka, Phys. Rev. B **55**, 726 (1997)
22. D. Sayers, in *Proceedings of Seventh International Conference on Amorphous Liquid Semiconductors*, ed. by W. Spear (University of Edinburgh, CICL, 1977), p. 61
23. A.V. Kolobov, P. Fons, J. Tominaga, in *Chalcogenide Alloys for Reconfigurable Electronics*, ed. by P.C. Taylor, A.V. Kolobov, A.H. Edwards, J. Maimon, Materials Research Society Symposium Proceedings, vol. 918, 2006, p. 181
24. B. Huang, J. Robertson, Phys. Rev. B **81**, 081204R (2010)

25. A.L. Ankudinov, B. Ravel, J.J. Rehr, S.D. Conradson, Phys. Rev. B **58**, 7565 (1998)
26. A.V. Kolobov, P. Fons, J. Tominaga, A.I. Frenkel, A.L. Ankudinov, S.N. Yannopoulos, K.S. Andrikopoulos, T. Uruga, Jpn. J. Appl. Phys. **44**, 3345 (2005)
27. P. Jovari, I. Kaban, J. Steiner, B. Beuneu, A. Schops, M.A. Webb, Phys. Rev. B **77**, 035202 (2008)
28. D.A. Baker, M.A. Paesler, G. Lucovsky, P.C. Taylor, J. Non-Cryst. Solids **352**, 1621 (2006)
29. M.A. Paesler, D.A. Baker, G. Lucovsky, A.E. Edwards, P.C. Taylor, J. Phys. Chem. Solids **68**, 873 (2007)
30. D.A. Baker, M.A. Paesler, G. Lucovsky, S.C. Agarwal, P.C. Taylor, Phys. Rev. Lett. **96**, 255501 (2006)
31. D.A. Baker, AIP Conf. Proc. **882**, 19 (2007)
32. T. Arai, M. Sato, N. Umesaki, J. Phys. Condens. Matter **19**, 335213 (2007)
33. M.F. Thorpe, J. Non-Cryst, Solids **57**, 355 (1983)
34. M. Wuttig, N. Yamada, Nature Mater. **6**, 824 (2007)
35. S. Kohara, K. Kato, S. Kimura, H. Tanaka, T. Usuki, K. Suzuya, H. Tanaka, Y. Moritomo, T. Matsunaga, N. Yamada, Y. Tanaka, H. Suematsu, M. Takata, Appl. Phys. Lett. **89**, 201910 (2006)
36. J. Akola, R.O. Jones, S. Kohara, S. Kimura, K. Kobayashi, M. Takata, T. Matsunaga, R. Kojima, N. Yamada, Phys. Rev. B **80**, 020201 (2009)
37. P.K. Khulbe, T. Hurst, M. Horie, M. Mansuripur, Appl. Opt. **41**, 6220 (2002)
38. S. Raoux, H.Y. Cheng, M.A. Caldwell, H.S.P. Wong, Appl. Phys. Lett. **95**, 071910 (2009)
39. M. Krbal, A.V. Kolobov, P. Fons, J. Tominaga, S.R. Elliott, J. Hegedus, T. Uruga, Phys. Rev. B **83**, 054203 (2011)
40. J. Akola, R.O. Jones, Phys. Rev. B **76**, 235201 (2007)
41. J. Hegedüs, S.R. Elliott, Nature Mater. **7**, 399 (2008)
42. B.S. Lee, G.W. Burr, R.M. Shelby, S. Raoux, C.T. Rettner, S.N. Bogle, K. Darmawikarta, S.G. Bishop, J.R. Abelson, Science **326**, 980 (2009)
43. G. Zhou, H. Borg, J. Rijper, M. Lankhorst, J. Honkx, Proc. SPIE **4090**, 108 (2000)
44. M. Naito, M. Ishimaru, Y. Hirotsu, M. Takashima, J. Appl. Phys. **95**, 8130 (2004)
45. J. Akola, J. Larrucea, R.O. Jones, Phys. Rev. B **83**, 094113 (2011)
46. M. Krbal, A.V. Kolobov, J. Haines, P. Fons, C. Levelut, R. Le Parc, M. Hanfland, J. Tominaga, A. Pradel, M. Ribes, Phys. Rev. Lett. **103**, 115502 (2009)
47. R. De Bastiani, E. Carria, S. Gibilisco, A. Mio, C. Bongiorno, F. Piccinelli, M. Bettinelli, A. Pennisi, M. Grimaldi, E. Rimini, J. Appl. Phys. **107**, 113521 (2010)
48. R. De Bastiani, A.M. Piro, M.G. Grimaldi, E. Rimini, G.A. Baratta, G. Strazzulla, Appl. Phys. Lett. **92**, 241925 (2008)
49. K.S. Andrikopoulos, S.N. Yannopoulos, A.V. Kolobov, P. Fons, J. Tominaga, J. Phys. Chem. Solids **68**, 1074 (2007)
50. R. Mazzarello, S. Caravati, S. Angioletti-Uberti, M. Bernasconi, M. Parrinello, Phys. Rev. Lett. **104**, 085503 (2010)
51. G.C. Sosso, S. Caravati, R. Mazzarello, M. Bernasconi, Phys. Rev. B **83**, 134201 (2011)
52. W. Welnic, A. Pamungkas, R. Detemple, C. Steimer, S. Blugel, M. Wuttig, Nature Mater. **5**, 56 (2005)
53. J. Robertson, K. Xiong, P.W. Peacock, Thin Solid Films **515**, 7538 (2007)
54. C. Lang, S.A. Song, D.N. Manh, D.J.H. Cockayne, Phys. Rev. B **76**, 054101 (2007)
55. J. Kalb, F. Spaepen, M. Wuttig, Appl. Phys. Lett. **84**, 5240 (2004)
56. J.L.F. Da Silva, A. Walsh, S.H. Wei, H. Lee, J. Appl. Phys. **106**, 113509 (2009)
57. J. Eom, Y. Yoon, C. Park, H. Lee, J. Im, D. Suh, J. Noh, Y. Khang, J. Ihm, Phys. Rev. B **73**, 214202 (2006)
58. S. Caravati, M. Bernasconi, T. Kühne, M. Krack, M. Parrinello, Appl. Phys. Lett. **91**, 171906 (2007)
59. J. Akola, R.O. Jones, J. Phys. Condens. Matter **20**, 465103 (2008)
60. A.V. Kolobov, A.S. Mishchenko, P. Fons, S.M. Yakubenya, J. Tominaga, J. Phys. Condens. Matter **19**, 455209 (2007)

61. W.A. Harrison, *Electronic Structure and the Properties of Solids: The Physics of the Chemical Bond* (Dover Publications, New York, 1989)
62. S. Caravati, M. Bernasconi, T. Kühne, M. Krack, M. Parrinello, J. Phys. Condens. Matter **21**, 255501 (2009)
63. M.H. Kwon, B.S. Lee, S.N. Bogle, L.N. Nittala, S.G. Bishop, J.R. Abelson, S. Raoux, B. Cheong, K.B. Kim, Appl. Phys. Lett. **90**, 021923 (2007)
64. Z. Sun, J. Zhou, A. Blomqvist, B. Johansson, R. Ahuja, Phys. Rev. Lett. **102**, 075504 (2009)
65. J. Akola, R.O. Jones, Phys. Rev. B **79**, 134118 (2009)
66. J.L.F. Da Silva, J. Appl. Phys. **109**, 023502 (2011)
67. L. Pauling, *The Nature of the Chemical Bond and the Structure of Molecules and Crystals: An Introduction to Modern Structural Chemistry* (Cornell University Press, Ithaca, 1960)
68. P.O. Lowdin (ed.), *Advances in Quantum Chemistry*, vol. 6 (Academic Press, New York, 1972)
69. M. Xu, Y. Cheng, H. Sheng, E. Ma, Phys. Rev. Lett. **103**, 195502 (2009)
70. N.F. Mott, Adv. Phys. **16**, 49 (1967)
71. T. Matsunaga, J. Akola, S. Kohara, T. Honma, K. Kobayashi, E. Ikenaga, R.O. Jones, N. Yamada, M. Takata, R. Kojima, Nature Mater. **10**, 129 (2011)
72. C. Steimer, V. Coulet, W. Welnic, D. Henning, R. Detemple, C. Bichara, B. Beuneu, J. Gaspard, M. Wuttig, Adv. Mater. **20**, 4535 (2008)
73. C. Bergman, C. Bichara, J.P. Gaspard, Y. Tsuchiya, Phys. Rev. B **67**, 104202 (2003)
74. M. Delheusy, J. Raty, R. Detemple, W. Welnic, M. Wuttig, J. Gaspard, Physica B **350**, 1055 (2004)
75. A.V. Kolobov, P. Fons, M. Krbal, R.E. Simpson, S. Hosokawa, T. Uruga, H. Tanida, J. Tominaga, Appl. Phys. Lett. **95**, 241902 (2009)
76. H. Endo, Y. Jinnai, R. Lan, M. Kuwahara, Y. Kobayashi, M. Susa, in 20th Symp. Phase Change Optical Information Storage (PCOS'08), Izu, 2008, p. 53
77. J.Y. Raty, V.V. Godlevsky, J.P. Gaspard, C. Bichara, M. Bionducci, R. Bellissent, R. Ceolin, J.R. Chelikowsky, Phys. Rev. B **65**, 115205 (2002)
78. C. Bichara, M. Johnson, J. Raty, Phys. Rev. Lett. **95**, 267801 (2005)
79. V.M. Glazov, S.N. Chizhevskaya, N.N. Glagoleva, *Liquid Semiconductors* (Plenium Press, New York, 1969)
80. J. Raty, V. Godlevsky, P. Ghosez, C. Bichara, J.P. Gaspard, J.R. Chelikowsky, Phys. Rev. Lett. **85**, 1950 (2000)
81. P. Fons, A.V. Kolobov, M. Krbal, J. Tominaga, K.S. Andrikopoulos, S.N. Yannopoulos, G.A. Voyiatzis, T. Uruga, Phys. Rev. B **82**, 155209 (2010)
82. T. Matsunaga, P. Fons, A.V. Kolobov, J. Tominaga, N. Yamada, App. Phys. Lett. **99**, 231907 (2011)
83. C. Bichara, M. Johnson, J.P. Gaspard, Phys. Rev. B **75**, 060201R (2007)
84. Z. Sun, J. Zhou, A. Blomqvist, L. Xu, R. Ahuja, J. Phys. Condens. Matter **20**, 205102 (2008)
85. Z. Sun, J. Zhou, R. Ahuja, Phys. Rev. Lett. **98**, 055505 (2007)
86. J. Akola, R.O. Jones, Appl. Phys. Lett. **94**, 251905 (2009)

61. W.A. Harrison, *Electronic Structure and the Properties of Solids: The Physics of the Chemical Bond* (Dover Publications, New York, 1989)
62. S. Caravati, M. Bernasconi, T. Kühne, M. Krack, M. Parrinello, J. Phys. Condens. Matter 21, 255501 (2009)
63. M.H. Kwon, B.S. Lee, S.N. Bogle, L.N. Nittala, S.G. Bishop, J.R. Abelson, S. Raoux, B. Cheong, K.B. Kim, Appl. Phys. Lett. 90, 021923 (2007)
[illegible]

Chapter 10
Pressure-Induced Transformations

10.1 Ge-Sb-Te Alloys

10.1.1 Pressure-Induced Amorphisation

Composition Dependence

Since pressure is also an important thermodynamical parameter, it can be surmised to have a strong effect on phase stability. The present section describes the available experimental and theoretical results [1–8]. We start the discussion of pressure-induced modification with the metastable crystal (cubic) phase of GST that is relevant to phase-change memories. While the local structure of metastable GST is distorted with Ge and Sb atoms displaced from the centre of the cell [9–11], its long-range order is best described as having the rock salt structure with a large thermal factor and in what follows pseudo-cubic peak indexing is used.

Figure 10.1 shows the pressure-induced evolution of the observed XRD pattern for cubic $Ge_2Sb_2Te_5$ and $GeSb_2Te_4$ [1, 2]. One can see that in the pressure range up to ca. 15 GPa, compression results in a simple shift of all peaks to larger angles indicating a decrease in the unit cell volume under compression. At higher pressures, the intensity of the GST peaks in the metastable cubic phases decreases, the peaks broaden and eventually are replaced by broad amorphous bands. It should be noted that pressures at which $Ge_2Sb_2Te_5$ amorphises is lower in $GeSb_2Te_4$. This decrease in the amorphisation threshold correlates with the concentration of vacancies which suggests their important role in the amorphisation process [2].

Pressure-induced amorphisation is *irreversible at room temperature*, i.e. the initial "cubic" crystalline structure is not restored upon decompression. At the same time, the stable trigonal phase of $Ge_2Sb_2Te_5$ that does not contain vacancies remains crystalline in this pressure range (cf. Sect. 10.1.2).

The observed amorphisation is rather atypical. It can, for example, be noted that GeTe that possesses a similar local structure does not amorphise upon compression.

A. V. Kolobov and J. Tominaga, *Chalcogenides*, Springer Series in Materials Science 164, DOI: 10.1007/978-3-642-28705-3_10,

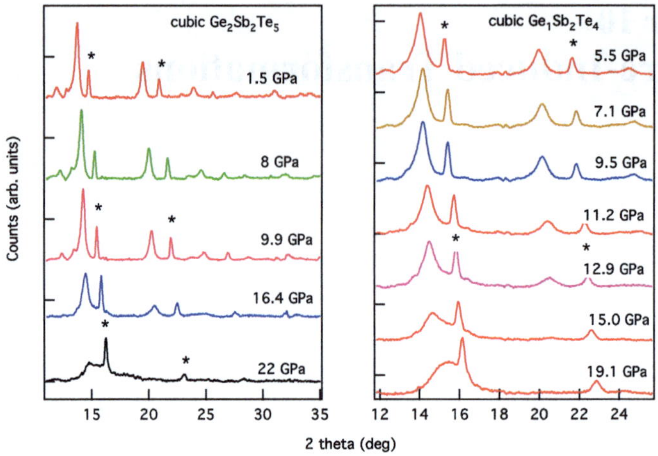

Fig. 10.1 Pressure-induced changes of the XRD pattern of $Ge_2Sb_2Te_5$ (*left*) and $GeSb_2Te_4$ (*right*) upon compression. The peaks marked by * (on some of the spectra) are the corresponding peaks of the NaCl marker. The pressures are shown at the right side of each plot. Reprinted with permission from Kolobov et al. [2]. Copyright 2007 by the American Institute of Physics

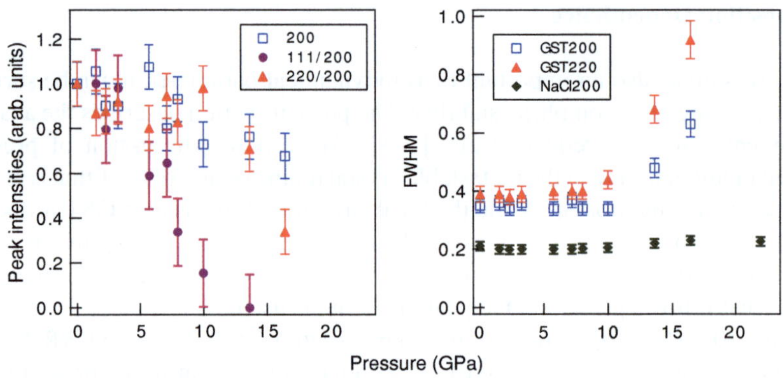

Fig. 10.2 Peak area (*left*) and peak FWHM (*right*) variation with pressure in $Ge_2Sb_2Te_5$. The unchanged NaCl peak width indicates that hydrostatic conditions are preserved throughout the pressure range used. Reprinted with permission from Kolobov et al. [1]. Copyright 2006 by the American Physical Society

Although pressure-induced amorphisation has been reported for numerous materials [12], it typically happens in corner-linked polyhedral compounds, hydrogen bonded materials and van-der-Waals solids. When a material with the compact rock salt structure (both ionically and/or covalently bonded) is subject to pressure it usually transforms to a CsCl-type structure [13, 14] since the latter possesses an increased packing density.

Even more unusual is the non-uniform decrease in the relative peak intensities with increasing pressure. Figure 10.2 shows the pressure dependences of the normalized

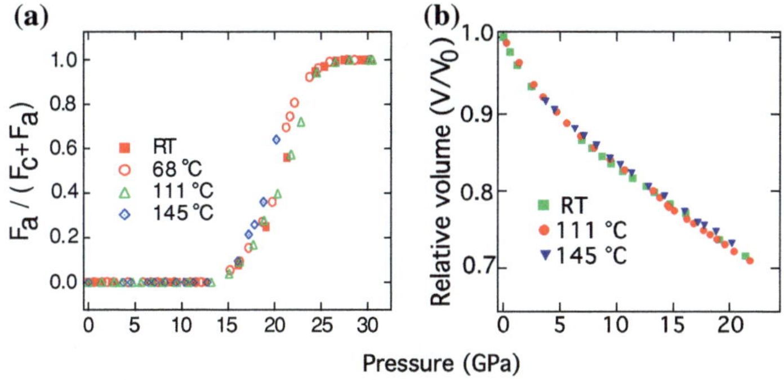

Fig. 10.3 Fraction of the amorphous phase as a function of pressure: the amorphous phase starts to grow at 14–15 GPa with the material becoming completely amorphous at 25 GPa *independent* of temperature (**a**). Pressure dependence of the normalised unit cell volume (**b**). The legends are indicated in the figure. Reprinted with permission from Krbal et al. [3]. Copyright 2008 by the American Institute of Physics

peak areas and full widths at half maximum (FWHM). The 111 peak disappears at pressures around 10 GPa. This is followed by the intensity decrease and subsequent disappearance of the 220 peak while the 200 peak disappears last. The width of the 200 and 220 peaks remains constant to about 10 GPa after which they rapidly broaden. At the same time, the NaCl pressure marker 200 peak width remains constant and narrow clearly demonstrating that the level of deviatoric stress remains low up to the maximum pressures used [1].

Temperature Dependence

In the temperature range from room temperature to ca. 150 °C, the behaviour of $Ge_2Sb_2Te_5$ upon compression is essentially identical. The cubic fcc phase starts to disappear and the amorphous phase grows from ca. 15 GPa. The amorphisation process continues until the crystalline fcc phase totally disappears at 25 GPa. The onset of the amorphisation and its completion *do not depend* on temperature in this temperature range (Fig. 10.3) [3].

The fact that the amorphisation pressure does not depend on the temperature (in the temperature range from room temparature to 145 °C) is of significant interest and indicates that the process is not limited by kinetics. This may be the major underlying reason for extreme speed of the amorphisation process in $Ge_2Sb_2Te_5$-based memory devices.

There are, however, substantial differences in the decompression behaviour. At lower temperatures (room temperature to 110 °C, the amorphous phase is stable upon decompression. At the same time, at 145 °C, i.e. just below the crystallisation temperature, the amorphous phase is stable up to (or, rather, down to) rather low

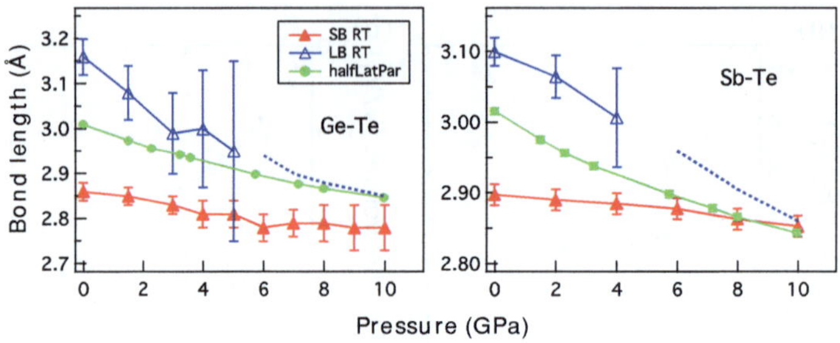

Fig. 10.4 Pressure-induced variations of the shorter (SB) and longer (LB) Ge–Te and Sb–Te bonds as well as the half-lattice parameter. The dotted lines show the variation of the longer bonds calculated based on the shorter bond lengths and the lattice parameter known from XRD. Reprinted with permission from Kolobov et al. [1]. Copyright 2006 by the American Physical Society

pressures but when the pressure is completely released the initial fcc phase is restored (mixed with a small fraction of the stable trigonal phase).

At temperatures close to the fcc-to-trigonal phase transition (220 °C in our experiment), when pressure is increased up to 5 GPa, the fcc phase is stable but at higher pressures the trigonal phase starts to form. This process is followed by the appearance of numerous new peaks suggesting decomposition of the material. Upon decompression, the trigonal form is obtained along with an additional phase which may correspond to an orthorhombic high-pressure form of GeTe [15].

Changes in the Local Structure

Extended X-ray absorption fine structure (EXAFS) measurements performed at Ge, Sb and Te K-edges provide insight into the structural modification on the local scale. The results (Fig. 10.4) show that the longer Ge(Sb)-Te bond lengths decrease faster than the half lattice parameter and at pressures around 10 GPa Sb occupies a position at the center of the cell while the shorter Ge–Te bonds remain somewhat shorter than half the lattice parameter [1].

Figure 10.5 shows XAFS spectra at Ge K-edge up to 25 GPa [2]. Because of the use of the diamond anvil cell in the experiment and strong diffraction glitches from the diamonds, the EXAFS range was limited to 7–8 Å^{-1} and conventional EXAFS analysis was complicated. The authors used an alternative approach analysing the position of the onset of EXAFS oscillations when the following relationship holds for the position of the onset of EXAFS and the first nearest neighbour distance [16]: $k \times R = \text{const}$, where k is the position of the onset of EXAFS oscillations and R is the first nearest neighbour distance. Knowing the Ge–Te distance at $P = 0$ GPa one can estimate the corresponding distance under pressure. The result is shown in Fig. 10.5 where variation of the Ge–Te distance upon compression and decompression is shown.

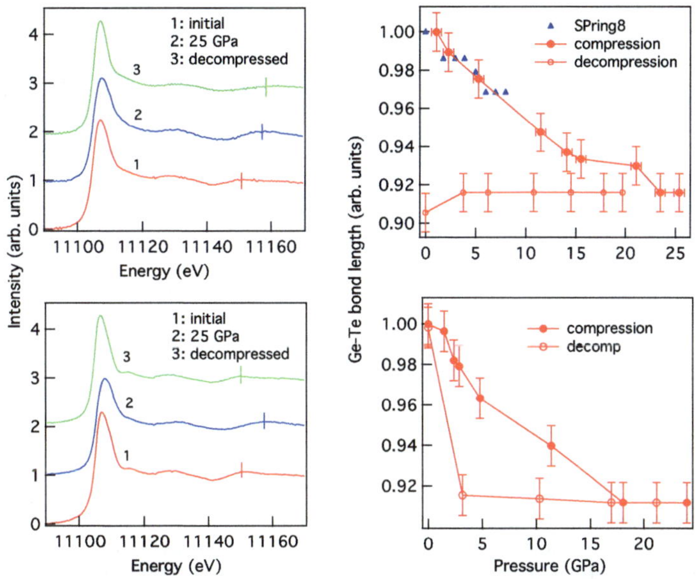

Fig. 10.5 Variations of XAFS spectra of cubic $Ge_2Sb_2Te_5$ at room temperature (*upper left*) and at 120 °C (*lower left*) upon compression and decompression as well as the corresponding changes in the Ge–Te bond lengths (*right-hand* panels). Closed and open symbols refer to the bond length changes upon compression and decompression, respectively. The vertical lines depict the error bars for the bond length determination. Reprinted with permission from Kolobov et al. [2]. Copyright 2007 by the American Institute of Physics. Triangles correspond to the results of the transmission measurements of Ge K-edge EXAFS using a large multi-anvil press [17]

One can see that upon compression the Ge–Te distance gradually decreases. In absolute terms (using the previously measured value of the Ge–Te bond length at ambient pressure [9]) the observed decrease in the bond length corresponds to a change from 2.83 to 2.61 Å, i.e. to the value corresponding to that in the as-deposited and laser-amorphised amorphous phase. Upon decompression at room temperature the Ge–Te distance remains unchanged. At the same time, decompression at a higher temperature (120 °C) results in the reversion of the Ge–Te bond length to its original value, i.e. recrystallisation takes place.

10.1.2 High-Pressure BCC Phase and Initial Structure Memory

Compression to higher pressures results in further modification of the structure [4]. The behaviours of the cubic and trigonal phases of $Ge_2Sb_2Te_5$ upon compression are different and the results are compared in Fig. 10.6. As already discussed above, the cubic fcc phase starts to disappear and the amorphous phase grows from about 15 GPa. The amorphisation process continues until the crystalline fcc phase totally disappears at 25 GPa (Fig. 10.6).

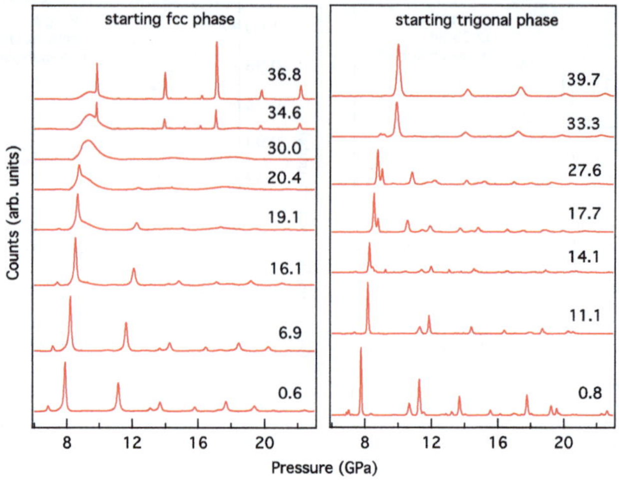

Fig. 10.6 Evolution of the X-ray diffraction patterns of the metastable fcc phase (*left*) and the stable trigonal phase (*right*) of $Ge_2Sb_2Te_5$ upon compression. The corresponding pressures are marked next to each curve. Reprinted with permission from [4]. Copyright 2009 by the American Physical Society

Upon further compression of the amorphous phase, a new crystalline phase starts to grow above 33 GPa, however some amorphous material remains up to the highest pressures reached [4]. The new crystalline phase is identified as body-centred cubic (bcc) with the space group $Im\overline{3}m$, $a = 3.38$ Å (at 35 GPa). The diffraction data are consistent with complete disorder on the 2a Wyckoff sites. In what follows we refer to this phase as bcc(C), with the C standing for the initial cubic phase.

The trigonal phase, on the other hand, remains crystalline in the pressure range studied. Upon compression at room temperature, the trigonal phase is transformed into an intermediate crystalline phase (orthorhombic GeS type [18]) and subsequently into bcc (Fig. 10.6, right-side panel). This phase is subsequently referred to as bcc(T).

It is interesting to note that the two high pressure phases—although both are bcc—possess different diffraction patterns (Fig. 10.7). While the bcc(C) possesses large crystallites, the starting trigonal phase exhibits fine-powder patterns in all three phases [4]. This difference may be due to the fact that the amorphous phase is uniform and continuous while the orthorhombic phase consists of small grains. It seems natural that larger crystallites can grow within a continuous uniform amorphous network than from a small individual grains when it may be possible that individual grains undergo the phase transition thus preserving the fine-grain structure.

The decompression behaviour is also very different for the two starting structures (Fig. 10.8). The bcc(T) phase changes into an intermediate crystalline phase and eventually the initial trigonal structure is restored, i.e. pressure-induced changes in the trigonal $Ge_2Sb_2Te_5$ are reversible. On the other hand, the bcc(C) phase gradually reverts into the amorphous phase with some remaining bcc(C) phase after complete decompression [4].

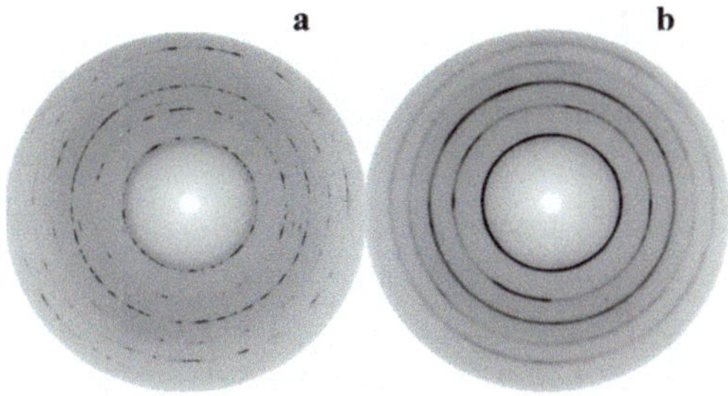

Fig. 10.7 X-ray diffraction patterns of bcc phases of $Ge_2Sb_2Te_5$ obtained starting with the fcc (*left*) and trigonal (*right*) phases. Reprinted with permission from Krbal et al. [4]. Copyright 2009 by the American Physical Society

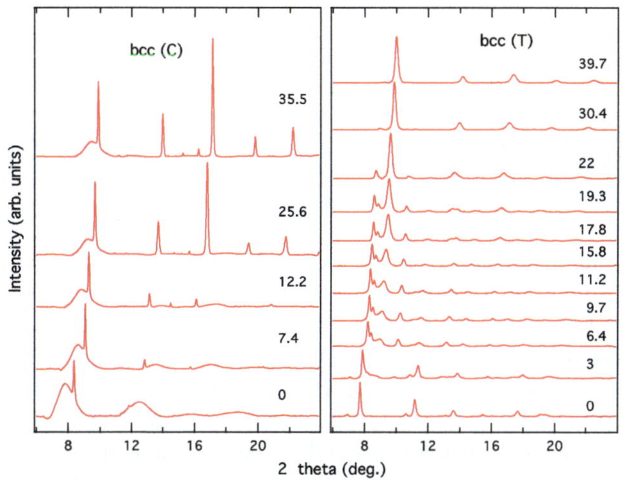

Fig. 10.8 Evolution of the X-ray diffraction patterns of the metastable fcc phase (*left*) and the stable trigonal phase (*right*) of $Ge_2Sb_2Te_5$ upon decompression. The corresponding pressures are marked next to each curve. Reprinted with permission from Krbal et al. [4]. Copyright 2009 by the American Physical Society

The different decompression behaviour suggests memory of the initial structure: the apparently similar (bcc) structures evolve differently depending on the structure of the initial phase used to generate the high-pressure bcc phase.

The unit cell volume change with pressure is shown in Fig. 10.9. The fcc $Ge_2Sb_2Te_5$ volume decrease upon initial compression (0 to 13 GPa) can be described by the Birch–Murnaghan equation of state (EOS) [19] with the bulk moduli $K_0 = 39 \pm 2$ GPa (Since K_0 and K_0' are strongly correlated and the number of experimental points is rather limited, we have chosen to limit the Birch–Murnaghan equation of state up to second order in elastic strain corresponding to $K_0' = 4$.) These values compare rather well with the known bulk moduli for GeTe ($K_0 = 49.9$ GPa) [20], $AgSbTe_2$ (45 GPa) [21], and Sb (39.7 GPa) [22]. At higher pressures (13 to

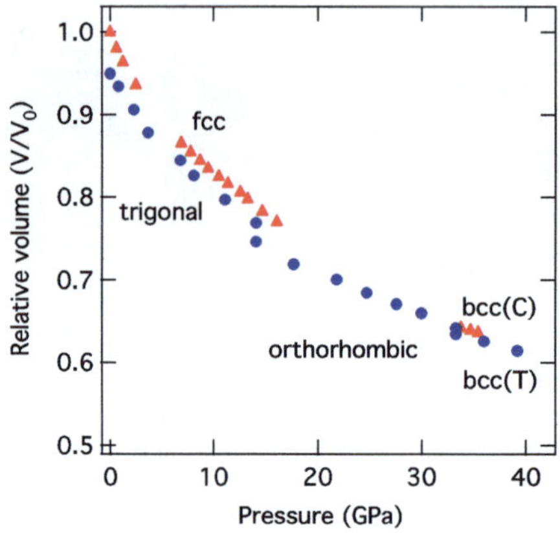

Fig. 10.9 Pressure dependence of the normalised volume per (Te) atom (assuming $Ge_2Sb_2Te_5$ stoichiometry throughout the complete cycle) upon compression and decompression for the metastable fcc and stable trigonal phases of $Ge_2Sb_2Te_5$. The error bars are smaller than the size of the symbols used. The triangles correspond to the starting fcc phase and the circles correspond to the starting trigonal phase. Reprinted with permission from Krbal et al. [4]. Copyright 2009 by the American Physical Society

22 GPa), the volume deviates from the EOS value at lower pressures with the material becoming softer due to partial conversion to the amorphous phase.

In the 33–38 GPa range the much less compressible bcc(C) phase appears. Fitting the data (over a rather limited range) to the EOS yields the following bulk moduli for the high-pressure phase: $K_0 = 61 \pm 10$ GPa and $V/V_0 = 0.88$. The bulk moduli and normalised volumes obtained for the starting trigonal $Ge_2Sb_2Te_5$ material are as follows (K_0' fixed at four in all cases). For the trigonal phase: $K_0 = 44 \pm 2$ GPa ($V/V_0 = 0.95$), for the orthorhombic phase: $K_0 = 48 \pm 6$ GPa ($V/V_0 = 0.91$), and, finally, for the bcc(T) phase: $K_0 = 57 \pm 10$ GPa ($V/V_0 = 0.87$).

The fact that the amorphous phase formed by compression of the fcc phase of $Ge_2Sb_2Te_5$ is actually an intermediate phase between the two crystalline phases deserves a special discussion. It strongly suggests that pressure-induced amorphisation cannot be treated as "melting" [12]. More likely it proceeds through nanophase separation [1]. Further experimental support of this hypothesis comes from the fact that at 30 GPa the volume of the bcc(C) phase is 1.5% larger than that of the bcc(T) phase. In the case of the bcc(C) the volume tends towards those of bcc Sb (40.6 Å^3) [22] and bcc Te (41.6 Å^3) [23] at 30 GPa. This is an indication that the bcc phase is richer in Sb and Te and that the remaining amorphous material is Ge-rich. It is interesting to note that the formation of an intermediate amorphous phase during compression has also been observed for $AgSbTe_2$ [21].

10.2 Pressure-Induced Changes in Ge-Free Phase-Change Alloys

10.2.1 Binary Sb_2Te

Upon compression, Sb_2Te exhibits three phase transitions (Fig. 10.10a, b) [5]. The initial crystalline structure of Sb_2Te is a trigonal phase and belongs to the space group R$\bar{3}$m1 with hexagonal lattice constants: $a = 4.274(1)$ Å and $c = 17.652(6)$ Å [24]. The first transition occurs within the pressure range of 7–9 GPa (Fig. 10.10a: 9.1 GPa) in which the starting phase becomes orthorhombic that can be described as the space group Pbnm (GeS-like) with lattice parameters: $a = 4.379(9)$ Å, $b = 11.309(6)$ Å and $c = 4.173(0)$ Å. A similar trigonal ⇒ orthorhombic transition was also observed in trigonal $Ge_2Sb_2Te_5$ [4] at 17 GPa. The next phase transformation occurred in the pressure range 12.3–15 GPa (Fig. 10.10a: 14.4 GPa) where a mixture of orthorhombic and body-centred-tetragonal BCT (SbII-like) phases can be detected. The BCT phase is an incommensurate composite phase comprising host (space group I4/mcm) and guest (space group I4/mmm) structures [22]. The refined lattice parameters of the BCT structure at 18.4 GPa were: $a_h = a_g = 7.951(4)$ Å, $c_h = 3.801(3)$ Å and $c_g = 3.082(7)$ Å, yielding a c_h/c_g ratio of 1.233 which is somewhat smaller than the values obtained in other materials with similar structures (1.404 for Sr–V at 56 GPa [25], 1.309 for Bi–III at 6.8 GPa [26] and 1.307 for Sb–II at 12 GPa [22]). Eventually, the body-centred-cubic (BCC) phase occurs at pressures above 28 GPa (Fig. 10.1a: 32 GPa). This is distinctly different from cubic Ge-Sb-Te alloys that exhibit the formation of an amorphous phase on their route to the high-pressure BCC phase.

Upon decompression (Fig. 10.10b), the BCC phase gradually retransforms to the BCT structure over the range of pressures from 21.5 to 15.3 GPa and remains stable down to ambient pressure.

10.2.2 Ag-In-Sb-Te Alloy

Incorporation of elements such as Ag and In into Sb_2Te leads to a change of the initial phase to an A7 structure (rhombohedral) that belongs to the space group R$\bar{3}$m1 with the lattice constants $a = 4.307(8)$ Å and $c = 11.214(6)$ Å. Over the studied pressure range of 0–40 GPa only one phase transition was observed at 12.4 GPa with the material adopting the BCT form (Fig. 10.10c: 12.4 GPa) with lattice parameters $a_h = a_g = 8.037(1)$ Å, $c_h = 3.866(4)$ Å and $c_g = 2.984(1)$ Å, giving a c_h/c_g ratio of 1.296. Upon decompression (see Fig. 10.10d), the AIST ($Ag_{11}In_4Sb_{57}Te_{28}$) sample reverted to the initial A7 structure with slightly different lattice constants: $a = 4.348(8)$ Å and $c = 10.975(7)$ Å [5].

The pressure dependences of the relative volume of Sb_2Te and AIST are shown in Fig. 10.11. The observed decrease in compressibility can be described by the Birch–Murnaghan EOS with the bulk moduli K_0 in units of GPa. The obtained values of EOS fitting are summarised in Table 10.1.

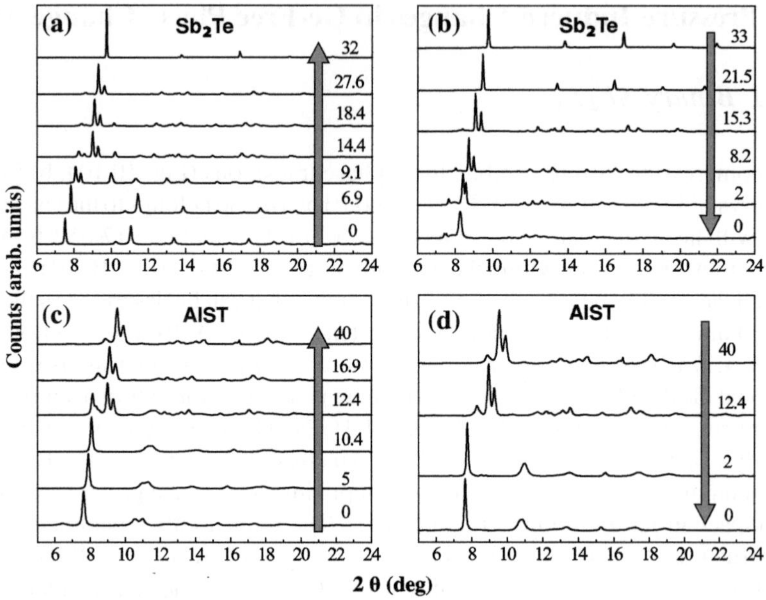

Fig. 10.10 The evolution of X-ray diffraction patterns of Sb_2Te (upper two plots) and AIST (lower two plots) upon compression and decompression. Arrows indicate compression (figures on the left) and decompression (figures on the right). Reprinted with permission from Krbal et al. [5]. Copyright 2011 by the American Physical Society

Table 10.1 Bulk moduli and the V/V_0 of respective phases observed in Sb_2Te and AIST upon compression

Sb_2Te Phase	V/V_0	K(GPa)	AIST Phase	V/V_0	K(GPa)
Trigonal	1	42.8 ± 0.33	A7	1	44.3 ± 1.5
Orthorhombic	0.842	43.7 ± 2	*Simple cubic	0.85	53.9 ± 2
BCT	0.782	61.2 ± 1	BCT	0.78	67.7 ± 2.3
BCC	0.744	70.7 ± 5.2	–	–	–

(* the simple cubic phase occurred when the A7 structure was exposed to pressures above 3 GPa for a few days.) Reprinted with permission from Krbal et al. [5]. Copyright 2011 by the American Physical Society

10.3 Ab Initio Studies of Pressure-Induced Changes

10.3.1 Pressure-Induced Amorphisation

Pressure-induced amorphisation has also been studied using *ab initio* simulations [6]. The crystalline order was found to disappear at 21 GPa in agreement with experiment. Furthermore, different kinetics for ⟨111⟩, ⟨220⟩ and ⟨200⟩ peaks was reproduced

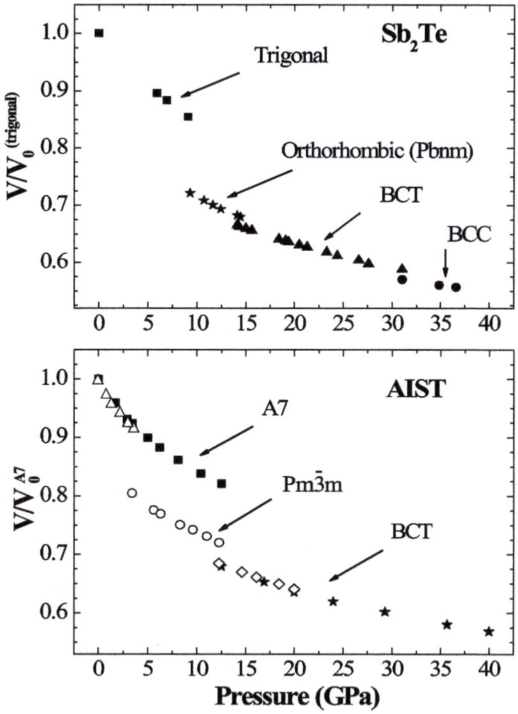

Fig. 10.11 The pressure dependences of the relative volume of Sb_2Te and AIST with denoted crystallographic symmetries observed using synchrotron radiation. In the case of AIST, open symbols represent the results obtained using a laboratory X-ray-based source. Reprinted with permission from Krbal et al. [5]. Copyright 2011 by the American Physical Society

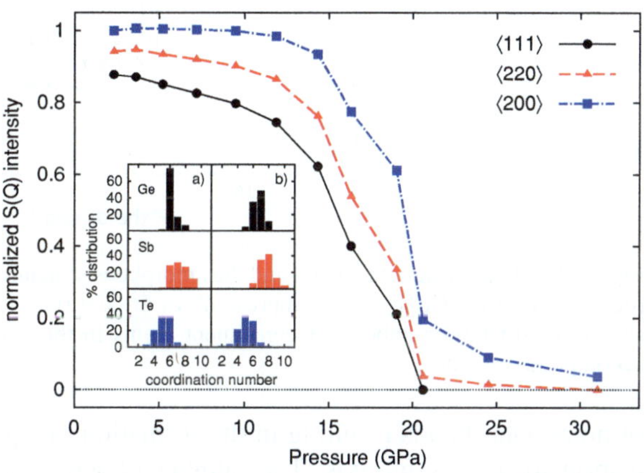

Fig. 10.12 Evolution with pressure of the intensity of the main peaks of the structure factor $S(Q)$ normalized to the intensity of the ⟨200⟩ peak at zero pressure. Reprinted with permission from Caravati et al. [6]. Copyright 2009 by the American Physical Society

(Fig. 10.12). The ⟨200⟩ peak that gives a measure of the alignment of square rings along ⟨100⟩ directions is the last to disappear.

By inspecting the atomic trajectories, the authors found that the initial change predominantly occurs in the vicinity of vacancies. Namely, it was found that Te atoms move towards the nearest vacancy (see Sect. 11.2: Fig. 11.6). This motion induces rupture of Te-X bonds opposite to the vacancy and subsequent formation

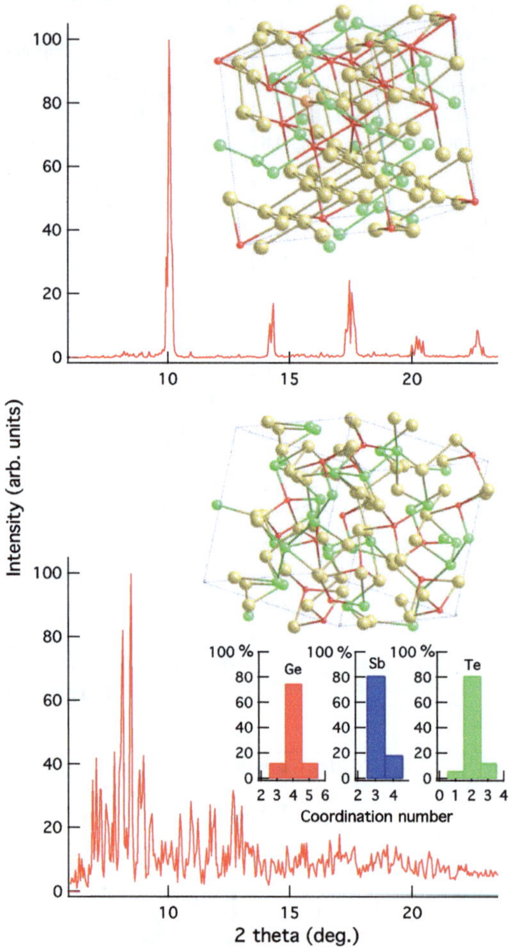

Fig. 10.13 The bcc structure (*top*) and the amorphous structure obtained along the decompression route (*bottom*) for $GeSb_2Te_4$ "compressed" to 46 GPa (Ge-*red*, Sb-*green*, Te-*yellow*). The inset shows coordination numbers for constituent atoms in the amorphous phase obtained along the decompression route

of homopolar bonds resulting in the formation of square rings rotated by 45° with respect to the crystal axis. The number of wrong bonds increases with pressure but only Te–Te dimers and trimers are formed according to [6]. The fraction of tetrahedrally coordinated Ge sites was found to be 17% [6], a value lower than that found in melt-quenched GST (33%) by the same authors [27].

It was also found that the density of the pressure-amorphised phase was higher than that of the melt-quenched phase (both refer to the phases obtained *in silico*) being 6.5 and 8.2% less dense, respectively, than the crystalline cubic phase. The concentration of wrong bonds was found to be the same for the two amorphous phases.

Table 10.2 Percentage of different bonds in the bcc phase of $GeSb_2Te_4$ at 46 GPa

Bond type	Ge–Te	Sb–Te	Ge–Sb	Ge–Ge	Sb–Sb	Te–Te
Percentage	22	37	8	1	7	25

10.3.2 High-Pressure BCC Phase

Upon further increase in pressure, the structure changes to bcc (Fig. 10.13, upper panel) in agreement with experimental observations [4]. The simulations clearly demonstrate an appearance of a large concentration of wrong Sb–Sb and especially Te–Te bonds, i.e. phase separation of $Ge_2Sb_2Te_5$ takes place upon strong compression in agreement with the experimental conclusion [4]. The distribution of bond types is shown in Table 10.2. The observed distribution is in good agreement with the conclusion drawn from the experimental data refinement, namely, that the bcc sites are randomly occupied by the three types of constituent atoms. A similar conclusion, namely, that the sites on the bcc lattice are randomly occupied has been independently reached in [7]. No vacant sites were found in the high-pressure phase.

The average coordination number in the structure was found to be $CN_{3.0} = 7.2$ for a bond length cut-off of 3.0 Å and $CN_{3.1} = 8.3$ for a cut-off of 3.1 Å. The obtained values are in good agreement with the crystal lattice parameter found from refinement of the experimental data. The first nearest neighbor distances in the generated structure vary in a range from 2.6 to 3.1 Å demonstrating a rather high degree of disorder remaining in the high-pressure phase. The high-temperature bcc phase was predicted to be metallic [7].

It was subsequently reported that the high-pressure bcc phase can be obtained by compressing as-deposited amorphous $Ge_2Sb_2Te_5$ [8]. The authors noted that while the as-deposited amorphous $Ge_2Sb_2Te_5$ remains amorphous up to pressures of ca. 20 GPa, an additional peak emerged in the XRD spectrum of the amorphous phase at pressures above 10 GPa alongside with a change in the relative peak intensities. From these observations the authors suggest that the structure of the amorphous phase changes upon compression even when the material remains amorphous [8].

10.3.3 Amorphisation Upon Decompression

Upon decompression of the bcc phase, the amorphous phase is obtained in agreement with experiment (Fig. 10.13, lower panel). The coordiantion numbers for the constituent atoms show that the amorphous phase obtained along the decompression route satisfy the 8-N rule. Other major differences between this phase and the amorphous phase obtained upon the compression route are a higher concentration of tetrahedrally coordinated Ge sites and presence of extended Te–Te chains. The mechanism of the amorphisation process upon decompression is discussed in Sect. 11.2.2.

References

1. A.V. Kolobov, J. Haines, A. Pradel, M. Ribes, P. Fons, J. Tominaga, Y. Katayama, T. Hammouda, T. Uruga, Phys. Rev. Lett. **97**, 035701 (2006)
2. A.V. Kolobov, J. Haines, A. Pradel, M. Ribes, P. Fons, J. Tominaga, C. Steimer, G. Aquilanti, S. Pascarelli, Appl. Phys. Lett. **91**, 021911 (2007)
3. M. Krbal, A.V. Kolobov, J. Haines, A. Pradel, M. Ribes, P. Fons, J. Tominaga, C. Levelut, R. Le Parc, M. Hanfland, Appl. Phys. Lett. **93**, 031918 (2008)
4. M. Krbal, A.V. Kolobov, J. Haines, P. Fons, C. Levelut, R. Le Parc, M. Hanfland, J. Tominaga, A. Pradel, M. Ribes, Phys. Rev. Lett. **103**, 115502 (2009)
5. M. Krbal, A.V. Kolobov, P. Fons, J. Haines, A. Pradel, M. Ribes, A.A. Piarristeguy, C. Levelut, R. Le Parc, V. Agafonov, M. Hanfland, Phys. Rev. B **83**, 024105 (2011)
6. S. Caravati, M. Bernasconi, T. Kuehne, M. Krack, M. Parrinello, Phys. Rev. Lett. **102**, 205502 (2009)
7. Y.Q. Cheng, M. Xu, H.W. Sheng, Y. Meng, X.D. Han, E. Ma, Appl. Phys. Lett. **95**, 131904 (2009)
8. M. Xu, Y. Meng, Y.Q. Cheng, H.W. Sheng, X.D. Han, E. Ma, J. Appl. Phys. **108**, 083519 (2010)
9. A.V. Kolobov, P. Fons, A.I. Frenkel, A.L. Ankudinov, J. Tominaga, T. Uruga, Nature Mater. **3**, 703 (2004)
10. S. Shamoto, N. Yamada, T. Matsunaga, T. Proffen, J.W. Richardson, J.H. Chung, T. Egami, Appl. Phys. Lett. **86**, 081904 (2005)
11. M. Wuttig, D. Lüsebrink, D. Wamwangi, W. Welnic, M. Gillessen, R. Dronskowski, Nature Mater. **6**, 122 (2007)
12. S. Sharma, S. Sikka, Progr. Mater. Sci. **40**, 1 (1996)
13. D. Zahn, S. Leoni, Phys. Rev. Lett. **92**, 250201 (2004)
14. P. Toledano, K. Knorr, L. Ehm, W. Depmeier, Phys. Rev. B **67**, 144106 (2003)
15. M. Shimada, F. Dachille, Inorg. Chem. **15**, 1729 (1976)
16. J. Stöhr, *NEXAFS Spectroscopy. Springer Series in Surface Science* (Springer, Berlin, 1992)
17. P. Fons, A.V. Kolobov, J. Tominaga, Y. Katayama, Nucl. Inst. Meth. B **238**, 160 (2005)
18. W. Zachariasen, Phys. Rev. **40**, 917 (1932)
19. F. Birch, Phys. Rev. **71**, 809 (1947)
20. A. Onodera, I. Sakamoto, Y. Fujii, N. Mo, S. Sugai, Phys. Rev. B **56**, 7935 (1997)
21. R.S. Kumar, A.L. Cornelius, E. Kim, Y. Shen, S. Yoneda, C. Chen, M.F. Nicol, Phys. Rev. B **72**, 060101 (2005)
22. O. Degtyareva, M.I. McMahon, R.J. Nelmes, Phys. Rev. B **70**, 184119 (2004)
23. G. Parthasarathy, W. Holzapfel, Phys. Rev. B **37**, 8499 (1988)
24. V. Agafonov, N. Rodier, R. Céolin, R. Bellissent, C. Bergman, J.P. Gaspard, Cryst. Struct. Commun. **47**, 1141 (1991)
25. M.I. McMahon, T. Bovornratanaraks, D.R. Allan, S.A. Belmonte, R.J. Nelmes, Phys. Rev. B **61**, 3135 (2000)
26. M.I. McMahon, O. Degtyareva, R.J. Nelmes, Phys. Rev. Lett. **85**, 4896 (2000)
27. S. Caravati, M. Bernasconi, T. Kühne, M. Krack, M. Parrinello, Appl. Phys. Lett. **91**, 171906 (2007)

Chapter 11
Mechanism of the Phase-Change Process

In early work, it was tacitly assumed that upon exposure of the crystalline phase to an intense laser or current pulse phase-change materials melt and are subsequently quenched into a (completely) disordered amorphous state. Recent years have witnessed significant progress in understanding the atomistic mechanism of the phase change process. This chapter presents a summary of these results.

11.1 Microscopic Models of Phase Change

11.1.1 Umbrella-Flip of Ge Atoms

The first model that attempted to provide an atomistic description of the phase-change process was the so-called umbrella flip model [1]. This model (cf. Sect. 9.1.2) was derived from X-ray absorption studies that demonstrated Ge-Te bond shortening in the amorphous phase accompanied by a significant change in XANES spectra that indicated a significant change in the local structure around Ge atoms, i.e. the number of the first nearest neighbours and the nature of bonding between the atoms. To account for the obtained results (and in particular for the bond shortening and decrease in the white-line intensity, a change in the slope immediately after the white line and the location of features at higher energies), it was suggested that Ge atoms in the amorphous phase are located on tetrahedral symmetry sites within the Te sublattice. In line with this, the underlying mechanism of amorphisation was argued to consist of rupture of the longer Ge-Te bonds and a subsequent transition of Ge atoms from octahedral to tetrahedral symmetry positions within the Te fcc lattice (cf. Fig. 9.5) flipping through a plane formed by three Te atoms (hence the name) with subsequent lattice relaxation. This provided a natural explanation for the large property contrast between the phases as being due to different bonding configurations of Ge species [1]. Furthermore, such a switch accounted for why the transition is very fast (just Ge atoms have to switch without significant atomic diffusion) and why the medium is very stable (the Te fcc sublattice is preserved).

A. V. Kolobov and J. Tominaga, *Chalcogenides*, Springer Series in Materials Science 164, DOI: 10.1007/978-3-642-28705-3_11,

It was further suggested [2, 3] that instability at the tetrahedral sites in the ordered phase leads to generation of disordered GST structures in which the Ge atoms are mostly fourfold coordinated with three short Ge-Te bond lengths. As the displacement has the lowest energy near intrinsic vacancy sites [3], the authors further suggested that a high degree of amorphisation can be achieved most easily when the system has a composition of $Ge_2Sb_2Te_5$, in which the number of Ge atoms and Te tetrahedral sites matches that of the vacancy sites.

Subsequent ab initio studies confirmed the existence of tetrahedral Ge atoms in the melt-quenched amorphous phase. At the same time, they also found that only ca. 30% of Ge atoms were tetrahedrally coordinated [4–7].

11.1.2 Four-Membered Rings Rearrangement

Ab initio studies performed by different groups have confirmed not only the existence of Ge atoms on tetrahedral symmetry sites (about 30% of the total number of all Ge atoms) but also found the presence of regular ABAB building blocks (A=Ge,Sb, B=Te) alternatively referred to as four-membered rings, square rings or even-membered rings [5–8] in amorphous GST (cf. Figs. 9.18 and 9.22). Such square rings were found even in the liquid phase and their concentration as well as the concentration of *connected square rings* increased upon annealing. These square rings were considered seeds for the crystallisation process. From these results it was proposed that the crystallisation / amorphisation process consisted of ordering / disordering of such building blocks. The presence of vacancies was argued to be favourable for spacial rearrangements [5, 6]. The fact that crystallisation could be explained by ordering of pre-existing four-membered rings was argued to be the reason for the experimentally observed high crystallisation rates.

11.1.3 Resonance Bonding

The proposal that the bonding in the crystalline state of GST is resonant while in the amorphous state it is purely covalent [9–11] implicitly sugested that the underlying mechanism of the phase transition consisted of establishing—or destruction—of resonance bonding between the covalently bonded fragments. In the crystalline phase, resonant bonding leads to ordering and alignment of p orbitals on adjacent molecular units. This alignment is lost in the amorphous phase, irrespective of whether the local coordination changes and this has a drastic effect on materials's properties [12].

The role of resonance bonding is well exemplified by considering bonding in trigonal Se and Te. In the crystalline phase of these elements, the spiral chains of Se and Te can be viewed as distortions of simple cubic structures, in which there is a strong ordering between bonds on adjacent chains so that the p-orbitals are aligned to form linear chains as in the A7 structure. However, on amorphising the trigonal structure, this order is lost; the structure becomes more molecular with shorter intrachain distances and no obvious correlations between chains. A similar mechanism can be considered for GST whose structure also contains shorter and longer bonds and the shorter bonds get even shorter in the amorphous phase.

There are two important aspects of resonance bonding that determine the behaviour of phase-change materials. First, as was mentioned in [9], long-range order is crucial for resonance bonding to exist. If the long-range order is lost, the system will be unable to achieve resonance bonding [9]. Another important aspect is that resonance bonding in IV–VI materials is formed through use of the backside lobes of the same p-orbitals that are used for the covalent backbone structure. Since *p*-orbitals have pronounced anisotropy, the bonds formed through their use are strongly directional. For this reason, for the formation of resonance bonding the participating atoms have to be aligned. Since there are subsets of shorter and longer bonds in IV–VI crystals, their structure can be approximated as a covalently bonded backbone with additional energy gain due to weaker (resonance) bonds. In what follows, important consequences of these aspects of resonance bonding are described for the case example of GeTe following the logic of Ref. [13].

Resonance Bonding in Solids with Bonding Energy Hierarchy

While for the case of classical resonance bonding (e.g. in benzene) all bonds are identical, the presence of short and long bonds in GeTe is likely to redistribute the electron charge density between the two kind of bonds. Indeed, the plot of the charge difference between the simulated GeTe model and isolated pseudo-atoms (Fig. 8.18) shows that the electron density is only significant along the short bonds. In contrast, the electron density pile-up along the long bonds is *significantly* lower demonstrating that there is a pronounced bonding energy hierarchy between the short and long bonds. The existence of a bonding energy hierarchy should manifest itself in the different response of the short and long bonds to thermal and/or electronic excitations.

To investigate the effect of the increased local distortion upon subsequent structural evolution, a series of DFT simulations has been performed for the GeTe binary alloy which is an end point of the quasi-binary $Sb_2Te_3 - GeTe$ tie line [13]. Starting with the relaxed GeTe structure representing the ground state of the system, the authors modelled the experimentally observed Ge(Sb)-Te bond shortening at higher temperature by introducing random local distortions following two algorithms. In both cases, the Ge-Te bond lengths were the same and equal to the experimental Ge-Te distance in the amorphous phase. In panel (a), the atomic misalignment was minimised by keeping the Te atoms at their original positions and displacing all Ge atoms along the pseudocubic $\langle 111 \rangle$ directions. In panel (b), on the other hand, the misalignment was maximised by randomly displacing both Te and Ge atoms. In both cases, the average structures retain the long-range order of the crystalline phase as evidenced by simulation of X-ray (Bragg) diffraction [13, 14].

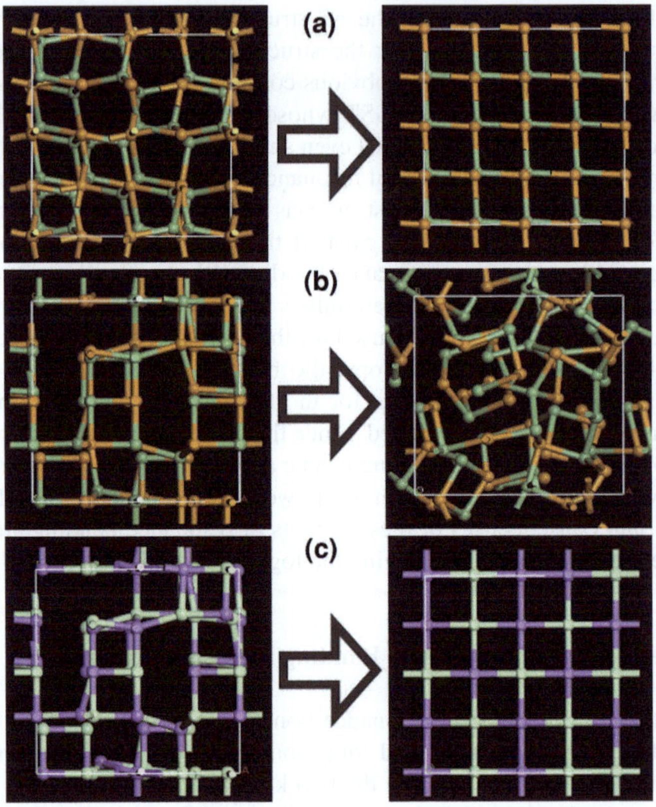

Fig. 11.1 Effect of the nature of distortion and bonding on the stability of the ordered phase. Resonantly bonded GeTe structure with distortions that preserve resonant bonding reverts to the crystalline phase **a** while that with distortions that break resonant bonding changes to the amorphous phase **b**. At the same time, ionically bonded NaCl with identical distortions readily reverts to the initial crystalline state **c**. Reprinted from Kolobov et al. [14] with pemission from Elsevier

The distorted structures were then allowed to relax at 0 K (Fig. 11.1). As expected, structure (a) in which the atomic misalignment was minimised relaxed back into the equilibrium structure. In contrast, structure (b) that possesses a similar degree of long-range order did not revert to the starting structure. Instead, a significant atomic relaxation took place leading to a complete loss of long-range order. At the same time, when halite (NaCl), an ideal rock salt structure in which all bonds have identical energy and length, was distorted in a similar way (c) (the same fractional coordinates with the appropriate unit cell volume), it readily relaxed to the perfectly ordered equilibrium state [13]. These results clearly demonstrate the importance of the nature of the bonding between the constituent atoms and the character of disorder for the ability of a materials to amorphise. Solids with resonance bonding that possess a bonding energy hierarchy are unique from this perspective [13].

While the weaker bonds may be broken due to thermal vibrations at elevated temperatures, electronic excitation of the system may strongly affect this process.

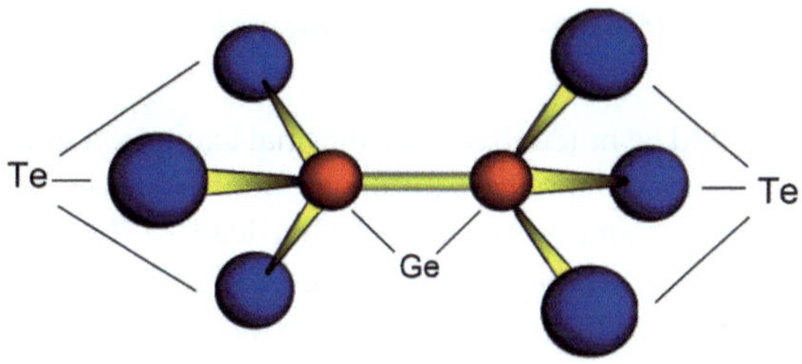

Fig. 11.2 The local structure proposed to account for Ge-Ge bonds in amorphous $Ge_2Sb_2Te_5$. Reprinted from Paesler et al. [15] with permission from Elsevier

Indeed, optical excitation (or carrier injection from the contact) generates a large concentration of non-equilibrium charge carriers that thermalise into tail states. Since these states are associated with the weakest bonds in the system, their population results in further weakening of these bonds which can be interpreted as dynamical lowering of the energy barrier between the two states that enables a transition to the new phase using thermal vibration energy at lower temperatures. In other words, *the system in the electronically excited state can melt at a significantly lower temperature*. We would like to reiterate that there exists abundant experimental evidence that various chalcogenide alloys, can be molten optically at temperatures significantly below their thermal melting point. Such processes have been described in Chap. 6.

11.1.3.1 Distortion-Triggered Collapse of the Ordered Phase

In contrast to *all previous* DFT studies when the amorphous phases was obtained by quenching the disordered melt obtained at high temperature, the authors of Ref. [13] succeeded in generating the amorphous phase *directly from the crystalline cubic phase*. A careful analysis of atomic trajectories reveals that the tetrahedrally coordinated Ge atoms in the amorphous phase *do not acquire this bonding geometry through the umbrella-flip process* proposed earlier [1]. While this observation does not negate the possibility of the umbrella flip for some particular configurations, it does demonstrate that there is a major alternative route, namely via rotation and distortion of "cubic" fragments with essentially preserved Ge(3):Te(3) local bonding geometry. The formation of tetrahedral sites in cases of GeTe may take place when two Ge atoms located at pyramidal apexes come close together during the lattice relaxation process, the resulting local structure being in agreement with the previous suggestion based on EXAFS data analysis (Fig. 11.2).

Finally, it is important that the *destabilisation of the subsystem of* the weaker bonds with the preserved covalent backbone not only allows for the low-temperature amorphisation of the material; the preserved covalent backbone also ensures the memory of the initial crystalline structure making the (reverse) crystallisation process fast and ensuring high cyclability of the crystallisation-amorphisation process.

It may be informative to remind the reader that *inter*molecular, rather than *intra*molecular, bond-breaking processes were also found predominant during the revrsible photstructural change in chalcogenide glasses [16], which underscores similarities between the processes in these two classes of chalcogenide alloys.

11.1.4 Photo-Assisted Amorphisation of $Ge_2Sb_2Te_5$

It should be noted that since thermal excitation of the system excites a broad spectrum of vibrational modes, a purely thermal process is not ideal from a switching efficiency viewpoint since only modes that lead to destruction of the atomic alignment are efficient in triggering the phase-change process. An important question is whether energy introduced into the system through a "non-thermal" channel can be efficiently coupled. Experimental observations of high efficiency of femtosecond laser pulses [17] and/or use of coherent phonons [18] indicate that this is possible.

11.1.4.1 In Situ Time-Resolved Sub-Nanosecond XAFS Studies

As described in the previous section, the presence of a bonding energy hierarchy in the resonantly bonded crystalline phase causes the crystalline phase to be unstable with respect to external stimuli which may lead to the formation of the amorphous phase. In addition, the random distribution of Ge and Sb atoms and vacancies in the metastable crystalline phase breaks translational symmetry and leads to a behaviour reminiscent of molecular crystals in which bonding and anti-bonding states are partially localised. This leads to a situation where the presence of a non-equilibrium electron distribution can lead to the selective destruction of the longer bonds and in turn the destabilisation of the *metastable* crystalline phase. Such electron distributions can exist in high concentrations either as a result of electron-hole pair generation or carrier injection from a contact. They necessarily modify the interatomic potentials and may effect the phase-change process as suggested, for example, in [2]. In order to obtain the details of the phase-change process and the role of electronic excitation, the structural transformation has to be studied in situ, i.e. a knowledge of the structural *evolution* is crucial. An optical pump/x-ray probe technique that is capable of the acquisition of XAFS spectra with sufficient time resolution makes such observations possible on sub-nanosecond time scales [19, 20].

Unlike EXAFS where oscillations quickly damp with temperature, XANES is insensitive to temperature as is exemplified in Fig. 11.3 (left). At the same time, it is highly sensitive to the three-dimensional structure present, in particular, the white-line intensity of $Ge_2Sb_2Te_5$ is significantly different among the crystalline, amorphous and liquid states [21, 22] as shown in Fig. 11.3 (right) which can serve as a signature of the structure present. A three-dimensional area map of the fluorescence intensity at the X-ray excitation energy corresponding to the white-line position of Ge is shown in Fig. 11.4 and demonstrates the usefulness of such an approach.

Figure 11.5 shows the time evolution of the white-line intensity of GST as a function of delay after the excitation pulse [20]. To exclude the possibility of a baseline shift, the corresponding intensity one electron orbit period before ($\sim 5\,\mu s$) the laser pulse, i.e. in the starting crystalline state, was simultaneously monitored; no change in baseline was observed. One can clearly see from the figure that following exposure to the laser pulse the white-line intensity first monotonically decreases

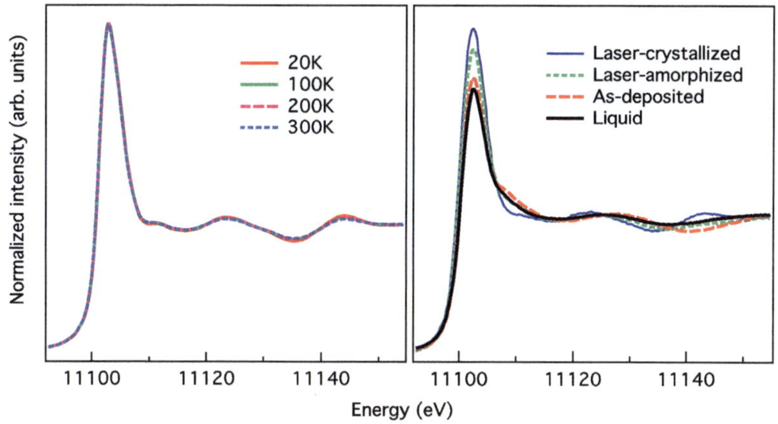

Fig. 11.3 Independence of XANES spectra of crystalline $Ge_2Sb_2Te_5$ on temperature (*left*) and pronounced differences among XANES spectra for laser-crystallized (*blue/thin solid line*), laser-amorphized (*green/fine dashed line*), as-deposited (*red/course dashed*), and liquid (*black/thick solid line*) $Ge_2Sb_2Te_5$ (*right*) make XANES a robust probe of the structure present. Reprinted with permission from P. Fons et al. [20]. Copyright 2010 by the Americal Physical Society

Fig. 11.4 A three-dimensional X-ray intensity map showing the crystallised band (*blue*), small amorphous marks (*green*), and as-deposited amorphous background (*red*). A ca. 1 μm diameter X-ray probe was used. Reprinted with permission from Fons et al. [20] Copyright 2010 by the Americal Physical Society

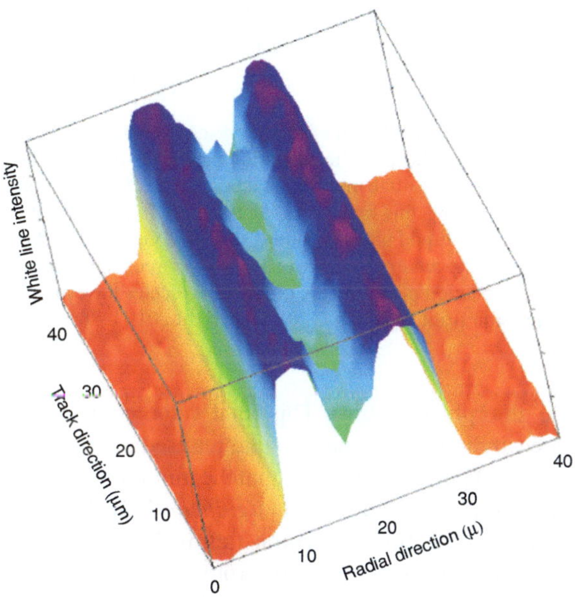

and reaches a minimum value in about 1 ns. This fast initial decrease in white-line intensity is followed by its partial recovery and within ~2 ns a new saturation value is reached. Scans, corresponding to longer time delays (not shown), confirm that the white-line intensity remains unchanged for delay times greater than 4 ns after

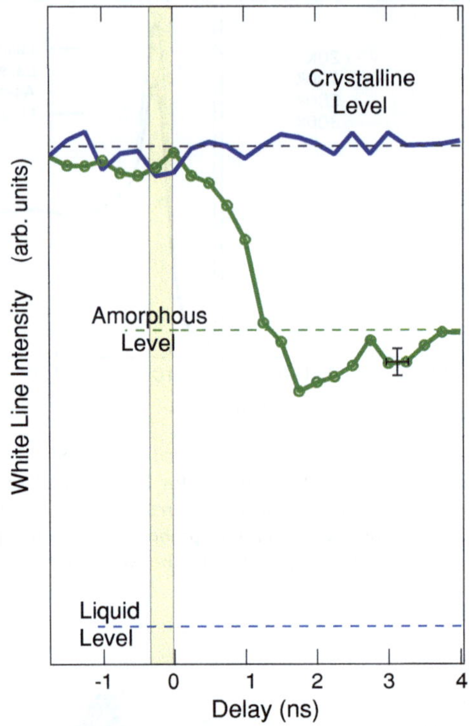

Fig. 11.5 Time evolution of the fluorescence intensity at the white line. A decrease in the white-line intensity following the excitation process is followed by a slower partial recovery. While the steady-state white-line intensity reaches exactly the static amorphous value, the minimum intensity acquired during this process is *significantly higher* than the corresponding white-line intensity value for the liquid phase demonstrating that $Ge_2Sb_2Te_5$ *does not melt* in a conventional sense on its way to the amorphous phase. The *yellow* rectangle indicates the duration and relative position of the excitation pulse and the cross bar indicates the estimated error on both axes. Reprinted with permission from Fons et al. [20]. Copyright 2010 by the Americal Physical Society

the amorphising pulse. This final value coincides with the value of the white-line intensity in the amorphous state realised after exposure to a laser pulse.

It is of particular interest that the minimum value of the white-line intensity during the amorphisation process is *significantly higher* than that corresponding to the static liquid state [22] clearly demonstrating that $Ge_2Sb_2Te_5$ does not melt in a conventional sense upon its transformation from the crystalline to the amorphous phase [20]. This result demonstrates that the amorphisation process in the sub-nanosecond excitation regime *does not* consist of conventional melting and subsequent quenching of the melt into the amorphous phase and may be a solid–solid process. Ultrafast amorphisation of Ge-Sb-Te alloys induced by femtosecond laser pulses [17, 23, 24] provides further support for this conclusion. At the same time, for cases when the change is induced by longer laser or current pulses thermal effect may play the dominant role.

The role of electronic excitation in the phase-change process has also been invetigated in [25] and it was concluded that electronic excitation considerably lowers the critical amorphisation temperature and reduces the atomic diffusion coefficient with respect to that of the corresponding liquid phase by at least one order of magnitude. In addition, the resulting structure has fewer wrong bonds, in line with the experimental EXAFS results (cf. Sect. 9.1.2). Nonthermal amorphisation induced by picosecond electric pulses has also been reported [26].

Why Intense Excitation is Needed?

Within the conventional thermal paradigm, the needed intense excitation level is naturally explained by the need to heat the material above the melting point. Within the framework of a non-thermal mechanism that considers electronic excitation, an explanation for the existence of the intensity threshold was proposed in [2]. The authors consider the covalently bonded fragment as a defect within a resonantly bonded crystalline phase. The crystalline phase of phase-change materials is a degenerate p-type semiconductor with the Fermi level located at the valence band (or slightly inside the valence band). Since the total energies of $Ge_2Sb_2Te_5$ containing Ge atoms in octahedral and tetrahedral symmetry positions are very close [27], the authors proposed that energy levels corresponding to individual tetrahedrally coordinated Ge atoms (and corresponding vacancies) to be located closer to the valence band.

At room temperature and at low-excitation levels, the states of the complex defects are located above the chemical potential and are hence unoccupied. In this situation, a single excitation event results in an unstable tetrahedral Ge site which readily decays back to the original structure. Under intense optical excitation, the chemical potential moves towards mid-gap. In this situation the defect states can become located below the chemical potential and hence occupied (stable). Once the concentration of these defects becomes sufficiently large and their wavefunctions overlap, the electronic structure changes to that of the 'amorphous' state.

11.2 Pressure-Induced Amorphisation

11.2.1 Amorphisation Upon Compression

The experimental observations of the lower amorphisation pressure in cubic $GeSb_2Te_4$ compared to $Ge_2Sb_2Te_5$ (the materials contain 25 and 20% vacancies, respectively) and absence of amorphisation in the denser bulk phase strongly suggest that vacancies play a crucial role in the studied process. It is quite natural that squeezing out vacancies under high pressure leads to randomisation of the atomic structure. While at ambient pressure the density of the amorphous phase is lower than

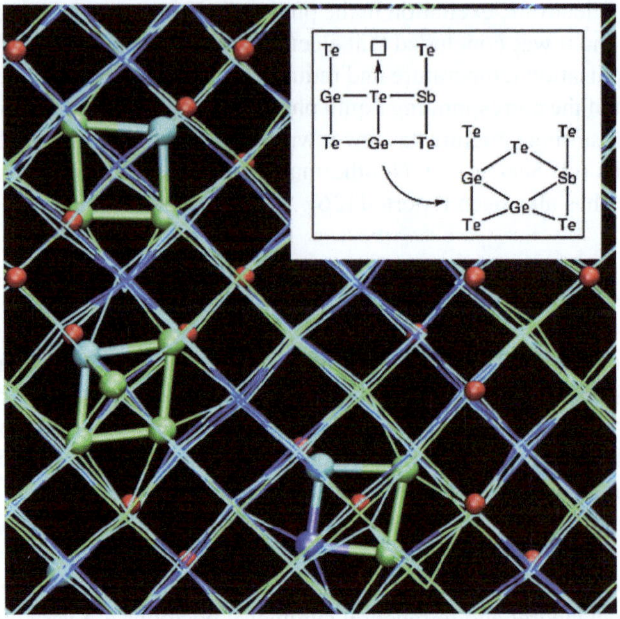

Fig. 11.6 Snapshot of the simulation at 14 GPa showing the appearance of squared rings formed by homopolar Ge/Sb bonds rotated by 45° with respect to the crystal axis. Large spheres indicate atoms involved in the formation of rotated rings (*blue* Ge; *green* Sb; *light blue* Te). Small red spheres indicate the position of Ge/Sb vacancies. *Inset* Sketch of the displacement of Te atom towards the vacancy giving rise to a rotated square ring with homopolar bonds. Reprinted with permission after Caravati et al. [28]. Copyright 2009 by the American Physical Society

that of the crystalline phase. It may be possible that the amorphous phase compresses more easily and at the higher pressures can become denser than the cubic crystalline phase.

Ab-initio simulations provided further insights into the atomistic mechanism of the amorphisation process under pressure [28]. It was found that upon compression Te atoms move towards the nearest vacancy. This process results in the destruction of the Te-X bonds opposite to the vacancy and the subsequent formation of Ge-Ge(Sb) bonds. This transformation gives rise to square rings rotated by 45 ° with respect to crystal axis (Fig. 11.6). The important role of vacancies was thus confirmed and their role was argued to be twofold. First, they increase the compressibility of the cubic phase and allow reaching a higher density at lower pressure via formation of rotated rings. Second, the rotated rings formed, thanks to vacancies, serve to destabilise the system once their concentration becomes sufficiently high.

We would like to note that this mechanism is essentially the same as the amorphisation mechanism described earlier in this chapter, namely, the amorphisation results from misalignment of the covalently bonded fragments of the crystalline "cubic" phase.

A question arises why GST becomes amorphous under pressure while the binary GeTe does not. In addition to an important role of vacancies discussed above, it may be that the presence of the second constituent (Sb_2Te_3) is needed. The different covalent radii of Ge and Sb give rise to a non-uniform strain distribution inside the material which may act as a driving force for the amorphisation via nanophase separation that is often a driving force for pressure-induced amorphisation [29] in cases where the total volume of the decomposed phases is smaller than the volume of the starting material. The nanophase separation into Ge-Te and Sb-Te phases was previously suggested to take place during photo-induced amorphisation [30].

It is also interesting to note that pressure-induced amorphisation is experimentally usually observed in solids with van-der-Waals [29], i.e. in materials that contain a sysbystem of week bonds.

11.2.2 Amorphisation Upon Decompression

The amorphous phase is also obtained upon decompression of the high-pressure bcc phase. The structure of this phase seems to be denser than that obtained on the compression route and the amorphisation mechanism can be the following.

In the high-pressure bcc phase each Ge atom has eight equidistant neighbours and this structure is stabilised by the applied pressure. As pressure is released, Ge atoms acquire their preferred fourfold coordinated configurations (or threefold if bonded through dative bonds formed using lone-pair electrons of neighbouring Te atoms) with other neighbours moving away. Since in the compressed fcc form Ge atoms are randomly bonded to either of the species present on GST, i.e. to Sb, Te or Ge that possess different atomic radii, during the structural relaxation the atomic structure becomes disordered.

11.3 Amorphisation of Ag-In-Sb-Te: Bond-Interchange Model

For AIST alloys it has been suggested that the transition can be presented as spatial rearrangement of the shorter and longer bonds (ordering or disordering in space of the subsystem of the shorter bonds) in the A7 structure (Fig. 11.7) [31]. The interchange between the shorter and longer bonds alters the orientation of the octahedra but does not require bond breaking or diffusion and the process is a sequential collective motion of Sb atoms, the principal component of AIST. This transition is reminiscent of either spin ordering in materials such as spin glasses or of an order–disorder transition in materials like GeTe (8.1.1). Destruction of the ordering leads to atomic misalignment and subsequent disappearance of the resonance bonding with the concomitant large optical contrast. At the same time, the very small atomic displacements account for the very fast transition rates. It is interesting to note that the authors consider the photon excitation as a possible cause for the alignment of the octahedra.

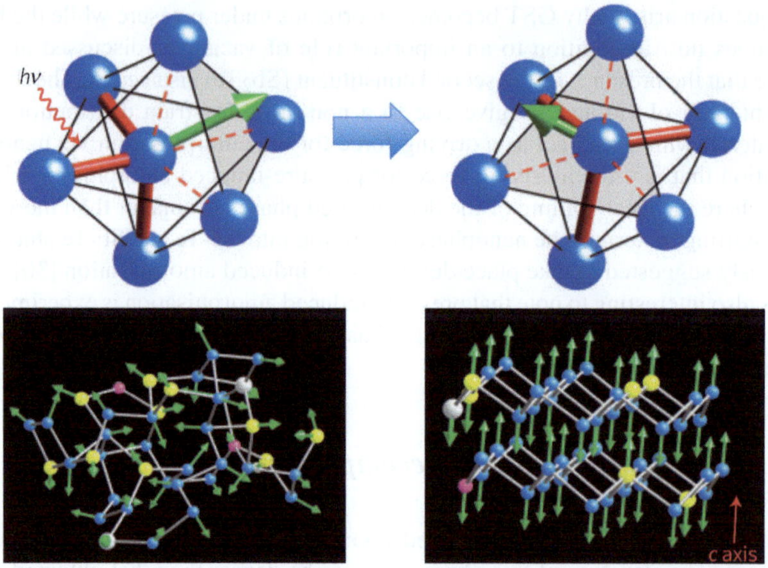

Fig. 11.7 Phase-change mechanism in AIST. The bonding electrons are excited by laser light, causing the atoms in the amorphous phase to move. Finally, the central atom with three short (*red*) and three long (*dashed*) bonds crosses the centre of the distorted octahedron, interchanging a short and a long bond. *Green* resultant vector (reminiscent of a pseudospin) of short bonds. Resonant bonding between periodic short and long bonds leads to the crystalline A7 network. The *grey* sticks (*lower right*) correspond to the *red* bonds (*upper right*). Ag, silver; In, magenta; Sb, *blue*; Te, *yellow*. Reprinted by permission from Macmillan Publishers Ltd.: Nature Mater. [31], copyright (2011)

One may ask a question why bond interchange in AIST results in disordering while a rather similar process in GeTe (the order-disorder ferroelectric transition) does not. The answer to this question is not clear but it could be that the differences between the short and long bonds is more significant in the A7 structure of AIST. As a consequence, the atomic displacement resulting from the bond interchange in AIST and subsequent lattice relaxation may be large enough to break the long-range order.

11.4 Origin of Optical and Electrical Contrast

11.4.1 Coordination Change

After the initial proposal of the umbrella flip model that attributed the change in optical properties to a change in the local coordination of Ge atoms [1], Welnic et al. [27, 32] performed theoretical studies of the effect of the change in the Ge coordination number on the optical properties of phase change materials (both GeTe and $GeSb_2Te_4$ were investigated). The amorphous state was modelled by putting all (or some of the) Ge atoms onto tetrahedral symmetry sites. For both materials, the experimentally observed changes in optical properties were qualitatively reproduced.

At the same time, the joint density of states alone in the two phases [32] could not explain the experimentally observed changes from which it was concluded that the optical contrast between the two phases can only be explained by significant changes in the transition matrix elements resulting from the changes in the number of bonds and in the local order upon the phase transition. A very similar conclusion was reached in [33].

11.4.2 Resonant Bonding

Effect of *p*-Orbital Alignment

Realisation that resonant interaction between the constituent atoms significantly modifies the electronic properties of a solid [9, 10] allowed the authors to come up with an alternative explanation that the optical contrast was due to the presence (absence) of resonance bonding in the crystalline (amorphous) state. This idea has been further developed in [12] where the authors model resonance interaction using trigonal and monoclinic selenium as a prototypical system.

The stable phase of selenium possesses a structure where atoms form ordered spiral chains. This form of selenium can also be viewed as a distorted simple cubic lattice (Fig. 1.12) with two short intrachain bonds and four longer interchain bonds so that the *p*-orbitals on each site align along the interchain bonds. In the monoclinic form, on the other hand, the *p*-orbitals do not align. The authors use monoclinic selenium to model the local order in the disordered phase and obtain different dielectric functions calculated for trigonal and monoclinic selenium.

The authors further use orthorhombic GeTe to model the amorphous phase of GeTe and also obtain reasonable agreement with experiment (Fig. 11.8). From these results they conclude that angular disorder within *p*-bonded polymorphs with the same coordination number can cause large differences in optical functions and argue that the key difference in bonding between the crystal and amorphous phases is not a reversion to the 8-N rule bonding but the loss of second-neighbour angular order. This conclusion has been subsequently confirmed in [33].

The authors implicitly argue that the reproduction of the experimental data by means of changing the coordination number following the octahedral-to-tetrahedral geometry change reported in [32] could be due to the resulting misalignment of the *p*-orbitals. It may be instructive to note that in both cases (the umbrella flip and destruction of resonance bonds with the preserved covalent skeleton of the material) the bonding nature changes from "resonant" to purely covalent which is likely to be the general underlying reason for the drastic change in optical properties.

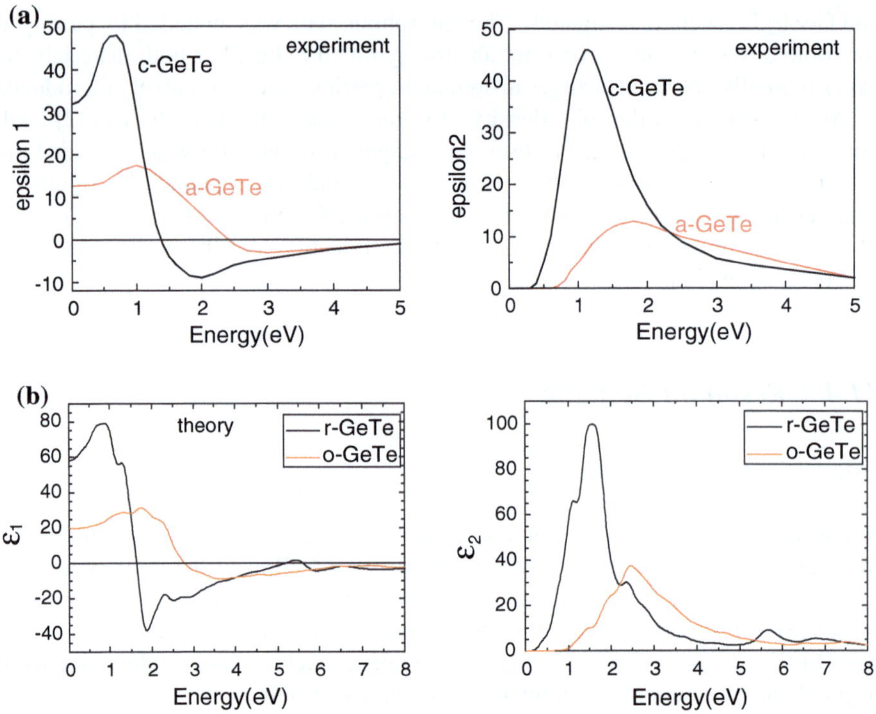

Fig. 11.8 **a** Experimental dielectric functions of crystalline and amorphous GeTe and **b** calculated dielectric functions rhombohedral and orthorhombic GeTe. Reprinted with permission from Huang and Robertson [12]. Copyright 2010 by the American Physical Society

Resonance Bonds versus Long-Range Order

Of interest is the observation that the principal change in the optical properties takes place during the initial stage of the phase-change process when resonance bonding is "turned off" but the atomic structure still preserves the average long-range order of the crystalline phase (Fig. 11.9). Subsequent relaxation and loss of the long-range order results in a much smaller change in the optical properties [13]. In other words, the destruction of "resonance" bonding causes a major change in optical properties, while subsequent disordering of the structure that does not alter the covalent nature of bonding results in a minor change in optical properties. The above results show that optical properties are primarily determined by the nature of the bonding between the atoms, i.e. by the *local* rather than the *average* structure. In other words, the change in properties is caused by a change in the local *bonding character* rather than being due to the loss of long-range order. It was subsequently suggested that dynamic change in bonding caued by laser-induced heating may account for the formation of transparent apertures in the excited state [34].

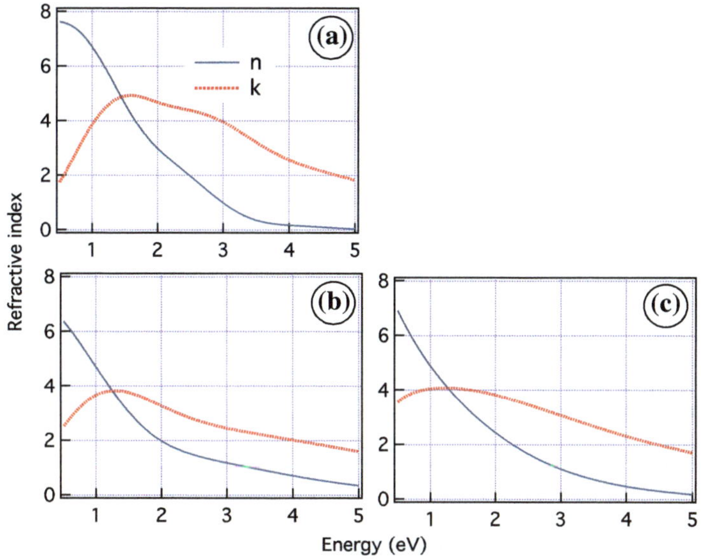

Fig. 11.9 Optical properties for equilibrated **a**, critically distorted **c** and amorphous **d** GeTe phases. Counterintuitively, the principal change occurs during the initial stage of the phase-change process when the system still possesses the *average structure of the crystalline phase*. Subsequent loss of long-range order produces only a minor change in optical properties. Reprinted by permission from Macmillan Publishers Ltd.: Nature Chem. [13], copyright (2011)

The results described above show that both a change in the *coordination number* and a change in the local *bonding nature* between the crystalline and amorphous phases cause strong (and similar) changes in optical properties. The latter factor is more general and includes the former since the tetrahedral Ge sites are characterised by purely covalent bonding. We would like to reiterate here that it is the nature of the bonding (strongly localised covalent bonds in the amorphous phase vs. less localised "resonance" bonds in the crystalline phase) that determines the optical properties of phase-change materials and the strong contrast between the two phases.

While the effect of bonding character on electrical properties of phase-change materials has not been investigated in sufficient detail, it may be suggested that the higher conductivity of the crystalline phase is also due not only to an increased degree of long-range order but also to the more extended nature of bonding electrons in the crystalline phase due to the formation of additional ("resonance") bonds. At the same time, the very strong bonding energy hierarchy, the resulting disorder and subsequent formation of tail states may be a reason why even in the crystalline phase the carriers undergo Anderson localisation [35].

The fact that the large contrast results from a change in the local structure and does not stem from the loss of long-range order suggests that certain crystal-crystal transitions may also be suitable for memory devices, such as the rhombohedral-to-orthorhombic transition in GeTe considered in [12]. Since such a transition is not associated with significant entropic losses that are unavoidable during a crystal-to-amorphous transition it may be beneficial for developing memories with lower energy consumption.

References

1. A.V. Kolobov, P. Fons, A.I. Frenkel, A.L. Ankudinov, J. Tominaga, T. Uruga, Nature Mater. **3**, 703 (2004)
2. A.V. Kolobov, A.S. Mishchenko, P. Fons, S.M. Yakubenya, J. Tominaga, J. Phys. Condens. Matter **19**, 455209 (2007)
3. J.L.F. Da Silva, A. Walsh, S.H. Wei, H. Lee, J. Appl. Phys. **106**, 113509 (2009)
4. S. Caravati, M. Bernasconi, T. Kühne, M. Krack, M. Parrinello, Appl. Phys. Lett. **91**, 171906 (2007)
5. J. Akola, R.O. Jones, J. Phys. Condens. Matter **20**, 465103 (2008)
6. J. Akola, R.O. Jones, Phys. Rev. B **76**, 235201 (2007)
7. J. Hegedüs, S.R. Elliott, Nature Mater. **7**, 399 (2008)
8. S. Kohara, K. Kato, S. Kimura, H. Tanaka, T. Usuki, K. Suzuya, H. Tanaka, Y. Moritomo, T. Matsunaga, N. Yamada, Y. Tanaka, H. Suematsu, M. Takata, Appl. Phys. Lett. **89**, 201910 (2006)
9. G. Lucovsky, R. White, Phys. Rev. B **8**, 660 (1973)
10. K. Shportko, S. Kremers, M. Woda, D. Lencer, J. Robertson, M. Wuttig, Nature Mater. **7**, 653 (2008)
11. J. Robertson, K. Xiong, P.W. Peacock, Thin Solid Films **515**, 7538 (2007)
12. B. Huang, J. Robertson, Phys. Rev. B **81**, 081204R (2010)
13. A.V. Kolobov, M. Krbal, P. Fons, J. Tominaga, T. Uruga, Nature Chem. **3**, 311 (2011)
14. A.V. Kolobov, P. Fons, M. Krbal, J. Tominaga, J. Non-Cryst. Solids (2011). doi:10.1016/j.noncrysol.2011.10.024
15. M.A. Paesler, D.A. Baker, G. Lucovsky, A.E. Edwards, P.C. Taylor, J. Phys. Chem. Solids **68**, 873 (2007)
16. S.R. Elliott, J. Non-Cryst. Solids **81**, 71 (1986)
17. H. Santoh, Y. Hongo, K. Tajima, M. Konishi, T. Saiki, in *Proceedings of the 2009 EPCOS Meeting*, (Aachen, Germany, 2009)
18. K. Makino, J. Tominaga, M. Hase, Opt. Express **19**, 1260 (2011)
19. P. Fons, A.V. Kolobov, T. Fukaya, M. Suzuki, T. Uruga, N. Kawamura, M. Takagaki, H. Osawa, H. Tanida, J. Tominaga, Jpn. J. Appl. Phys. **46**, 3711 (2007)
20. P. Fons, H. Osawa, A.V. Kolobov, T. Fukaya, M. Suzuki, T. Uruga, N. Kawamura, H. Tanida, J. Tominaga, Phys. Rev. B **82**, 041203 (2010)
21. A. Kolobov, Nature Mater. **7**, 351 (2008)
22. A.V. Kolobov, P. Fons, M. Krbal, R.E. Simpson, S. Hosokawa, T. Uruga, H. Tanida, J. Tominaga, Appl. Phys. Lett. **95**, 241902 (2009)
23. T. Ohta, J. Optoelectron. Adv. Mater. **3**, 609 (2001)
24. M. Konishi, H. Santo, Y. Hongo, K. Tajima, M. Hosoi, T. Saiki, Appl. Opt. **49**, 3470 (2010)
25. X.B. Li, X.Q. Liu, X. Liu, D. Han, Z. Zhang, X.D. Han, H.B. Sun, S.B. Zhang, Phys. Rev. Lett. **107**, 015501 (2011)
26. D.Q. Huang, X.S. Miao, Z. Li, J.J. Sheng, J.J. Sun, J.H. Peng, J.H. Wang, Y. Chen, X.M. Long, Appl. Phys. Lett. **98**, 242106 (2011)
27. W. Welnic, A. Pamungkas, R. Detemple, C. Steimer, S. Blugel, M. Wuttig, Nature Mater. **5**, 56 (2005)
28. S. Caravati, M. Bernasconi, T. Kuehne, M. Krack, M. Parrinello, Phys. Rev. Lett. **102**, 205502 (2009)

29. S. Sharma, S. Sikka, Prog. Mater. Sci. **40**, 1 (1996)
30. A.V. Kolobov, P. Fons, J. Tominaga, A.I. Frenkel, A.L. Ankudinov, S.N. Yannopoulos, K.S. Andrikopoulos, T. Uruga, Jpn. J. Appl. Phys. **44**, 3345 (2005)
31. T. Matsunaga, J. Akola, S. Kohara, T. Honma, K. Kobayashi, E. Ikenaga, R.O. Jones, N. Yamada, M. Takata, R. Kojima, Nature Mater. **10**, 129 (2011)
32. W. Welnic, S. Botti, L. Reining, M. Wuttig, Phys. Rev. Lett. **98**, 236403 (2007)
33. S. Caravati, M. Bernasconi, M. Parrinello, J. Phys. Condens. Matter **22**, 315801 (2010)
34. R.E. Simpson, P. Fons, X. Wang, A.V. Kolobov, T. Fukaya, J. Tominaga, Appl. Phys. Lett. **97**, 161906 (2010)
35. T. Siegrist, P. Jost, H. Volker, M. Woda, P. Merkelbach, C. Schlockermann, M. Wuttig, Nature Mater. **10**, 202 (2011)

Part IV
Applications

Chapter 12
Memory Devices

While chalcogenide glasses have numerous applications, most of them rely on their optical properties, such as strong optical nonlinearity. The interested reader is referred to the following recent sources [1–3] where applications of chalcogenide glasses are described. In this chapter we concentrate on phase-change materials whose application in memory devices has a strong effect on modern society. Some aspects of phase-change memory technology have been recently reviewed in [4–6].

12.1 Optical Memories

Optical memory comprises compact discs (CDs), digital versatile discs (DVDs) and Blu-ray discs (BDs). Since compact discs were first released in 1982, optical discs have spread worldwide quickly and household vinil record players were replaced by CD players within a decade. Nowadays, various kinds of optical discs are not only available as music and film sources, but also as distribution media for computer software. Phase-change materials are consumer friendly and became suitable for re-writable media in CD and DVD families because a reflectivity difference between the recorded and non-recorded states can be detected using simpler systems than magneto-optical (MO) materials that need an additional polarising plate to discriminate a Kerr-rotation angle of less than a degree. Presently, the same phase-change alloys that are used in rewritable optical media are also the leading candidate for non-volatile phase-change random access memory (PCRAM).

Below, we first review the background of the development of the CD and DVD, and then step forward to the recording mechanism, switching speed, write-erase cyclability and optical near-field characteristics in phase-change optical discs. This chapter is concluded by a review of non-volatile solid-state memory.

A. V. Kolobov and J. Tominaga, *Chalcogenides*, Springer Series in Materials Science 164, DOI: 10.1007/978-3-642-28705-3_12,

Fig. 12.1 A gramophone (phonographic) disc

12.1.1 Compact Disc and Digital Versatile Disc Families

Predecessors

Before 1982, the year the CD was released, popular music media were gramophone records (commonly known as phonographic records in American English) and magnetic tapes. The record is an analogue sound storage medium consisting of a flat disc (vinyl has been the material most commonly used after about 1950) with an inscribed, modulated spiral groove. The gramophone system utilises a diamond or sapphire tip that vibrates in accordance with the shape of the modulated groove, and these vibrations are transformed into a weak electrical current that is subsequently transferred to an amplifier connected to a speaker system.

The way to the vinyl disc started in the mid-1850s. The ealiest recording of sound dates back to 1857 and is associated with the phonautograph, patented by Léon Scott de Martinville, that used a vibrating diaphragm and stylus to graphically record sound waves as tracings on sheets of paper, purely for visual analysis and without any idea of playing them back. The first system capabale of both recording and reproducing the sound was the phonograph invented in 1877 by Thomas Edison. The recording medium of a phonograph was a cylinder covered with a tin foil subsequently replaced by a wax cylinder. Soon after this, the cylinder was superseded by a flat disc thanks to Emile Berliner who subsequently founded *The Gramophone*. Berliner's gramophone records were more convenient to handle, cheaper to make and louder in reproduction. For almost a century, the gramophone records (Fig. 12.1) have been loved by millions of musicians and music lovers worldwide.

Magnetic tape recording, on the other hand, was first demonstrated in 1898 by Valdemar Poulsen and was called telegraphone. For historical fairness it should be noted that originally the idea was introduced in 1888 by Oberin Smith, although he could not complete his system. A magnetic tape for recording and readout was made of steel with a 3.0 mm width, and the drum was rotating at a speed of 1.2 m/s. The metal tape was later replaced by a plastic film coated with magnetic powder (metal or ferrite), rolled up and placed into a small box. One of the advantages of magnetic tapes over gramophone was that the magnetic system had a potential of not only being a music tape recorder but also was a step towards a video tape recorder, VTR. In the early 1950s, several companies started to develop video tape recorders, and the first VTR system was developed in 1953 in United States where TV broadcasting companies badly needed such systems in news program business because of the time difference between the East and West coasts.

As time went on and noise reduction systems had been invented, listeners of music gradually required systems with more realistic sound of higher quality similar to that of the source. In addition, both gramophone discs and music tapes were degrading under wear or abrasion due to the contact with a diamond tip or a magnetic head, with concomitant sound quality degradation and increasing noise after numerous playbacks. Furthermore, a magnetic tape could become tangled, while a gramophone record often twisted or bent even at relatively low temperature and humidity.

Characteristics of CDs and DVDs

Nobody could think that the exisiting analogue recording media and devices would be replaced by digital ones when the laser, in particular a semiconductor laser with a wavelength near 830 nm or shorter, was developed by General Electric in 1962.[1] Consumers had to wait for almost thirty years before witnessing the sound revolution. It was in 1982 that an alliance including Sony and Philips released a digital disc music player to the music market. The disc diameter is 12 cm, which is smaller and easier to handle than a gramophone disc. It is intersting to note that CD's capacity (with the corresponding audio recording duration of 74 min) was determined by the need to hold on a singe disc Beethoven's 9th symphony. This optical disc size has been subsequently superseded by digital versatile disc (DVD) and Blu-ray disc (BD).

Figure 12.2 shows schematic of the fundamental read-write optical system used in a CD drive [7]. Thanks to a semiconductor laser diode and a focusing lens with a high numerical aperture (NA), physically contact-less recording and readout was realised. In far-field optics, the spot size ϕ of the focused laser beam is determined by a simple equation due to the diffraction of light [8]:

$$\phi = \lambda/(2\mathrm{NA}) \tag{12.1}$$

where λ is the wavelength of the light used and NA stands for the numerical aperture of the lens. Combinations of the laser wavelength and NA used in CD, DVD and Blu-ray systems are listed in Table 12.1. The shorter the wavelength of the laser, the higher the recording and storage density. However, the success of the story does not always follow one's expectation, although a shorter wavelength laser (beyond

[1] Zhores I. Alferov and Herbert Kroemer were awarded the Nobel Prize in physics in 2000 "for developing semiconductor heterostructures used in high-speed- and optoelectronics" that form the basis of semiconductor lasers.

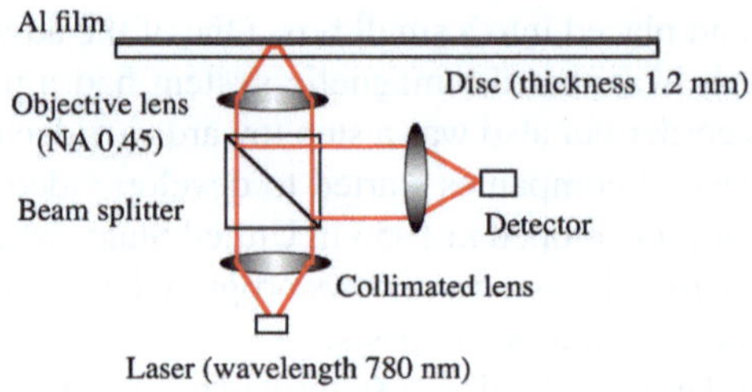

Fig. 12.2 A schematic representation of the optical pickup system of a CD player. The laser beam is collimated and introduced into a beam splitter, then focused by an objective lens. The beam is focused through a transparent disc substrate (1.2 mm thick) on an Al reflector deposited on the back surface of the disc. On the surface, pre-patterned data for music are fabricated as pit sequences with a depth of about 500 nm

Table 12.1 Specifications of CD, DVD and Blu-ray discs

Disc type	Wavelength (nm)	NA	Spot size (μm)	Minimum pit length (μm)	Release year
CD	780	0.45	1.0	0.83	1982
DVD	650 (635)	0.60	0.6	0.40	1996
Blu-ray	405	0.85	0.3	0.14	2003

400 nm) has recently been developed. First, the light absorption of a polycarbonate disc increases as the wavelength gets shorter, and secondly the focus depth becomes shorter for larger NA in order to obtain a higher resolution and density. For example, while the 1.2 mm thick polycarbonate substrate used in CD is tough and rigid to realise easy handling, a 0.1 mm thick BD disc is much softer and easily bends by hand. As a consequence, an additional 1.1 mm dummy plate is required to make the BD disc strong and tough. In DVD, two 0.6 mm disc substrates are glued together. Thirdly, the operating distance between the bottom surface of a higher NA lens and a recording disc surface gets shorter, resulting in an increased possibility of the lens touching the rotating disc resulting in a crash. To avoid such failures and to increase the density further, several technologies have been evaluated and developed. We discuss some of them in subsequent sections.

In CD, DVD and Blu-ray drives, the semiconductor laser that is necessary to read and write the data is installed in a small module (optical pick-up). The beam profile at the exit window of the laser is not circular and has to be reshaped into a circle by a cylindrical lense. The collimated beam is next delivered to an objective lens and focused on the storage medium after passing through a beam splitter and a quarter-wavelength plate. The objective lens is supported and focused by a two- or three-dimensional electro-magnetic actuator that is used to compensate for distance deviation from the focal depth due to the disc rotation. It should be noted that in case of a CD a laser beam focused by the objective lens makes a spot on the recorded data pit stream through a 1.2 mm thick polycarbonate disc as mentioned earlier [9]. Therefore, the spot size on the polycarbonate disc is much larger than that of the focused spot on the pit stream, which is stamped on the backside surface. If dust particles or oil spots (e.g. finger prints) are deposited on the polycarbonate disc surface, the beam can be focused on the pit stream even though a part of the light is scattered on the surface. This is a great advantage of CD and DVD systems over near-field optical data storage, AFM or STM storage using a needle-like tip or probe [8].

A read-only memory (ROM) disc-like CD is made of a 12 cm diameter and 1.2 mm thick transparent polycarbonate disc with a thin Al alloy film deposited atop the protection coating. The incident laser beam is to a large extent reflected (>80%) by the film if the surface is flat. The back surface of the polycarbonate disc has an information pit stream embossed upon it from the central area (starting 23.0 mm from the disc centre) outward towards the edge area (ending at 58.5 mm from the disc centre) in a spiral shape. The width between adjacent pit streams in the radial direction is stipulated to be 1.6 μm for the CD format [9]. Thus, the total length of the stream is roughly 5 km, and 20,000 tracks can be pre-recorded or stamped onto a single CD. The maximum storage capacity of a CD is 700 MB. In a DVD, the optical readout method and system are essentially the same as in CD except for the laser wavelength of 650 nm and a 0.60 NA lens. Thus, the laser spot size is further reduced and the storage capacity consequently increases to 4.7 GB. In Blu-ray discs, use of a laser with $\lambda = 405$ nm and a 0.85 NA lens enables an increase in the storage capacity to 25 GB [10].

Disc substrates of CD, DVD and Blu-ray discs are commonly produced from a molten polycarbonate polymer using a mould-injection method. To make a master disc with the pit streams of music information, first a glass disc is prepared. After polishing the surface to a few nanometer roughness and depositing a photoresist film by spin coating, the glass disc is transferred to a laser-cutting processing, where the pit streams are recorded along with the music programs by an ultra-violet laser beam. After development, the glass disc is dipped into a chemical bath containing Ni ions for electrical plating. Finally, the Ni film grown on the glass is separated from the master glass. This Ni plate is usually called *mother*. A replica made from *mother* is used in the same process to fabricate numerous *children*, especially when millions of copies of a CD are required for record sellers.

Figure 12.3 shows SEM and AFM images of a CD substrate surface after removing the Al reflective layer. The sound track contains nine different pit sizes. Currently, in CD families, the laser beam does not read pits or marks of identical size, but the data are actually addressed at the edges (front and tail) of the pits. This readout method is called edge-position readout, while the method to read a pit position is called pit-position readout. The former method allows one to increase the recording density by 1.5 compared to the latter method.

The discs described above are read-only memory (ROM) discs. As music sounds or a movie are premastered on the disc substrate surface, no one can touch or change the source. When the first format and specification for the CD were issued in 1982, no one expected recordable or erasable CDs. However, as the number of CD users increased explosively, garage band musicians and some hardcore users urgently demanded a recordable CD for spreading their original sounds and music into the local

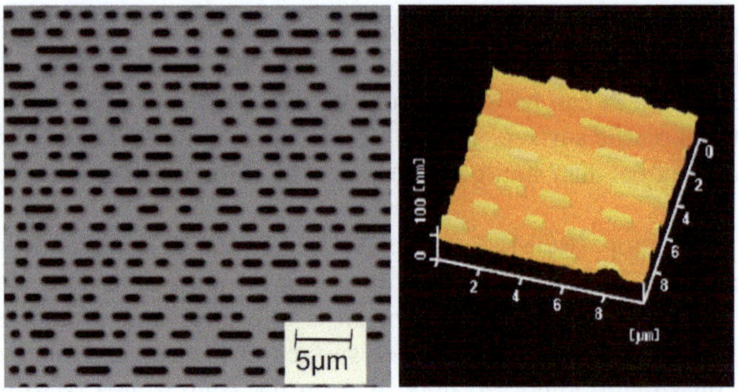

Fig. 12.3 SEM (*left*) and AFM (*right*) images of a CD disc surface after removing the Al layer. Z-height is inversed to make pits appear clear. The CD track contains nine different pit lengths

community. Finally, in 1987 the CD alliance introduced a recordable CD (CD-R), although research and development (R&D) divisions in several companies had been investigating this possibility since the early 1980s. In the CD format and specifications, referred to as the Red Book, there was one technical difficulty for recordable CDs: more than 70% reflectivity from the disc surface was specified, and the backward compatibility to the already spread huge number of CD drives on the market had to be insured. Before CD-R, several recordable optical discs were invented, but all of them except for CD-R could not satisfy the required specification for the reflectivity [9]. Therefore, these discs had been pushed out of the CD family market. Optical phase-change (PC) rewritable discs (reflectivity $< 40\%$) were not an exception at that moment.

12.1.2 Recordable Discs (CD-R and DVD-R)

The CD-R is an optical storage disc using an organic dye as the recording layer and is fully compatible with CD [11]. The disc structure is very simple and is comprised of two layers: one is a cyanine dye (Fig. 12.4) layer absorbing light with a wavelength of around 780 nm, and the other is a highly reflective metallic layer such as silver or gold (presently, Al alloys are used due to cost considerations). These two layers are deposited over a polycarbonate substrate, on which a spiral-shaped guide track for the laser is cut from the innermost track outwards with a track pitch the same as that of CD, i.e. 1.6 μm. A gold or silver film on top is used to obtain the high reflectivity required for the CD specification.

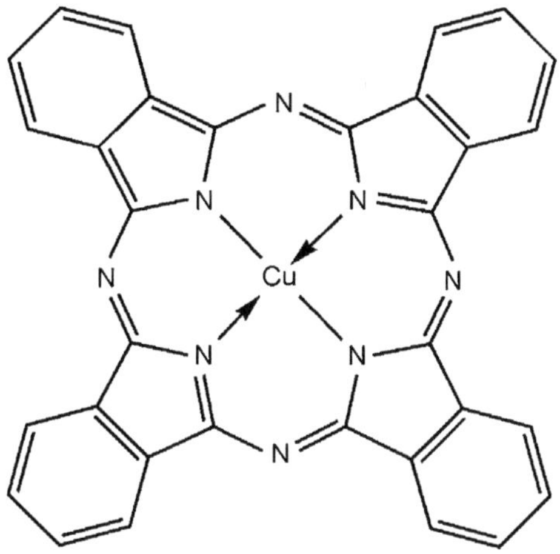

Fig. 12.4 A typical organic dye of Cu phthalocyanine. π electrons are conjugated and form a flat sheet molecular structure

12.1.3 Erasable Discs (CD-RW, DVD-RW and DVD-RAM)

Besides write once recordable discs, the demand for erasable discs is also high because of a need for recovery of a recording mistake or temporary backup of data. Technically, an erasable disc is more difficult to fabricate and design than CD-R and DVD-R. The recording material used in the disc must be able to recover its initial characteristics after an erasing procedure. Since the early 1980s, various approaches have been investigated to realise an erasable disc. Until the early 1990s, the development of magneto-optical (MO) discs was given priority and some MO systems were commercialised. In a MO disc system, Kerr-rotation or Faraday rotation effects were used as the principle of recording and erasing [12]. For example, an amorphous Tb-Fe-Co alloy shows a relatively large Kerr rotation (ca. 1 °) angle in magnetisation and by applying an external magnetic field perpendiculat to the film surface one can turn the magnetisation direction up and down. To read the recorded data, a linearly polarised laser beam is focused on the recorded track and the reflected beam (transmitted beam in systems that utilise the Faraday effect) is observed through a linear polariser. An optical detector can monitor the Kerr rotation as a reflectivity change at the up-magnetised recorded spot and the down-magnetised non-recorded area. The Kerr rotation effect of the material is relatively large for longer wavelengths (780 or 650 nm) but becomes smaller for the shorter wavelength range of less than 500 nm needed for denser storge. As the development of optical discs had shifted towards a higher storage density using shorter wavelengths (400 nm instead of 780 nm), this disadvantage became fatal, and finally MO discs disappeared from the market. Instead, attention has gradually turned to the phase-change system since the mid-1990s because of their good compatibility with CD and DVD optical systems.

The R&D of phase-change optical storage has a long history starting with the discovery of the memory effect (a reversible change in resistivity) in some chalcogenide alloys by Ovshinsky [13]. In the late 1960s Hitachi began to study

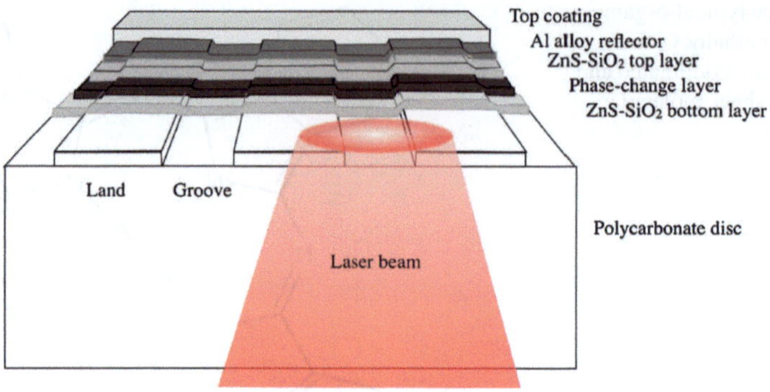

Fig. 12.5 A typical structure of a phase-change erasable optical disc

phase-change materials and in the early 1970s Matsushita Electric (Panasonic) initiated a similar research [6]. Matsushita Electric tested numerous compositions and eventually discovered in the late 1980s the promising Ge-Sb-Te alloys, and especially, the $Ge_2Sb_2Te_5$ composition [14, 15]. It is worth mentioning that the series of annual meetings dedicated to phase change memories started in Japan as early as 1985. This work resulted in commercialisation of DVD-RAM by Hitachi and Matsushita Electric in 1998 [16, 17]. At about the same time, Ricoh, Toray Chemicals, TDK and other major Japanese electronics companies joined the search for other compositions and alloys [18–21]. In the early 1990s, Ricoh developed a Sb-rich alloy without Ge, namely Ag-In-Sb-Te (AIST) alloy [22] that is currently used in DVD-RW [23]. The interested reader may find additional information in [24].

In phase-change erasable optical discs, neither an optical polariser nor an external magnet to rotate the magnetisation is required. Therefore, the optical pickup system that is a module equipped with a semiconductor laser, collimating lens, beam splitter, objective lens and corresponding actuator, and a four-sector differential photo detector can be simply designed to be compatible with CD and DVD systems. The $Ge_2Sb_2Te_5$alloy and the Sb-rich alloy both generate about a 30% difference in reflectivity by optical interference in combination with two $ZnS\text{-}SiO_2$ films and Al-alloy films, resulting in more than 60% signal contrast. Due to the appearance of the phase-change discs, the reflectivity contrast specification was eventually modified.

Figure 12.5 shows the structure of a phase-change disc. All phase-change discs have the same structure: the bottom $ZnS\text{-}SiO_2$ layer used to control the maximum reflectivity level, a phase-change layer erasable recording, the top $ZnS\text{-}SiO_2$ film to enhance the optical contrast between the crystalline and amorphous states and an Al alloy layer to release heat accumulated in the phase-change film as quickly as possible.

Using such a structure, the reflectivity contrast between the amorphous and crystalline state is significantly enhanced. The reflectivity change can be easily observed by an optical microscope as shown in Fig. 12.6. The uppermost layer has the additional role in determining the total reflectivity level. The bottom layer thickness is

Fig. 12.6 Optical contrast between the amorphous and crystalline states observed by a high-resolution laser microscope. Recorded marks (*black*) are in the amorphous state, the surrounding area is crystalline

determined by the laser wavelength, and becomes thinner as the wavelength becomes shorter. The phase-change layer thickness is fixed around 15–20 nm in DVD-RAM and -RW, and becomes thinner as well when using a shorter wavelength laser. The top $ZnS\text{-}SiO_2$ film must be thin enough. Otherwise, if its thickness is larger than 40 nm, the heat generated during the recording process is not released properly, resulting in a decrease in cyclability [25]. TEM images of recorded marks are shown in Fig. 12.7. As one can see the crystal grain size is larger in Ge-Sb-Te than in AIST alloys.

The phase-change recording and erasing cycle is somewhat similar to a thermodynamical cycle. In every record/erase cycle, the system (a phase-change mark whose size is about one micron or less) does work. The work is associated with the volume change of the recorded bit (about 6–10%) and a large part of the input energy is wasted as entropy [26]. From rough estimates, the entropic losses can consume as much as 95% of the input energy. This issue is also very important in phase-change non-volatile solid-state memory as described below in this chapter.

12.1.4 Super-RENS Discs

Optical systems that use far-field optics when light propagates between an object and the detector suffer from the optical diffraction limit described by Eq. 12.1. One of the solutions to go beyond this limitation is to use optical near field. The best known example is a tip of a scanning near-field optical microscope. When light is squeezed through an aperture with a diameter smaller than the wavelength of the light, it cannot propagate further in the z direction. At the same time, a certain fraction of the electromagnetic field, called optical near field, can soak through the aperture; its intensity decreases rapidly away from the aperture as shown in Fig. 12.8. The properties of the near-field light can be obtained by solving the Maxwell equation for

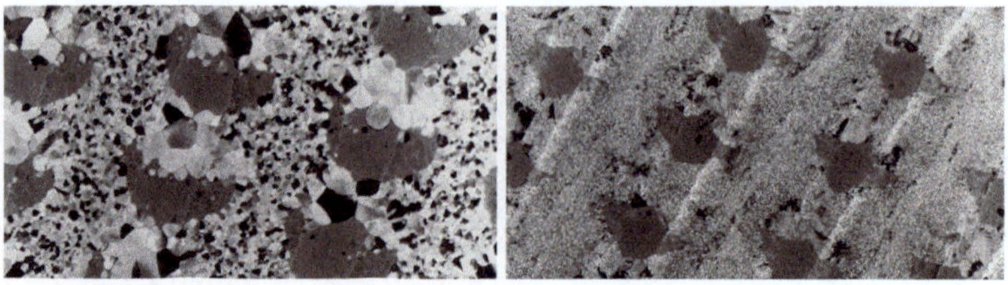

Fig. 12.7 The shortest amorphous marks written in DVD-RAM (*left*) and in DVD-RW (*right*). Even when using different materials (GST in -RAM and AIST in -RW discs), the mark shapes are similar due to control of the cooling rate. In general, GST alloys are made of small crystalline grains around amorphous marks, while AIST alloys have relatively large grains; the edge shape of the marks is clearer in AIST than in the case of GeSbTe because the crystallisation process is growth limited. (Courtesy of TDK Corporation)

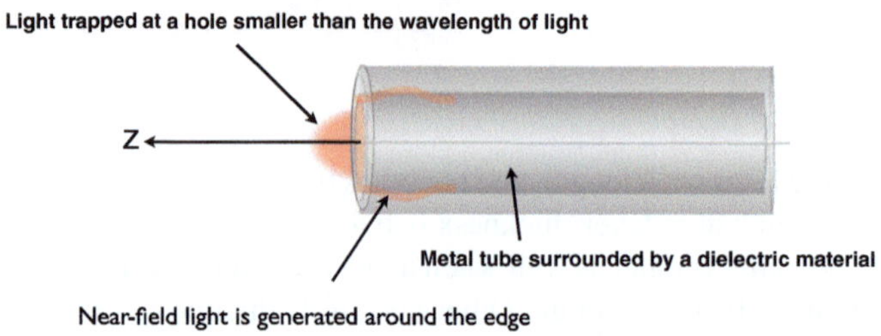

Fig. 12.8 Light cannot propagate in a tunnel with a diameter smaller than the wavelength. However, it can penetrate through the wall as near-field light

an electromagnetic wave with the appropriate boundary conditions [27]. The light trapped in the aperture has been described in terms of localised surface plasmon polaritons, or localised surface plasmons [8].

The super-resolution near-field structure (Super-RENS) optical disc is a development at the cutting edge of modern optics using local plasmons. Super-RENS was invented with the objective to overcome the optical diffraction limit and to increase the storage density of optical discs [28, 29].

One big technological issue that must be cleared in order to apply local plasmonics to optical data storage is the data transfer rate with a typical requirement being more than ten million bits per second for high-definition video reproduction. Scanning near-field optical microscopes (SNOM) with a spatial resolution beyond the optical diffraction limit potentially enable ultra-high density recording [30]. Almost all commercial SNOMs, however, depend on a piezoelectric-scanning drive as shown in Fig. 12.9 [31]. Therefore, the scanning area is limited to a maximum of $100 \times 100\,\mu m$ square or a much smaller area, and the typical speed of less than 0.1 m/s. Even for a SNOM probe aperture size of 50 nm, the maximum data transfer speed is about 2 Mbps, a value that cannot satisfy the requirements for video reproduction. In addition, it is simply impossible to record and reproduce a two-hour video. The fatal issue using a SNOM is intrinsic to the SNOM system itself, i.e. using a piezoelectric stage

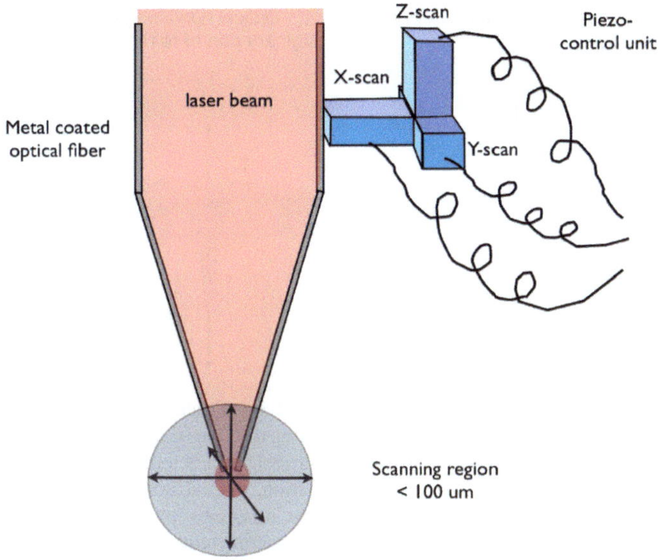

Fig. 12.9 An optical fibre or a metal-coated tip used in SNOM cannot be used for optical recording or data reproduction because spatial and positional controls of the tip depend on a piezo-control unit that is rather slow while the required data transfer rate in real optical storage systems is much higher (>10 Mbps)

and controlling the distance between the SNOM tip and a recording medium surface by the weak force (atomic, electrical and shear forces) generated between the two materials. The weak forces are precisely fed back and controlled in order to maintain a separation distance within 100 nm or less. As mentioned earlier, the intensity of the electrical field falls exponentially from the surface. To keep the light intensity at the recording medium constant at the high speed required for video reproduction is at the moment impossible using a SNOM-like system.

In the development of Super-RENS, a new innovation was required to overcome the conventional concept when an optical recording source is installed in a recording unit or module (like an optical pickup unit), and the light reaches the recording medium going through a space between the light source and the medium, i.e. the optical recording module was always separated from the recording medium. In super-RENS, the recording medium is a part of the recording module: an aperture, that can be opened and closed using a laser and far-field optics, is incorporated into the disc itself as discussed below.

The spatial separation between the optical module and the recording medium in an optical read-out system is typically about four orders of magnitude larger that that used in SNOM. To ensure a constant distance even at high rotation speed, an additional optically non-linear (chalcogenide) layer is used. When the laser beam from an optical pickup used in DVD or Blu-ray systems is focused on a chalcogenide film under the far-field (i.e. diffraction limit) conditions, the minimum optically

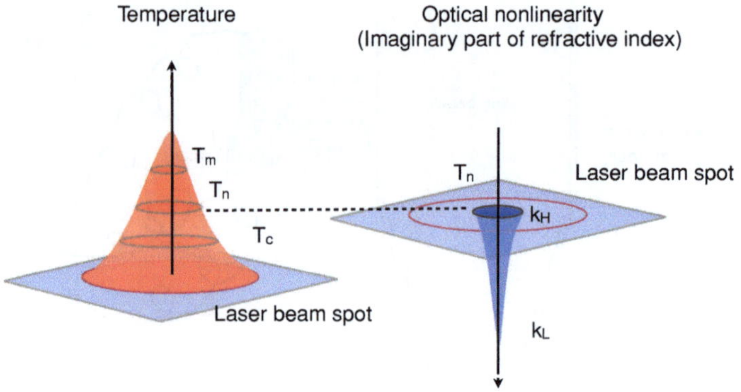

Fig. 12.10 The temperature profile and threshold switching of a chalcogenide film. The temperature of the chalcogenide film has a Gaussian profile reflecting the Gaussian distribution of the laser beam intensity (focused by an objective lens). Due to intrinsic characteristics of the phase transition, three phenomena are important, namely, crystallisation, optical nonlinearity and melting. Once the temperature reaches the threshold point (T_n), the imaginary part of the refractive index, k_H steeply drops to k_L. (T_c and T_m are crystallisation and melting temperatures, respectively.) The spot diameter can be one-fifth of the laser spot size or even smaller. Using the optical nonlinearity of the chalcogenide thin film enables optical recording and consequently the reproduction of data beyond the far-field diffraction limit

determined spot using a laser with a wavelength of 635 nm, is about 530 nm. Can a recorded mark be smaller than this?

The answer is 'yes'. The additional chalcogenide alloy film on top of the medium layer plays the decisive role. Some chalcogenide alloys, such as $Ge_2Sb_2Te_5$and Sb_2Te show a large optically nonlinear response that depends on temperature (achieved due to laser-induced heating). The temperature profile usually has the same shape (typically Gaussian) as that of the beam intensity profile. That is, the centre of the spot has the highest temperature and at the edge the temperature is lower. The film exhibits an intrinsic optical response, in particular, a large change in the refractive index in response to the temperature change, as shown in Fig. 12.10. While the exact temperatures vary from one chalcogenide alloy to another, $Ge_2Sb_2Te_5$and Sb_2Te exhibit a large change in the refractive index with a very steep threshold temperature at around 400 °C, resulting in the formation of an optically transparent window with a diameter of less than 50 nm. This window size is much smaller than the laser spot defined by the optical diffraction limit. Such a window cannot pass or transmit the far-field light to the backside of the film, instead local plasmons are generated there [8].

Because the transparent window is a dielectric while the medium surrounding it is metallic, at the edge of the window, a strong optical near field is present. As the film dynamic window is placed on a recording layer with a transparent spacer layer (<50 nm thickness), strong electric fields and high spatial frequency components are scattered if small pit or marks pass through the region. This is illustrated in Fig. 12.11. Thanks to the window with a steep circle boundary in the laser spot, super-RENS

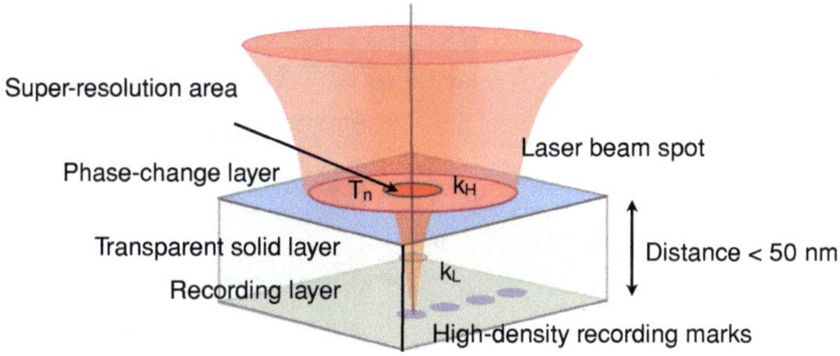

Fig. 12.11 Principle of a super-RENS optical storage disc. When a phase-change layer is illuminated by a laser beam, its temperature rises. Once the temperature exceeds T_n, an optical window is formed due to nonlinearity of the refractive index. Although propagating light cannot pass through the window, a non-propagating light field, such as local plasmons soak through the aperture to a depth of about 50 nm. The distance between the aperture and the recording layer is determined by the thickness of a dielectric film. If the laser power is precisely controlled, the window size is fixed and recording marks are clearly formed with a precise diameter. When the laser power is decreased below T_n or turned off the window closes. The window can catch up with the speed even when the structure is rotated at the same speed as a DVD or Blu-ray disc

media can resolve individual 75 nm pits, and playback a 50 GB movie at the same speed as that of a Blu-ray system.

In actual super-RENS media, the disc is a multilayer structure that consists of the following layers: ZnS-SiO_2 (60–100 nm)/PtO_x (4 nm)/ZnS-SiO_2 (20–40 nm)/PC layer (10–15 nm)/ ZnS-SiO_2 (20–40 nm)/Reflector. The phase-change layer is made of $Ge_2Sb_2Te_5$, or Sb_2, or In-doped Te, or Sb alloys. The PtO_x layer is used as a recording layer and the recording mechanism consists of the formation of Pt nanoparticles within nano-bubble pits (Fig. 12.12) due to laser-Induced heating through the following reaction [32]:

$$PtO_x \rightarrow Pt + O_2 \uparrow \tag{12.2}$$

The optimum thickness of each layer is determined using a computer simulation for each particular laser wavelength in an optical system. Figure 12.13 shows a typical super resolution effect observed from a super-RENS disc designed for a Blu-ray system. At 1.0 mW of laser power, the signal (carrier-to-noise ratio, CNR on a log scale) steeply drops at the optical resolution limit (ca. 156 nm), while the CNR of 100 nm pit pattern is preserved at 40 dB at a laser power of 4.0 mW power due to the super resolution effect.

Figure 12.14 shows a comparison of the super resolution effect using different optics. The effect is normalised to the wavelength and numerical aperture of the objective lens used. As can be seen from in Fig. 6.14, the resolution is determined only by the wavelength and the lens numerical aperture. This provides clear evidence that the super resolution phenomenon is not related to the optics but to a thermal threshold mechanism due to the beam focusing.

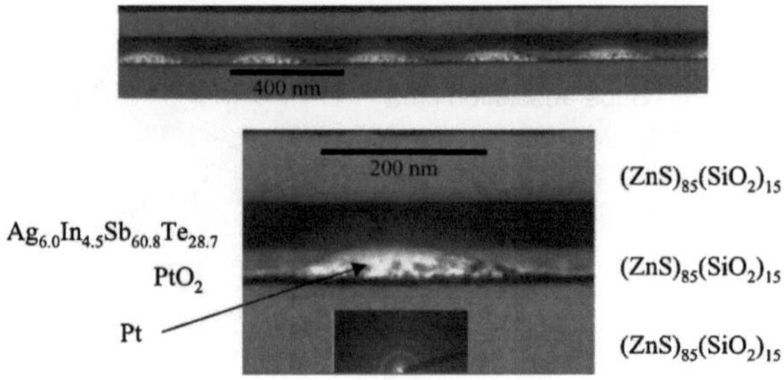

Fig. 12.12 TEM bright field images of as-recorded pits at 10 mW (*top*) and the magnified image of a recorded pit (*bottom*). The inset shows selected area electron diffraction pattern around recorded pits. Reprinted from Kikukawa et al. [32]. Copyright 2002 by the Amercian Institute of Physics

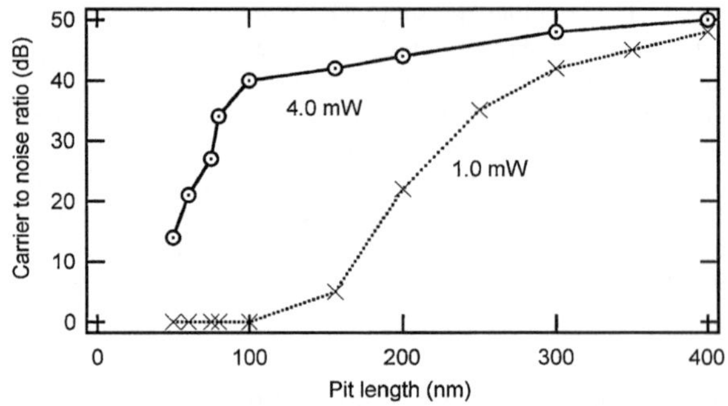

Fig. 12.13 Super resolution effect of a PtO_x super-RENS disc measured by a system using a 405 nm wavelength and a 0.65 NA lens. The optical resolution limit is 156 nm

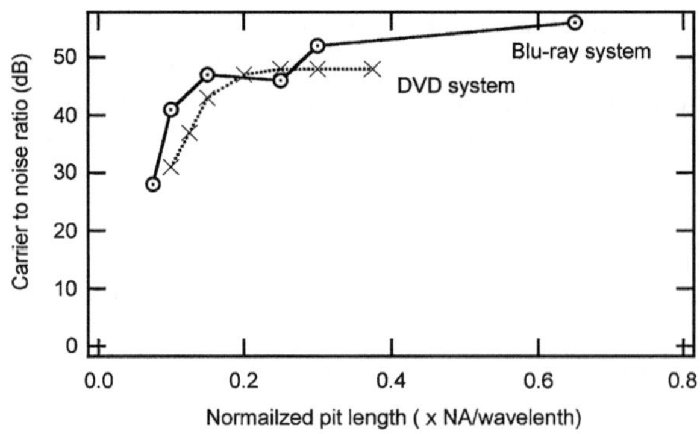

Fig. 12.14 Signal intensity drops against normalized pit length (d × NA/λ) in DVD and Blu-ray systems

InSb was shown to be another promising material for super-resolution recording [33, 34]. In 2009, a super-resolution disc with an InSb active layer with a capacity of 50 GB (Blu-ray × 2) was used to demonstrate video playback of a HDTV signal (Fig. 12.15) [29].

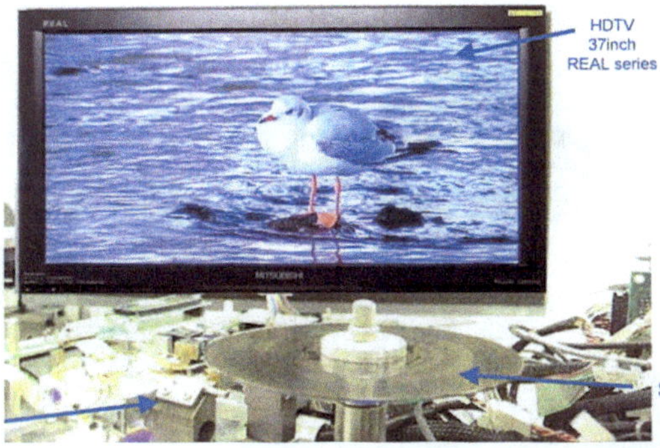

Fig. 12.15 A snapshot of a HDTV video recorded using a Super-RENS disc. Reprinted with permission by Japan Society of Applied Physics from Nakai et al. [29]

12.2 Non-Volatile Electronic Memories

Non-volatile electrical memory (NVM) has recently attracted increased attention because of its higher storage density, easier handling and especially because it is free of an energy consuming rotating optical drive or a hard disk drive equipped with a rotating motor. In the last decade, the storage capacity of non-volatile USB memory has exceeded 1 GB with a low price compatible with that of a recordable DVD medium. At the moment of writing 256 GB USB memory stickes were commercially available at ca. $600 and 32 GB at ca. $10. Now, NVMs are essential components of mobile and smart phones, and e-book devices. Although almost all NVMs are at the moment based on electron trapping (the so-called Flash Memory), charge-based storage is facing serious problems with further size reduction in order to increase the data capacity. As the device becomes smaller, the cell size to trap electrons must be as small as possible. The size reduction, on the other hand, reduces the number of electrons trapped in the cell, even if a high voltage is applied because electrons repel each other due to Coulomb forces. This has a negative effect on device reliability and write/erase cyclability ($<10^6$). As alternative devices, phase-change RAM (PCRAM or PRAM), magnetoresistive RAM (MRAM), and resistive RAM (RRAM) are being developed with the goal of replacing Flash. Excellent scaling characteristics of phase-change alloys [35–37] makes them a very attractive candidate for future memory devices (Fig. 12.6). In February 2012, Samsung announced 20 nm 1.8 V 8 Gb PC-RAM with 40 MB/s program bandwidth [38].

12.2.1 Phase-Change Random Access Memory

PCRAM cell is a two-terminal device whose resistence can be reversibly switched. The cell operation is based on the current-pulse-induced changes in the active

Fig. 12.16 Samsung's 512 Mb (64 MB) PRAM cells using 60 nm technology (reproduced from http://www.samsung.com with permission from Samsung)

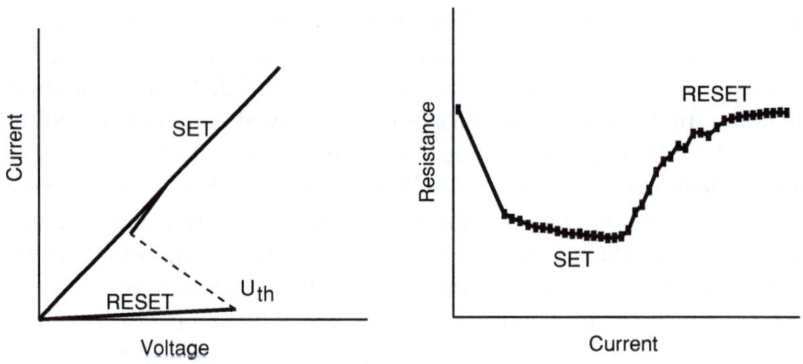

Fig. 12.17 Current-voltage *I–V* (*left*) and programming (*R–I*) (*right*) curves for a PCRAM cell

chaclogenide layer. From the amorphous RESET state the device can be turned into the crystalline SET state by application of a current pulse that is sufficient to heat the active layer above the crystallisation temperature. As the applied voltage reaches the threshold value of V_{SET}, the device switches into the low-resistance crystalline SET state. A typical *I–V* curve of a PCRAM device is shown in Fig. 12.17 (left panel). On the other hand, application of a short intense pulse that can supply enough energy through Joule heating can melt the chalcogenide alloy that is subsequently quenched into the amorphous (high-resistance) RESET state.[2]

Device performance is usually also characterised by the so-called programming (*R–I*) curves that show current values necessary to switch the cell between the two

[2] Recently there has been growing evidence that the amorphisation process can also be athermal under certain conditions [39–43].

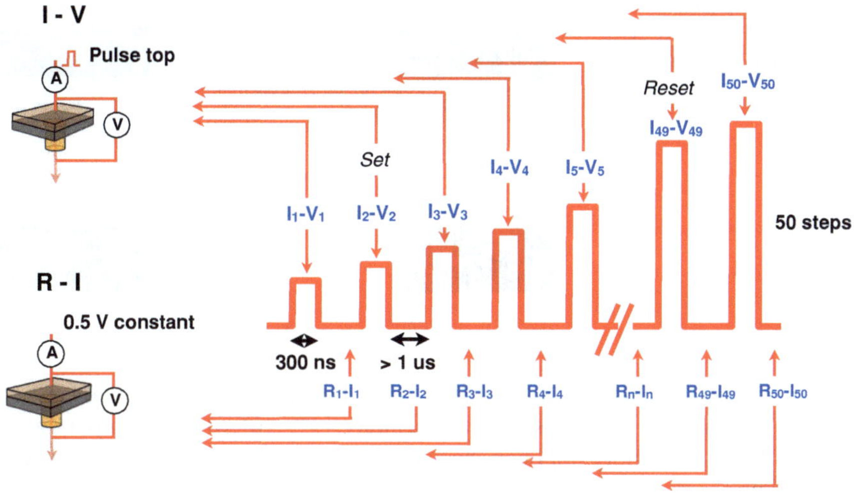

Fig. 12.18 The pulse pattern used in *I–V* and *R–I* measurements. While the *I–V* curve is measured *during the pulses* with gradually increasing intensity, the *R–I* curve was measured *between the pulses* at a constant voltage of 0.5 V

states. An example of an *R–V* curve is shown in Fig. 12.17, right panel. The *I–V* and *R–V* curves are both important characteristics of the PCRAM cells.

The *I–V* and *R–I* curves shown in Fig. 12.17 are measured using the following standard technique. A pulse train consisting of 50 pulses of gradually increasing voltage and duration of 300 ns each is applied to the sample. The intervals between the pulses are kept larger than 1 μs to allow for cooling to occur between pulses (Fig. 12.18). The *I–V* plots are measured at the centre of each pulse. As the applied voltage (and hence the Joule heating and sample temperature) increases, the materials first crystallises and then melts and reverts to the amorphous phase. The *R–I* plots, on the other hand, are measured between pulses in the cooled condition after each pulse using a constant voltage of 0.5 V. It should be stressed that the *R–I* plots represent the resistance of the sample *after the application of the pulse that generated current I* in the sample during the switching process and the shown *R* values *are not* a result of calculation ($R = V/I$) of the sample resistance *during the switching pulses*.

Properties such as high speed switching and a large number of SET/RESET cycles exceeding those of Flash are essentially satisfied in PCRAM using the same materials as commercial rewritable optical discs (DVD-RAM). The prototypical alloy is, $Ge_2Sb_2Te_5$ and it has the SET speed of about 50 to 70 ns. In addition, the SET temperature of a thin film, deposited by physical deposition technique such as sputtering, is around 155–160 °C [15]. The SET temperature can be controlled (higher or lower) by changing the GeTe/Sb_2Te_3 content ratio. With increasing GeTe content, the SET temperature rises, while it gradually drops with decreasing the GeTe content. The effect depends on microscopic interactive phase balance and stresses in the nanoscale grains. GeTe has the crystallisation temperature of around 190–200 °C, while that

Fig. 12.19 Schematic of a mushroom cell in the RESET (*left*) and SET (*right*) states

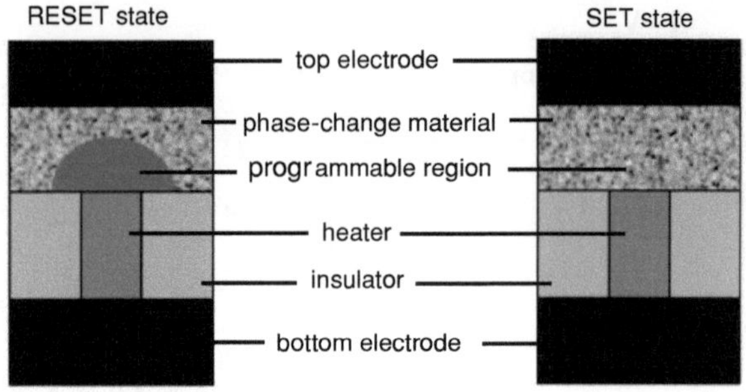

of Sb_2Te_3 is less than 100 °C [44]. Doping with elements such as nitrogen has been shown to increase the crystallisation temperature [45, 46] which is important for certain embedded memory applicaitons such as in the automotive industry when a chip is located in the immediate proximity of a hot engine.

The SET/RESET cyclability using a $Ge_2Sb_2Te_5$ film in PCRAM can be in the range of 10^6, which is of the same order of magnitude as that in DVD-RAM. There are indications that cyclability can be further improved by downsizing a switching cell size. Thus, applying a 45 nm size technologies is expected to increase the lifetime up to 10^9 or more [47].

Cell Designs

There are two major cell designs. In the so-called mushroom cell shown in Fig. 12.19, the temperature of the phase-change layer is increased using heat generated in the heater. In the high-conducting SET state the phase-change layer is crystalline. Upon application of a short intense pulse, the phase-change layer melts and forms a mushroom-shaped amorphous region (hence the name) with high resistivity (RESET state). A longer less intense pulse leads to crystallisation of the amorphous dome and reverts the system to the SET state.

An alternative design called a 'line cell' is shown in Fig. 12.20. It has been argued [48] that the the line cell's simplicity promises relatively low manufacturing costs because of the limited number of processing steps and rather low programming currents have been reported.

12.2.2 Phase-Change Memory for the Future

Cognitive Computing

It has always been a human dream to build a cognitive computer working similar to nerve cells. Figure 12.21 shows how nerve cells operate biologically. The incoming

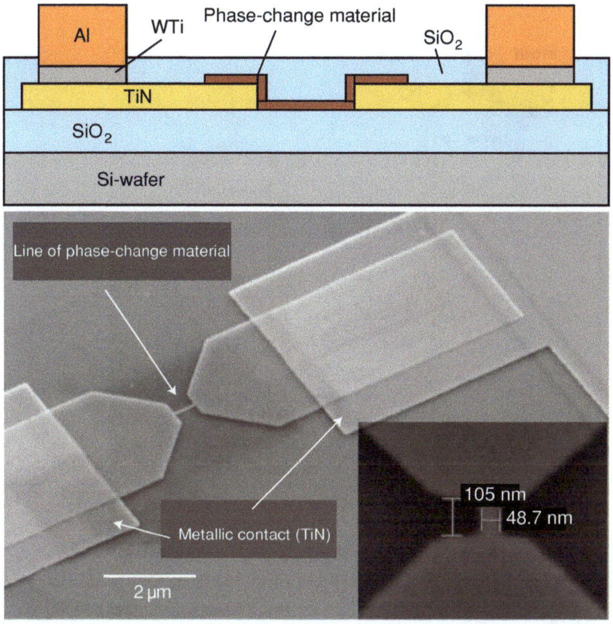

Fig. 12.20 Schematic cross-section of a line-concept memory cell with TiN contacts and Al bond-pads processed on a silicon wafer (*top*) and scanning electron micrograph of such a cell (length 500 nm, width 50 nm) made after structuring of the phase-change layer, which is done by electron-beam lithography. TiN contacts are structured by contact lithography. Inset, detail of similar cell with approximate dimensions 100 nm by 50 nm (*bottom*). Reprinted by permission from Macmillan Publishers Ltd.: Nature Mater. [48], copyright (2005)

information comes through the synapse and when summed in energy to a threshold condition, the nerve cell fires and sends information to other cells.

In 2004 Ovshinsky proposed that phase-change materials can be used for cognitive computing [49–51]. This proposal is based on the fact that with properly selected pulse intensity and duration, the first several pulses do not cause any detectable change in reflectivity (Fig. 12.22), i.e. the system can "learn" and act only after the appropriate combination of pulses has been input. Operation of an Ovonic cognitive device is illustrated in Fig. 12.23.

It should be noted that (i) an Ovonic congnitive device combines both a memory and a logic element in a singe device and (ii) the memory can be distributed and not separate function. Inputs can be optical, electrical or others, and outputs can be directed to various types of human interfaces. Early applications include pattern recognition and are expected to progress to more sophisticated functions. The device can be used to factor large numbers and therefore be applied in coding/decoding. The synaptic pulses can be optical or electrical. They operate in the amorphous state where the information is very secure and cannot be forensically investigated. Therefore,

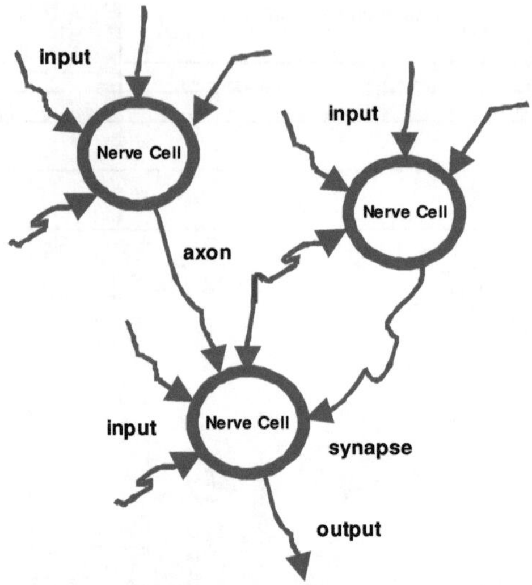

Fig. 12.21 Schematics illustration of a cognitive process in a biological system. An axon is a long, slender projection of a nerve cell, or neuron, that conducts electrical impulses away from the neuron's cell body. Axons make contact with other cells at junctions called synapses. Reprinted with permission after Ovshinsky [50]

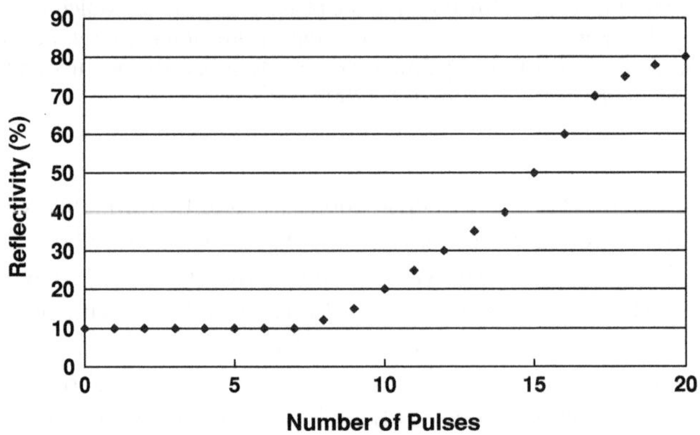

Fig. 12.22 Optical cognitive behaviour (the system "learns" during the initial pilses) of an Ovonic phase change material. Reprinted with permission after Ovshinsky [50]

they are perfect for encryption. Another major issue that has to be mentioned is that Ovonic cognitive computer will be able to operate at room temperature [51].

The arithmetic functionality of phase-change materials and their ability to "remember" previous excitations have been recently experimentally demonstrated in [52]. Use of phase-change materials as synapse for ultra-dense neuromorphic systems have also been investigated in [53].

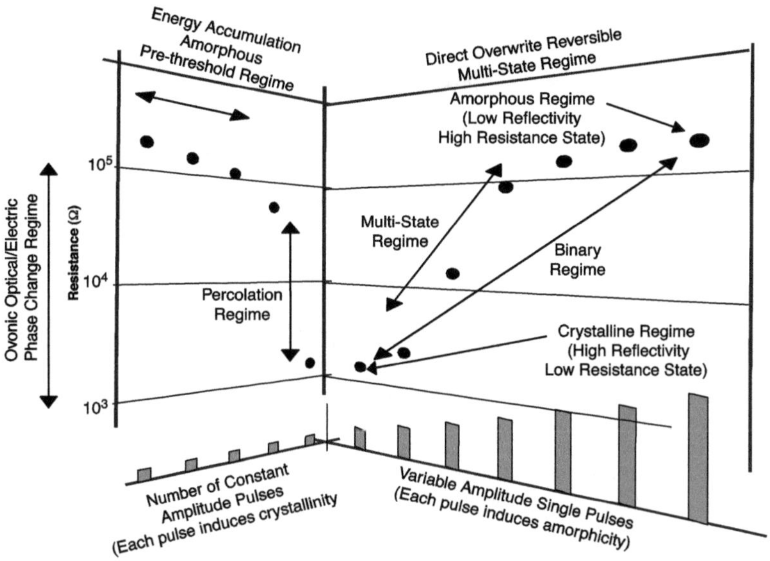

Fig. 12.23 Resistance characteristics of an Ovonic cognitive device. The cognitive amorphous pre-threshold synaptic regime (*left side*) culminates in a percolative transition to crystalline material, the equivalent of neurosynaptic switching. The resistance change accompanying the transition to the crystalline regime can provide readout and transferring of a completed signal to other devices. The leftmost and rightmost data points of the figure (the high resistance endpoints) both correspond to material that is substantially amorphous and the material becomes increasingly crystalline towards the centre of the figure, with the lowest resistance states having the greatest crystallinity. The right side is the multi-state crystalline cognitive regime. One can look upon the left side as being either standalone when the crystalline sums up the synaptic information or united with the activities of the right side. Reprinted with permission after Ovshinsky [50]

Interfacial Phase-Change Memory

As mentioned above, phase-change memory technology relies on the electrical and optical properties of certain materials changing substantially when the atomic structure of the material is altered. The energy needed to switch between the two phases depends on the intrinsic properties of the phase-change material and the device architecture.

While in composite GST Ge atoms switch stochastically resulting in large entropic losses and somewhat different local surrounding around each Ge atom giving rise to a broad distribution of the needed switching energies, in an engineered layered structure such as interfacial phase-change memory (*i*PCM) cell all Ge atoms should switch along the same direction and have unique local structure. As a result, one may expect significantly lower switching currents and improved reproducibility [54]. In the somewhat random local atomic structure of composite GST the Ge atoms can occupy both threefold coordinated (pyramidal) and/or tetrahedral sites [55, 56] with

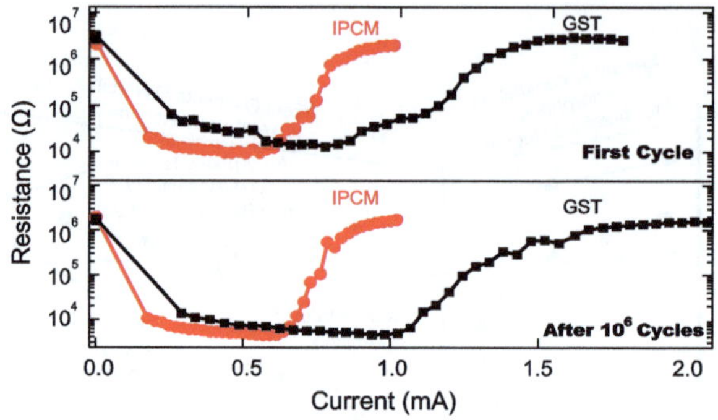

Fig. 12.24 Electrical switching characteristics of *i*PCM devices. Resistance-Current plots for phase change random access memory (PCRAM) devices on the first cycle (*upper panel*) and after 10^6 cycles (*lower panel*). Filled *squares* are from a device fabricated from a single GST target while filled *circles* are for a device containing a $(GeTe)_4(Sb_2Te_3)_2$ *i*PCM. The SET pulse lengths were 50 and 100 ns for the *i*PCM and GST materials, respectively. The RESET pulse length was fixed at 50 ns for both the *i*PCM- and GST- based devices. Reprinted by permission from Macmillan Publishers Ltd.: Nature Nanotech. [54], copyright (2011)

principally covalent bonding; the phase technologically known as the RESET state. In contrast, the cubic crystalline phase is assembled from approximately octahedral subunits which are usually described as being 'resonantly' bonded [57–59]; this bonding phase is technologically known as the SET state. Given the absence of significant long-range order in nanoscale structures it is appropriate to discuss the different phases of the *i*PCM in terms of the bonding nature rather than as being amorphous or crystalline. Therefore, the covalent (RESET) and resonance (SET) bonding states are used below when referring to these two phases.

The advantages of *i*PCM device are illustrated in Fig. 12.24 where the resistance R of *i*PCM- and GST-based PCRAM cells is plotted against the applied electrical current I [54]. The plot clearly shows that the currents necessary to reversibly switch *i*PCM-based devices between the SET and RESET states are substantially lower than those required for identical devices based on GST. Indeed, the electrical energy used to SET the GST and *i*PCM devices was 90 and 11 pJ respectively. Furthermore, due to the homogeneity of the Ge switching environment, the switch (or 'current snap') between the SET and RESET states was substantially more abrupt for the *i*PCM cells. This produced a tighter distribution of device characteristics facilitating the accurate determination of the SET and RESET states.

The efficiency improvements observed for the *i*PCM have been explained using entropy arguments [54]. The crystallisation process in GST is three dimensional, that is atomic movements occur stochastically in all directions [60]. The covalent state is composed of a range of local atomic configurations [55] and therefore it has a relatively large entropy. In contrast, the *i*PCM structures are designed to minimise the change in configurational entropy between SET and RESET states. The ultra-thin, uniform strata of Sb_2Te_3 and GeTe lower the entropy of the covalent state by

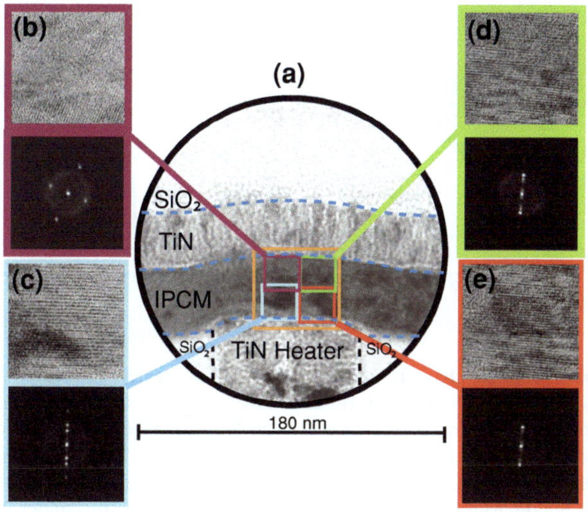

Fig. 12.25 TEM analysis of the *i*PCM RESET state. **a** A TEM image of a $(GeTe)_2(Sb_2Te_3)_4$ structure in the RESET state after 10^3 SET-RESET cycles. In contrast to GST, the main TEM image shows that there is no amorphous region surrounding the TiN heating electrode. The coloured *squares* indicate the areas chosen for further TEM inspection. The associated high resolution TEM images and SAD patterns are displayed in **b–e** and are bordered by a colour corresponding to the area of inspection. Reprinted by permission from Macmillan Publishers Ltd.: Nature Nanotech. [54], copyright (2011)

restricting the number of atomic configurations that can exist. For the SET operation, the *i*PCM- based devices use just 12% of the energy required by similar GST-based devices; an improvement consistent with the theoretical efficiency enhancement [61]. This increase in efficiency means that less energy is wasted during the phase transition, which in turn leads [4] to more than an order of magnitude improvement in the SET-RESET cycleability.

Reducing the entropic losses has the further ramification of producing device characteristics that are highly repeatable: the resistance of the *i*PCM during the first SET-RESET cycle is identical to that of the millionth cycle (see Fig. 12.24a). This implies that the *i*PCM structure is sustained for at least one million write-erase cycles.

The structure of the *i*PCM device in the RESET state after extended cycling is shown in Fig. 12.25a. The amorphous dome-shaped region is absent and, more importantly, a closer inspection of the *i*PCM directly above the heating electrode, Fig. 12.25b–e, reveals the structure to remain ordered in layers. The selective area diffraction (SAD) patterns associated with these images similarly display superlattice reflectivity thus demonstrating that even after cycling, the layer structure is present. The SAD pattern, which was collected from the whole *i*PCM structure, Fig. 12.25f, confirms the overall crystallinity of the cell.

On the basis of these results the authors argued that the most efficient and fastest phase-change memory devices will be developed using nanostructured materials that

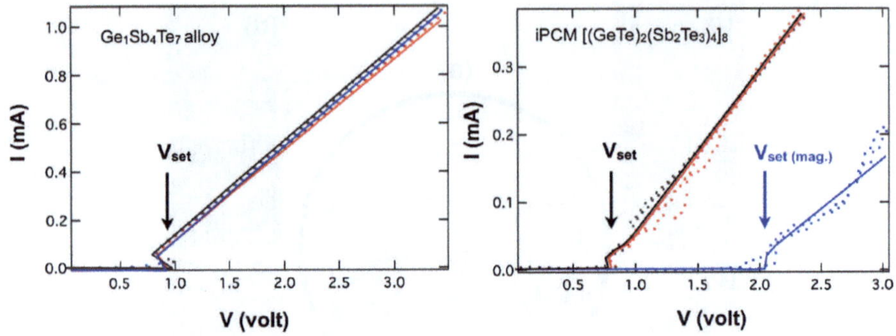

Fig. 12.26 Current-Voltage characteristics of a PCRAM device fabricated using composite $Ge_1Sb_4Te_7$ (*left*) and an identical composition *i*PCM device (*right*). As the voltage reaches the threshold value of V_{set}, the device switches into the low-resistance state. Device characteristics prior to application of a magnetic field, with an external magnetic field of 0.1 T applied, and after removal of the field are shown. The composite device (*left*) did not show any change among the three states, while the iPCM device (*right*) clearly showed a voltage shift under the magnetic field, indicated as V_{set}^{mag}. A sequence of 300 ns pulses was used for the measurements. Two scans for each state are shown to demonstrate reproducibility. *Dots* are experimental data and *solid lines* are guides for the eye. Reprinted from Tominaga et al. [62]. Copyright 2011 by the American Institute of Physics

take advantage of entropy controlled switching and, furthermore, exhibit effective electrical properties that are unattainable in composite GST [54].

Recently it was demonstrated that *i*PCM structures exhibit a strong response to an applied magnetic field (Fig. 12.26) despite all constituent elements being non-magnetic [62]. Since this effect was only observed in the layered structures and not in the composite material of the same average composition, the authors concluded that it arises from interfaces and interpreted it as due to topological insulator properties of *i*PCM. The observed effect suggests that use of a magnetic field as an extra degree of freedom may pave the way to conceptually novel memory devices combining the merits of both phase-change and magnetic memories [62, 63].

References

1. R. Fairman, B. Ushkov, *Semiconducting Chalcogenide Glass III: Applications of Chalcogenide Glasses* (Elsevier Science, Amsterdam, 2004)
2. A. Zakery, S.R. Elliott, *Optical Nonlinearities in Chalcogenide Glasses and their Applications* (Springer, Berlin, 2007)
3. B. Eggleton, B. Luther-Davies, K. Richardson, Nature Photonics **5**, 141 (2011)
4. G.W. Burr, M.J. Breitwisch, M. Franceschini, D. Garetto, K. Gopalakrishnan, B. Jackson, B. Kurdi, C. Lam, L.A. Lastras, A. Padilla, B. Rajendran, S. Raoux, R.S. Shenoy, J. Vac. Sci. Technol. B **28**, 223 (2010)
5. S. Raoux, W. Welnic, D. Ielmini, Chem. Rev. **110**, 240 (2010)
6. M. Terao, T. Morikawa, T. Ohta, Jpn. J. Appl. Phys. **48**, 080001 (2009)

7. V. Capel, *Audio and Hi-Fi Engineer's Pocket Book* (Heinmann Professional Publishing, London, 1988)
8. J. Tominaga, T. Nakano, *Optical Near-Field Recording-Science and Technology* (Springer, Heidelberg, 2005)
9. *Philips Consumer Electronics B.V.: Orange Book*, Part II, ver. 2.0 (1994)
10. O. Kawakubo, in *15th Symposium Phase Change Optical Information Storage* (PCOS'2003), Atami, 2003
11. E. Hamada, Y. Takagishi, T. Yoshizawa, T. Fujii, R. Negisgi, T. Nakajima, Jpn. J. Appl. Phys. **39**, 785 (2000)
12. M. Mansuripur, *The Physical Principles of Magneto-Optical Recording* (Cambridge University Press, Cambridge, 1995)
13. S.R. Ovshinsky, Phys. Rev. Lett. **21**, 1450 (1968)
14. N. Yamada, E. Ohno, N. Akahira, K. Nishiuchi, K. Nagata, M. Takao, Jpn. J. Appl. Phys **26**, 61 (1987)
15. N. Yamada, E. Ohno, K. Nishiuchi, N. Akahira, M. Takao, J. Appl. Phys. **69**, 2849 (1991)
16. H. Kitaura, E. Ohno, K. Nishiuchi, N. Yamada, in *11th Symposium Phase Change Optical Information Storage* (PCOS'99) (Mishima, Japan, 1999), p. 89
17. T. Ohta, J. Optoelectron. Adv. Mater. **3**, 609 (2001)
18. M. Terao, N. Nishida, Y. Miyauchi, S. Horigome, T. Kaku, N. Ohta, in *Proceedings of the SPIE*, vol. 695 (1986), p. 105
19. T. Kikukawa, in *13 Symposium Phase-Change Optical Information Storage* (PCOS2001) (Japan, 2001)
20. M. Horie, T. Ohno, N. Nobukuni, K. Kiyono, T. Hashizume, M. Miuno, in *Proceedings of the SPIE*, vol. 4342 (2001), p. 77
21. M. Terao, K. Shigematsu, M. Ojima, Y. Taniguchi, S. Horigome, S. Yonezawa, J. Appl. Phys. **50**, 6881 (1979)
22. H. Iwasaki, Y. Kageyama, M. Harigaya, Y. Ide, Jpn. J. Appl. Phys. **35**, 500 (1996)
23. K. Ito, M. Harigaya, M. Kinoshita, T. Shibaguchi, E. Suzuki, M. Shinotsuka, Y. Kageyama, in *10th Symposium Phase Change Optical Information Storage* (PCOS'98) (Mishima, 1998), p. 91
24. M. Okuda (ed.), *Technology and Materials for Future Optical Memories* (in Japanese) (CMC, Japan, 2004)
25. M. Terada, K. Furuya, T. Okamura, T. Morimoto, M. Nakao, Jpn. J. Appl. Phys. **32**, 5219 (1993)
26. P.W. Atkins, *Physical Chemistry*, 6th edn. (Oxford University Press, Oxford, 1998)
27. H. Raether, *Surface Plasmons on Smooth and Rough Surfaces and on Gratings* (Springer, Heidelberg, 1988)
28. J. Tominaga, T. Nakano, N. Atoda, Appl. Phys. Lett. **73**, 2078 (1998)
29. K. Nakai, M. Ohmaki, N. Takeshita, M. Shinoda, I. Hwang, Y. Lee, H. Zhao, J. Kim, B. Hyot, B. Andre, L. Poupinet, N.T. Shima, T.J. Tominaga, Jpn. J. Appl. Phys. **49**, 08KE02 (2010)
30. S. Hosaka, T. Shintani, M. Miyamoto, A. Hirotsune, M. Terao, M. Yoshida, K. Fujita, S. Kammer, Jpn. J. Appl. Phys. **35**, 443 (1996)
31. J.M. Vigoureux, F. Depasse, D. Courjon, Appl. Opt. **31**, 3036 (1992)
32. T. Kikukawa, T. Nakano, T. Shima, J. Tominaga, Appl. Phys. Lett. **81**, 4697 (2002)
33. B. Hyot, X. Biquard, F. Laulagnet, in *European Phase Change and Ovonics Science Symposium* (Zermatt, 2007)
34. M. Krbal, A.V. Kolobov, B. Hyot, B. Andre, P. Fons, R.E. Simpson, T. Uruga, H. Tanida, J. Tominaga, J. Appl. Phys. **108** (2010)
35. T. Uno, K. Nagata, N. Yamada, in *11th Symposium Phase Change Optical Information Storage* (PCOS'99) (Mishima, 1999), p. 83
36. S. Raoux, J.L. Jordan-Sweet, A.J. Kellock, J. Appl. Phys. **103**, 114310 (2008)
37. R.E. Simpson, M. Krbal, P. Fons, A.V. Kolobov, J. Tominaga, T. Uruga, H. Tanida, Nano. Lett. **10**, 414 (2010)

38. Y. Choi, I. Song, M. Park, H. Chung, S. Chang, B. Cho, J. Kim, Y. Oh, D. Kwon, J. Sunwoo, J. Shin, Y. Rho C. Lee, M.G. Kang, J. Lee, Y. Kwon, S. Kim, J. Kim, Y. Lee, Q. Wang, S. Cha, S. Ahn, H. Horii, J. Lee, K. Kim, H. Joo, K. Lee, Y.-T. Lee, J. Yoo, G. Jeong, *IEEE Solid State Conference Digest of Technical Papers*, (San Francisco, 2012), p. 46
39. H. Santoh, Y. Hongo, K. Tajima, M. Konishi, T. Saiki, in *Proceedings of the 2009 EPCOS Meeting*, (Aachen, Germany, 2009)
40. P. Fons, H. Osawa, A.V. Kolobov, T. Fukaya, M. Suzuki, T. Uruga, N. Kawamura, H. Tanida, J. Tominaga, Phys. Rev. B **82**, 041203 (2010)
41. A.V. Kolobov, M. Krbal, P. Fons, J. Tominaga, T. Uruga, Nature Chem. **3**, 311 (2011)
42. X.B. Li, X.Q. Liu, X. Liu, D. Han, Z. Zhang, X.D. Han, H.B. Sun, S.B. Zhang, Phys. Rev. Lett. **107**, 015501 (2011)
43. D.Q. Huang, X.S. Miao, Z. Li, J.J. Sheng, J.J. Sun, J.H. Peng, J.H. Wang, Y. Chen, X.M. Long, Appl. Phys. Lett. **98**, 242106 (2011)
44. J. Tominaga, P. Fons, A.V. Kolobov, T. Shima, T.C. Chong, R. Zhao, H.K. Lee, L. Shi, Jpn. J. Appl. Phys. **47**, 5763 (2008)
45. R. Kojima, S. Okabayashi, T. Kashihara, K. Horai, T. Matsunaga, E. Ohno, N. Yamada, T. Ohta, Jpn. J. Appl. Phys. **37**, 2098 (1998)
46. V. Sousa, L. Perniola, A. Fantini, G.B. Beneventi, E. Gourvest, S. Loubriat, A. Bastard, A. Roule, A. Persico, H. Feldis, A. Toffoli, D. Blachier, S. Maitrejean, B. Hyot, J.F. Nodin, C. Jahan, G. Reimbold, T. Billon, B. Andre, B.D. Salvo, F. Boulanger, S. Lhostis, P. Mazoyer, D. Bensahel, P. Zuliani, R. Annunziata, in *European Phase Change and Ovonics Science Symposium* (Milano, 2010)
47. S. Lai, in *Tech. Digest Int, Electron Devices Meeting*, 2003
48. M. Lankhorst, B. Ketelaars, R. Wolters, Nature Mater. **4**, 347 (2005)
49. S.R. Ovshinsky, B. Pashmakov, Mater. Res. Soc. Proc. **803**, 49 (2004)
50. S.R. Ovshinsky, in *European Phase Change and Ovonics Science Symposium*, The Ovonic cognitive computer—a new paradigm (Balzers, 2004)
51. S.R. Ovshinsky, Jpn. J. Appl. Phys. **43**, 4695 (2004)
52. C.D. Wright, Y. Liu, K.I. Kohary, M.M. Aziz, R.J. Hicken, Adv. Mater. **23**, 3408 (2011)
53. M. Suri, O. Bichler, D. Querlioz, O. Cueto, L. Perniola, V. Sousa, D. Vuillaume, C. Gamrat, B. De Salvo, *IEEE International Electron Devices Meeting Technical Digest*, (Washington, 2011), p. 79
54. R.E. Simpson, P. Fons, A.V. Kolobov, T. Fukaya, M. Krbal, T. Yagi, J. Tominaga, Nature Nanotech. **6**, 501 (2011)
55. M. Krbal, A.V. Kolobov, P. Fons, J. Tominaga, S.R. Elliott, J. Hegedus, T. Uruga, Phys. Rev. B **83**, 054203 (2011)
56. J. Akola, R.O. Jones, S. Kohara, S. Kimura, K. Kobayashi, M. Takata, T. Matsunaga, R. Kojima, N. Yamada, Phys. Rev. B **80**, 020201 (2009)
57. K. Shportko, S. Kremers, M. Woda, D. Lencer, J. Robertson, M. Wuttig, Nature Mater. **7**, 653 (2008)
58. B. Huang, J. Robertson, Phys. Rev. B **81**, 081204R (2010)
59. R.E. Simpson, P. Fons, X. Wang, A.V. Kolobov, T. Fukaya, J. Tominaga, Appl. Phys. Lett. **97**, 161906 (2010)
60. J. Hegedüs, S.R. Elliott, Nature Mater. **7**, 399 (2008)
61. J. Tominaga, R. Simpson, P. Fons, A. Kolobov, in *European Phase Change and Ovonics Science Symposium*, Milano, 2010
62. J. Tominaga, R.E. Simpson, P. Fons, A.V. Kolobov, Appl. Phys. Lett. **99**, 152105 (2011)
63. M. Segal, Nature Nanotech. **6**, 689 (2011)

Index

A. V. Kolobov and J. Tominaga, *Chalcogenides*, Springer Series in Materials Science 164, DOI: 10.1007/978-3-642-28705-3,